Workshop Machining

Workshop Machining is a comprehensive textbook that explains the fundamental principles of manually operating machinery to form shapes in a variety of materials. It bridges the gap between people who have traditional toolmaking skills and those who have been trained in programming and operation of CNC machines in a focused production environment, rather than general machine shop.

Using a subject-based approach, David Harrison intuitively guides readers and supplies practical skills. The chapters cover everything from the basic machine controls to advanced cutting operations using a wide range of tooling and work-holding devices. Theory and practice are shown via a mixture of diagrams, text and illustrated worked examples, as well as through exercises.

The book is ideal for students and lecturing staff who participate in, or lead, practical machining sessions, and for those who wish to further develop their machining skills. It also serves as an excellent reference to understand the principles and limitations of producing shapes with cutters that move in a limited combination of linear and radial paths.

David Harrison is a chartered engineer and lecturer at Newcastle College University Centre. He started his career as an MoD apprentice, and over his thirty years of hands-on experience, he has seen the progression of manufacturing from lines of manual and semi-automatic metal cutting machines to computer numeric controlled machines and additive manufacturing. David is focused on developing higher education engineering subjects to ensure that students upon graduation have the required skills to go into work.

Workshop Machining

A Comprehensive Guide to Manual Operation

David Harrison

Routledge
Taylor & Francis Group

LONDON AND NEW YORK

First published 2022
by Routledge
2 Park Square, Milton Park, Abingdon, Oxon OX14 4RN

and by Routledge
605 Third Avenue, New York, NY 10158

Routledge is an imprint of the Taylor & Francis Group, an informa business

© 2022 John David Harrison

British Library Cataloguing-in-Publication Data
A catalogue record for this book is available from the British Library

Library of Congress Cataloging-in-Publication Data
Names: Harrison, David (Chartered Engineer), author.
Title: Workshop machining : a comprehensive guide to manual
    operation / David Harrison.
Description: Abingdon, Oxon ; New York, NY : Routledge, 2022. |
    Includes bibliographical references and index.
Identifiers: LCCN 2021020906 (print) | LCCN 2021020907 (ebook) |
    ISBN 9780367278403 (hbk) | ISBN 9780367278397 (pbk) |
    ISBN 9780429298196 (ebk)
Subjects: LCSH: Machining.
Classification: LCC TJ1185 .H195 2022 (print) | LCC TJ1185 (ebook) |
    DDC 671.3/5—dc23
LC record available at https://lccn.loc.gov/2021020906
LC ebook record available at https://lccn.loc.gov/2021020907

ISBN: 978-0-367-27840-3 (hbk)
ISBN: 978-0-367-27839-7 (pbk)
ISBN: 978-0-429-29819-6 (ebk)

DOI: 10.1201/9780429298196

Typeset in Sabon
by Apex CoVantage, LLC

Contents

4 Milling   125

5   Surface grinding   281

6   Cylindrical grinding   323

7   Drilling

Acknowledgements

The images for this book would not have been possible without the support of Newcastle College and especially the kind assistance of Martin Garrett and Rob Spiers.

Chapter 1

Introduction

As technology continually advances through all walks of life manufacturing machining has also undergone a similar step change in capability. The introduction of computer controlled machinery has allowed profiles to be machined that were previously obtainable other than through a series of protracted operations and a degree of hand finishing. With the introduction of computer numeric controlled (CNC) machines productivity had a step change increase as has precision and repeatability.

However, with this improvement, have we lost anything? The answer to this is a qualified yes. Modern manufacturing often involves the creation of a computer program either remotely or directly entered into a machine. The closest machinists come to the process of machining is observing metal cutting happening inside a cloud of coolant through the closed door of a CNC or a tool path graphic on the machine display. The close connection to seeing and feeling exactly what is happening at the tool point, and the direct relationship between a machine operator directly controlling how metal is cut, is being eroded.

The intention of this book is to bridge the widening gap in knowledge between people who once learned skills through traditional four-year tool-making apprenticeships, and those who are trained to program and operate CNC machines in a production environment. It is intended to support apprentices, students, and lecturing staff who would like to participate in, or lead, practical sessions, and for those who want to develop an understanding of the principles and limitations of producing shapes with cutters that move in a limited combination of linear paths. It is also intended to provide a starting point for those who wish to broaden their skills from one machining discipline to another, and assist the hobbyist in taking their first steps along a path from basic turning and milling, to machining more complex components.

Workshop Machining is a comprehensive textbook that explains the basic principles of manually operating machinery to form shapes in a variety of materials through a subject based approach for those who wish to develop

DOI: 10.1201/9780429298196-1

both practical skills. The theory and practice will be shown through a mixture of diagrams, text and illustrated worked examples.

The contents of the book are broken down into five sections covering turning, milling, surface grinding, cylindrical grinding, and drilling. Each of those sections is structured in a way that introduces the reader to the machine and how it operates. The tooling that is used on it and the method of basic operation. There are further chapters in each section that identify how to operate more complex tooling, work holding techniques and alternative approaches. Common problems encountered when undertaking each type of machining are identified and have suggestions for resolution.

Located in the appendices are a number of simple and more complex machining exercises that are intended to introduce the reader to machining shapes in basic materials on a lathe and milling machine. Two simple tasks are intended to provide the reader with drawings and a step by step guide to machining the object, to give the operator confidence in using a lathe and a milling machine.

It is hoped that the following pages provide a useful reference for experienced machine operators, a 'how to' reference source for those wishing to deepen their knowledge of aspects of machining, and an introduction to material removal techniques for those taking their first steps in the wider world of the general engineering workshop.

Chapter 2

Degrees of freedom

Every machine tool has the capability to undertake a combination of translational movements and rotary movements, and it uses these to allow a cutter to be applied to a workpiece in order to shape it. The term that describes this is *kinematics*, or the study of motion without reference to force or mass. While there are a number of equations that can be employed to solve a problem, these tend, in a manufacturing environment, to be more of concern to robotics. However, its relevance to machining is that the combination of rotary movement and translational movement is what allows us to undertake the reductive process that is machining. Figure 2.1 identifies the six possible translational and rotational movements known as 'degrees of freedom'.

Each type of machine will have constraints imposed on it by the number of degrees of freedom, and that is why we often use multiple types of machines to produce a desired component. Modern computer numeric controlled (CNC) machines are capable of utilising five degrees of freedom simultaneously and are capable of producing complex developed shapes, that are either impossible to produce on manual machinery, or extremely time consuming especially to the level of accuracy achievable on a five-axis CNC machine.

Machine axis are defined on a common basis. Typically a milling machine will have the x-axis making translational movements to the left and right of the operator. The y-axis makes translational movements away from, and towards the operator, and the z-axis relates to translational movements made in the vertical plane. However on a lathe this differs with the z-axis making translational movements to the left and right of the operator along the axis of the machine, and the x-axis applying the cut moving away and towards the operator.

DOI: 10.1201/9780429298196-2

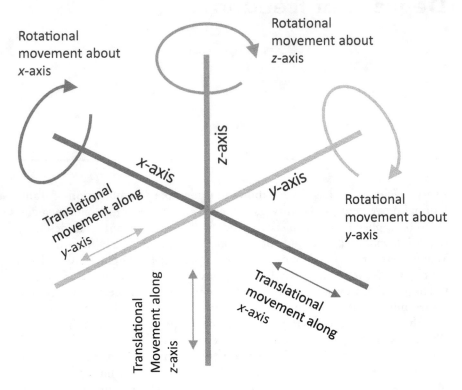

Figure 2.1 Six degrees of freedom

Chapter 3

Turning

Turning as a process is almost exclusively undertaken on a centre lathe. Turning differs from other types of machining as it is the workpiece that rotates rather than the cutter. Its function is to produce workpieces that are shaped into round components, generally, although not exclusively sharing the same central axis.

The lathe is one of the oldest tools that craftsmen have utilised to produce items, with some evidence to suggest turning on a lathe could date back as far as 1300 BC. While there is some evidence for this, and much stronger evidence to suggest that turning as a process was being undertaken in 400 BC, the centre lathe that engineers would recognise came into being in the 18th century. This type of machine has three degrees of freedom with a translational movements provided along the x- and z-axis and a single rotational movement along the z-axis. Figure 3.1 shows the notation of lathe axis on a centre lathe.

All machine shop lathes have the same basic configuration which consists of:

- **Headstock.** This consists of the chuck or principal work holding device, motor, gearbox and central spindle. It also has all of the ancillary controls such as speed selection, coolant on/off switch, gear selectors for configuring screw cutting settings, and feed speeds.
- **Bed.** This is a ground trackway which forms the principal chassis of the machine to allow the tool to be moved along the axis of the machine. It is generally manufactured from cast iron and is required to be very stiff to accommodate the loads imparted to it when machining.
- **Saddle.** This is mounted upon the ground trackway of the bed and is used to move the cross slide, compound slide, and tool post along the axis of the machine. Attached to the saddle are the controls for manually moving the saddle along with levers to operate the automatic feed mechanism and engage the lead screw where fitted.
- **Cross slide.** This sits on top of the saddle mounting the compound slide and tool post, and is used to move the tool across the axis of the machine.

DOI: 10.1201/9780429298196-3

- **Compound slide.** This mounts the tool post and is capable of being rotated to allow generation of features at any angle across the plane of the machine bed. This overcomes the single angle restrictions of the saddle and cross slide. A significant feature of the compound slide is a circular base engraved with a scale most commonly using whole degrees.
- **Tool Post.** This is the clamping device for holding the cutting tool. They come in a number of designs, which mount a single tool, or are of a more complex design such as those that allow for four tools to be mounted at 90 degrees to each other and can be indexed round using a simple lever, or are of the 'quick release' type allowing multiple positions of tools but are principally designed to allow quick changes of a tool carrier to facilitate changes of tool types for differing operations. They all have their strengths and weaknesses but the essential function is the same: to mount a cutting tool on centre height so that it can be addressed in the correct configuration to the workpiece.
- **Tailstock.** This is mounted on the ground slide way of the bed of the machine at the opposite end to the headstock. It consists of a cast body, containing an extending cylinder or quill which can be extended from the body using a hand wheel. Its function is to mount a separate chuck to facilitate drilling, or to mount a live or dead centre to support long workpieces. It is able to be moved along the bed to a desired position and locked in place. The centre height of the tailstock cylinder is exactly that of the axis of the machine, and it also has adjusting screws built into the body to allow transverse adjustment to ensure the centre is completely aligned in two planes. The significance of having a tailstock properly aligned with machine axis, is that if improperly set up then when machining material supported by a centre the result is a tapered workpiece.

While centre lathes come in all shapes and sizes from a small bench top modelling lathe to substantial machines with beds that are many meters long, they all operate using the same principles. While size makes a difference for the depth of cut that can be undertaken in a single pass, the length of bed constrains the longest component that can be machined and the height of the axis above the bed which limits the 'throw' of the machine, it is the accuracy of the machine and the surface finish that can be obtained which is most affected by size.

The accuracy of all lathes is a function of the precision to which they were manufactured and the combination of slide way screw pitches, and marking of the scales on control wheels. Small cheap machines tend to be less robust, and the diameter of the control screws and hand wheels tend to be less precise than larger machines. That said there are some extremely precise and commensurately expensive, machines available, however, these tend to be more directed at specialist markets than general engineering. The size of the machine is also often key to accuracy. A very large machine such as that used for turning ships propellor shafts is often poor at machining small diameter

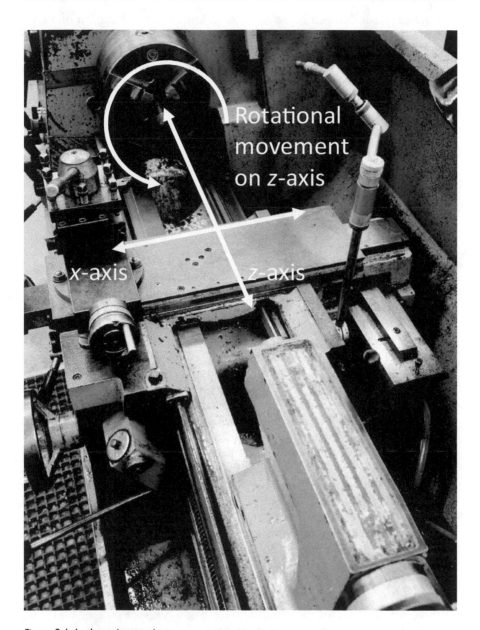

Figure 3.1 Lathe axis notation

components, so the size of machine used needs to be matched within a sensible parameter to achieve the accuracy required. A 'standard' workshop centre lathe in reasonable condition with a bed of approximately 750–1000mm and a throw of approximately 200mm should have little problem in achieving an accuracy of 0.02mm or better.

Surface finish that can be achieved on a lathe is a combination of a considerable number of factors all of which have a direct impact on the surface finish achieved. However, assuming that all factors have been optimised then achieving a surface finish of between 12.5–1.6μm with 3.2μm being considered a smooth turned finish. Factors that can (adversely) affect surface finish include the following:

- rotational speed;
- tool type;
- feed;
- material type;
- tool holding rigidity;
- work holding configuration;
- machine stiffness;
- use of coolant; and
- swarf clearance.

Tool and speed selection is affected by the workpiece material and tool being used. Rotational speed is normally calculated as a product of the material surface m/min rate and these rates vary for each broad type. In general terms, if high-speed steel tooling is used to machine mild steel it should be cut at around 27 surface m/min, whereas carbide tooling would use a cutting rate of around 90 surface m/min. This surface machining rate is calculated as the most efficient speed for the tooling and material that maintains tool life, efficiently cuts material and provides a decent surface finish. Speed selection is also constrained by the machine gearbox. Modern lathes will allow the operator to set a surface cutting rate (i.e. as the tool progresses towards the centre of the machine axis the speed automatically increases up to the maximum speed for the machine motor). Older or less expensive machinery tends to have a gearbox with a series of settings from comparatively low (i.e. 27 rpm) to quite high speeds (i.e. 2500 rpm). In this situation the closest speed is selected, and for work where speed is critical this speed may be adjusted between cuts to ensure close correlation with a surface machining rate as the tool progresses towards the machine axis.

3.1 WORKPIECE HOLDING – THE THREE-JAW CHUCK

A three-jaw chuck is the most common form of workpiece holding and lends itself well to holding hexagonal bar with three faces clamped within the jaws.

The workpiece is placed with the chuck and is securely tightened. As the three-jaw chuck utilises an internal scroll to tighten all three jaws simultaneously, it is good practice to tighten the chuck in two places using the correct chuck key properly engaged in the square holes to ensure that all three jaws have gripped the workpiece securely. It is also worth noting that some chuck keys have a sprung collar to eject the key preventing it from remaining in the periphery of the chuck unless pressure is applied. While these types of key are relatively uncommon, and are more often of a hexagonal type for use with four-jaw chucks, they ensure that the key cannot be left in the chuck to be ejected on machine start causing injury to the operator.

A constraint for the length of metal protruding from the chuck is the internal diameter of the spindle to allow the bar to slide through its centre and also the amount of material that be accommodated within the central diameter of the chuck itself. Typically chucks have a larger hole through them than the internal spindle diameter providing a pocket to accommodate the workpiece. The amount the workpiece can extend from the jaws of the chuck is slightly more complex. The correct extension is subjective and relative to the size of the bar being machined. A rough guide is that after 4–6 diameters workpieces tend to need support from a centre or rest to prevent it bending away from the tool, providing both a poor surface finish and, potentially a tapered workpiece. This tends to be more evident with a smaller diameter workpiece than a larger diameter workpiece which is generally stiffer and less prone to deflection away from the cutting tool. Figures 3.2 and 3.3 show a workpiece inserted in a chuck of a diameter that allows it to enter the machine spindle, and a larger workpiece that is retained externally within the chuck jaws.

Figure 3.2 Bar mounted in three-jaw chuck

Figure 3.3 Large bar mounted in three-jaw chuck

The amount of engagement in the jaws, and the length protruding from the chuck jaws is important. Longer bars can pass through the chuck and down the spindle of the lathe, however, care must be taken to ensure that excessive lengths do not protrude beyond the spindle, as some health and safety legislation requires any rotating shaft to be guarded, but more importantly thinner bars of metal droop which when rotated by the machine operation often bend and then form a high speed rotating cone of metal.

Mounting workpieces that are too large to pass into the central well of the chuck can still be securely held either by widening the gap between the jaws and pressing the workpiece up against the front face of the chuck. However, if the diameter of the workpiece is such that more than 25% of the jaws protrude beyond the external diameter of the chuck, the jaws can be removed and turned around providing a seating for the workpiece. The depth of step in the jaws is not great and unless a very short workpiece is being machined a supporting centre will need to be used or it is likely the workpiece will be ejected from the chuck when the tool is cutting. It is considered good practice for the full length of jaw to be used for clamping the workpiece unless supported by a centre.

3.2 TOOLING

Turning tools have over many years come in a number of configurations, with them being left-handed, right-handed, bent cranked and straight to assist the operator with presenting a cutting edge to the workpiece without interfering with another part. Modern practice has seen the wide adoption of cemented carbide inserts, which has improved cutting performance markedly and removed the requirement for tools to be regularly re-sharpened. While the shape and size of the cutting tools can significantly vary, the actual cutting edge geometry has relatively little variation and this variation tends to be driven by the material the cutting tool is to be used on than the configuration of the tool holder. One characteristic of turning tools compared to the vast majority of tools used in milling, drilling or grinding operations is that cutting is a single point operation and there is only one cutting edge in use during machining.

For many years solid high-speed steel (HSS) tools were the mainstay of turning tooling. They were, and still are, one piece of HSS with a tip that is ground to a series of angles, that are referred to as the side cutting angle, end cutting angle, side and end relief angles, side and back rake. These angles form the essential geometry of a cutting tool and vary according to the material being turned. More friable materials such as brass that have a tendency to generate swarf chips in short broken sections, will have different cutting and rake angles to those more ductile materials which tend to produce long continuous chips. In general all of the tool tip geometry issues that apply to HSS tooling also apply to carbide inserts, and while there are minor variations the general principles apply to both material types.

The greatest enemy of tooling is heat. Much of the power exerted by the machine in shearing the workpiece material is converted into heat and can rapidly erode the cutting edge. While the application of coolant is one method of removing heat, the angles ground onto the tool and the angle the tool is presented to the workpiece will have a significant effect on the loads and heat transmitted to the cutting edge of the tool, and the life of a tool before re-sharpening is required. Positive and negative lead angles and rake angles are also a factor in machinability and tool life. Tooling with a positive rake angle are more triangular in profile and turn the sheared chip through less of an angle than negative rake tools. This assists the smooth formation and thickness of the swarf chip, reducing the load on the tool and the power required to cut the material. Reducing the power required to cut allows the operator to adjust other parameters such as the feed of depth of cut to enhance the material removal rate. However, the cutting edge has less support and is more vulnerable to wear and erosion. Negative rate tooling is more substantial with a more 'wedge shaped' profile and heavier tool tip. This allows the tool to accept much higher loads, extends the life of the tool, and allows a heavier cut to be taken and it is not unusual to see this tool profile used on harder materials such as cast iron. However, it also requires significantly more power from the machine. Changing the rake of the tool, and its clearance angles allows the operator to reduce or increase chip thickness, feed rate, type of swarf chip produced and material removal rates. There is no one size fits all approach, for instance the cutting force required to remove a depth of material will reduce by roughly 1% for every degree increase in the side rake angle, but only to a point where the tool tip becomes weak, and that the length of the cutting edge is significantly reduced. The nose radius of the cutting tool is also an important factor. It forms a critical part of the cutting edge, and while a very small nose radius, will provide a suitable square corner the tip of the tool is weak, and a poor surface finish can be experienced. A larger nose radius will provide a much improved surface finish, allow for improved feed rates, and have a significantly longer tool life, but if a large tool nose radius is used chatter can occur.

Tooling manufacturers have extensive information on their websites and publications to assist the operator with selecting the correct tool profile for the material and machine being used, for both HSS tooling and carbide inserts. Some of the finer details of tool type may be of less importance to the general workshop operator, as the nature of the work probably removes any advantage in choosing one particular tool tip over another as it gives 0.01% more tool life. However, choosing the correct tool that is matched to the power of the machines within the workshop, and the workpiece material type, such that a high standard of surface finish is achieved with a satisfactory depth of cut and feed rate achieved.

There are approximately 14 differing types of carbide insert holder which provide for a wide selection of tool tip geometries and applications. On top of this there are a substantial number of insert materials and configurations and the option for right-handed or left-handed tools. While there is perhaps much less choice for HSS tooling, the operator has easy access to tooling to efficiently undertake a task. While the number of configurations are seemingly endless the generic types of tooling used for turning are as follows:

- side cutting tools;
- facing tools;
- parting tools;
- boring tools;
- knurling tools;
- threading tools; and
- form tools.

Side cutting tools are without doubt the most common type, and tools with a positive rake and left-hand crank can be used for both edge cutting and facing. Facing tools are designed specifically for facing, and commonly have a more rounded nose radius to improve surface finish.

Parting tools come in a number of configurations manufactured from either HSS or using a carbide insert. Two specific types are found, with a solid HSS shank and relatively short cutting point being less common in modern workshops, and a blade type of parting tool. Blade type tools using either HSS or carbide inserts. The solid parting tools with a short tool point have relatively little to commend themselves, as the depth of material that can be parted is restricted to the length of the tool point which in itself is limited by the depth of the tool bar, as once the length of the tool point is greater than the depth of the tool bar, the stiffness of the tool rapidly reduces and breakage is common. Blade types, especially those with a carbide insert, can accept large overhangs, and normally have a narrow width, with 3mm being common. This allows for a significant diameter of material that can be parted off on quite modest machines, and the narrow width ensures that less material wastage is incurred.

There is a rule of thumb when selecting boring tools that the largest bar should be used, especially when boring into deep cavities. Larger boring bars are stiffer and therefore less prone to chatter when a significant tool overhang is required when boring deep cavities. This overhang can restrict depths of cut, and as the tool cutting edge is on the front face of the tool allows for not only opening up the diameter of the cavity but facing of a cavity closed end.

Knurling tools come in two generic types, those that use wheels that are clamped vertically from opposing sides of the workpiece, and those that are pushed into the workpiece from one side. Vertically clamping types tend to only have one set of knurling wheels (i.e. coarse, medium, or fine), whereas the side type normally have a rotating carousel of pairs of knurling wheel and provide all three types. Vertical clamping knurling tools are less prone to distortion of the workpiece, especially if it is long and thin, however, multiple wheel sets need to be changed to alter the type of knurling being undertaken, or multiple tools with each type obtained. Side entry knurling tools are flexible, and provide choice, but hard materials require a significant transverse load to create the indent and can lead to workpiece distortion on smaller diameter workpiece sections.

Threading tools are those that have a tool point ground to the form of a thread profile. They can be sourced as a carbide insert or be operator ground using a solid HSS tool. The can be used to either generate a thread profile by running the tool down the trailing flank of the thread and only cut on one face or to sequentially press the tool into the workpiece forming both leading and trailing flanks at the same time.

Form tooling again tends to be of two types. Tool inserts that are formed at standard angles are widely available and commonly used for forming chamfers, without having to resort to rotating the compound slide to create a generated chamfer. However, the second type tends to be more bespoke. While computer numeric controlled (CNC) machines can easily generate profiles, on manual machines complex curved shapes require tooling that matches the profile of the feature to be created. This type of tooling is predominantly made from HSS given the expense of having a carbide tool ground. In fact the cost of obtaining any form tool of any size would probably make outsourcing the work to a workshop with CNC capacity more economic. However, the grinding of a bespoke HSS tool bar into a form tool is a simple, albeit sometimes lengthy task. It is common for the tool to be roughed out on an off-hand grinder, then finished on a surface grinder fitted with a profile dressing tool allowing and extremely accurate profile to be generated. That said, hand grinding when undertaken with care can be quite precise, and it is the dimensional tolerance on the component drawing that may decide the approach taken. One issue that operators should be aware of is the length of cutting edge being employed at any one time. As form tools are pushed into a workpiece to take up the form of the cutting edge the entire length of the cutting tool edge is in contact with the workpiece, and as such is subject to high loads, heat build up and wear. Operators should be circumspect when selecting the size of machine to undertake the task on the depth of cut, and the cutting speed to ensure that the tool cutting edge is preserved, the surface finish is satisfactory and no chatter experienced.

3.3 INITIAL TOOL SET-UP

Turning down the outside diameter of a round or hexagonal piece of bar is the simplest form of lathe use. The output of this operation is a round diameter, irrespective of the initial shape of the workpiece material.

Prior to commencing any machining operation there are a few checks to undertake which relate to safety and to effective operation of the machining process. Firstly check that the area around the machine is free of debris and loose items to ensure that the operator has a secure footing. Check that all guards are in good condition and operate correctly and any vision panels are clean enough to see through. Interlocked guards are a good feature to have on any machine tool as this provides a second layer of safety in preventing inadvertent operation of the machine while the operators hands are anywhere near a part that rotates. Where interlocked guards are not present, isolating the machine at the main switch is good practice.

The next check is to ensure that the correct tool is properly and firmly inserted in the tool post and that it is at the correct height (i.e. the point in contact with the workpiece is set at exactly the same height as the centre axis of the machine). There are a number of methods of checking for centre height, some of which are rather tedious and some quick but require care.

The significance of having a tool on centre height is that it ensures that the designed tool rake and clearance angles are correctly presented to the workpiece and that any chip breaker formed into the tool works effectively. Having a tool set at the correct height also reduces the force required to cut into the workpiece, and anyone who has tried to use a parting tool set above or below centre height would recognise this. Finally a tool with a sharp cutting edge of the correct geometry for the material and operation, set on centre height and using the correct feed and speed rate will provide a fine surface finish. Tools set below or above centre height will always have a compromised surface finish.

Method 1 – Use of dead/live centre

Rotate the tool post such that the cutting point of the tool can be compared with the point of a live or dead centre as shown in Figure 3.4. Depending on whether the tool point is observed to be above or below centre height will drive adjustment of the tool holder. In quick release tool posts the tool is firmly clamped in a carrier and a height is adjusted by rotating an adjusting screw as shown in Figure 3.5. Operating the tool post locking clamp secures the carrier and the tool point can be checked against the centre inserted in the tailstock.

Where a fixed or indexing tool post is used the height of the tool is adjusted by raising or lowering the tool in its holder by inserting or removing shims. To do this the tool locking screws need to be loosened such that a selection of shims can be inserted or removed as shown in Figure 3.6. The accuracy that can be achieved by this method is constrained by the thickness of shim used (i.e. the thinnest shim is the smallest adjustment that can be made); however, care needs to be taken to have a variety of thicknesses as the least number of shims used generally provides the most stability. It is also important to firmly and evenly clamp the tool down onto its pack of shims when assessing centre height. Failure to do so, either leaves it below centre following final tightening of the tool clamping screws, or secure tightening at one end, particularly the cutting end while leaving the other loose, results in the tool moving above centre when the remaining screws are tightened.

If errors remain the process is repeated until it is assessed that the tool is on centre. On completion of this exercise the tool post is rotated back to its normal position ready for use.

Figure 3.4 Tool point rotated to be aligned with tailstock centre

Figure 3.5 Adjustment of quick release tool holder

Figure 3.6 Tool height adjustment using shims

Method 2 – Using a trial cut

Another approach is to have a look at the tool and see if it appears to be on centre. If so it can be used to face off a scrap piece of material and identify if a small 'pip' is left in the centre of the bar when the cut has passed from the periphery to the centre. The diameter of the 'pip' gives a clear indication of how much a tool is under centre as seen in Figure 3.7. Comparison of the tool point to the upper surface of the 'pip' or observed centre point where it has been torn off identifies the degree that the tool is over centre height. The method of setting the height is exactly the same as for using a live/dead centre where shims, or tool holder adjusting screws are manipulated to raise or lower the tool height until it is on the centre axis of the machine.

Figure 3.7 'Pip' left on workpiece when tool under centre

Method 3 – Use of a straight edge or rule

This method is among the quickest and most accurate, however, has the potential to damage the tool point (particularly carbide tool inserts) if excessive force is used.

A piece of round bar, or the workpiece to be used is inserted into the chuck. With the machine isolated, and the tool securely mounted in its tool holder it is then wound in towards the workpiece using the cross slide wheel. As it approaches the work piece the operator places a thin flat piece of material, typically the operators 150mm rule, between the workpiece and the tool point as can be seen in Figure 3.8. The cross slide continues to be wound in until it lightly grips the rule. The rule is then observed from the end of the machine.

Figure 3.8 Steel rule used to indicate tool on centre

If the upper part of the rule or flat bar leans away from the operator then the tool is above centre height. Conversely if the upper part leans towards the operator then the tool is below centre height. Adjustment is applied using shims or the tool post adjusting screw and the process repeated until the rule or flat bar is plumb vertical.

It is important that the bar used is not bent, and is required to be flat. The longer the bar is the better, however, if it is heavy due to thickness or width then the force required to have it stay in place raises the possibility of crushing or breaking the tool point. While a 300mm long standard engineers rule can be used, a 150mm lightweight rule is often the best solution having a suitable balance of reasonable length, flatness and least mass.

Following insertion of the workpiece in the chuck machining can commence. When a bar of material is inserted in a three- or four-jaw chuck they are rarely held exactly centrally due to the way that the jaws tighten onto the workpiece. The result of this is that the first cuts taken are rarely critical.

3.4 FACING OFF

While the vast majority operations on a lathe relate to reducing a diameter, sections of diameter, screw cutting or knurling; it is normal practice to face off a bar of metal to remove what is often a rough sawn surface, or the deformed end of a rolled bar. The reason for this is to provide an end face that is square to the axis of the machine, provides a pleasing end finish and more importantly provides a plane surface that can be drilled with a centre drill should a live or dead centre be required. Figure 3.9 shows an example of a bar that has been faced off.

While specific facing tools can be used, a standard tool used for external turning generally provides a perfectly adequate surface finish and rarely damages the tool due to the light nature of the cut. Following the insertion of the workpiece in the chuck and checking of tool height and setting of the correct speed, the hand wheel of the saddle and cross slide are rotated to bring the tool close to the corner of the workpiece. The machine is started, a small depth of cut (<0.75mm) set using the saddle hand wheel and the cross slide hand wheel used to pass the tool across the face of the material until the centre is reached. The tool is then removed and machine switched off. If the surface is uniformly smooth then the operation is complete. If rough or distorted material is observed the process is repeated until smooth across its entire surface.

Light cuts are utilised in this operation as its rarely required to take heavy facing cuts. Where a significant amount of material is required to be removed for a length of bar its often significantly faster to either 'part off'

Figure 3.9 Facing off

the material using a parting tool or parallel turn to length. However, use of a robust facing tool is often employed where heavier sections such as castings require machining. In these situations cuts of 0.75mm would be slow and attempts to take heavier cuts with a standard turning tool lead to the tool point digging in and can form a cone of material behind the cutting face leading to an unsatisfactory cut. Figure 3.9 shows facing off being undertaken.

3.5 PARALLEL EXTERNAL TURNING

Parallel external turning is the simplest, and probably most common operation performed on a lathe. It involves applying a tool point to a rotating workpiece to reduce its diameter. This can also be to produce a rounded section of a workpiece formed from rectangular, polygonal or other irregular shaped section. The key point is that almost without exception, it is normal to be producing a rounded section, or reducing an existing diameter, to a known finished size from a piece of bar be it round or irregularly shaped. The finished shape will always be a plain round diameter, and the centre of this diameter will always be that of the machines central rotating axis.

Establishing the initial size and then reducing a diameter to a known size is where some skill is required.

When a piece of material is inserted into the chuck of the lathe generally it is slightly off-centre, either due to wear, or some asymmetry, in the manufacture and operation of a three-jaw chuck. Alternatively the difficulty or precision encountered with getting a workpiece perfectly located in a four-jaw chuck. While collet chucks generally provide more precise centring, and turning between centres is accurate only if the initial centre holes have been precisely drilled, some error is likely to be seen and the surface presented to the tool will not be uniformly round. This is further compounded by roundness tolerances in the bar stock being used. While ground bar stock is generally quite excellent with respect to its roundness, many other materials will have a degree of asymmetry relating to its manufacture (i.e. hot rolled, cold rolled steels, and cast or extruded non-ferrous materials).

The surface of materials such as steels that have been hot rolled will have a hard and rough scale on the exterior. This can sometimes be difficult to machine requiring more power and a depth of cut to remove and where possible removed in one cut.

While in some machines the combination of work holding errors and roundness tolerances of the workpiece bar stock may be only 0.1mm, however, hot rolled bar and worn work holding equipment means that selection of workpiece material must be sized such that it allows for errors. Ideally this does not require substantial amounts of material to be removed to achieve the required diameter, however, it can occasionally be an issue. The result of this is that an allowance must be made for cutting material to the finished size and any selection of bar stock must be sized to allow for a minimum 'clean up' cut to be taken to establish roundness with respect to the machine z-axis.

Once the workpiece has been inserted into the machine it is ready for its initial cut. The depth of this cut is arbitrary with it being linked to the roundness of the workpiece stock, machine size, tooling size, tooling type (i.e. carbide or high-speed steel), surface finish and difference between the stock size and the finished size. Generally this initial cut is light enough to clean up the outside in one cut, and also establish a datum size before undertaking successive deeper and heavier cuts.

Having established a plain cylindrical diameter of a known size metal reduction to a desired size can be undertaken. Depths of cut are directly related to the stiffness and power of the machine being used and the material being worked. High carbon alloy steels, cast irons and stainless materials will have a lower depth of cut than softer materials such as many non-ferrous metals and plastics. The size and strength of tooling is also a factor as a small tool in a very large powerful machine would break before the machine if subjected to an excessive load. However, it can often be found that a peck depth of 2.5mm is used by many people when using any sized machines larger than those considered as 'hobbyist' where a depth of 1mm is often used especially if the machine motor has a duty cycle.

Example 1

Having set the machine speed to the closest setting for the diameter and material being cut, or on more advanced machines having set the surface metres/min rate, start the machine. Bringing the tool into contact with workpiece can be approached in a number of ways. Merely moving the saddle along the z-axis until the tool is perpendicular to the workpiece and then advancing the cross slide until the tool touches the workpiece and an annular ring is produced is probably the most common method. The saddle is then moved towards the tailstock away from the end of the workpiece and an initial cut set using the scale on the cross slide wheel, or x-axis on any digital readout. The tool is then moved along the z-axis using either the saddle hand wheel or automatic feed until close to the chuck, any depth mark on the workpiece or to a known length. The machine is then measured using a mechanical device such as a micrometer, digital/dial/vernier calliper, manual calliper and rule or the machine digital read out (DRO) if a machine zero datum has been set. Figure 3.10 shows a workpiece that has been marked with an annular ring.

Figure 3.10 Marking an annular ring

While machines may be capable of greater cut depths than those commonly used, removing material to a finished cut size in one pass is often not desirable as any errors in accuracy related to machine load cannot be assessed with the risk of cutting undersize. Also if the machine is not utilising significant amounts of coolant heavy cutting will potentially heat treat steels, but also burn both tooling and the workpiece leading to high levels of tool replacement and discolouration or deterioration of the workpiece material. As previously discussed the correct depth of cut is subject to many variables, however, the experienced machinist will be listening to the cutting action to determine if anything seems distressed, observing the chip formation i.e. a continuous or small chip production that is free of burning or blueing and looking at the surface finish being produced. Generally when a good surface finish is seen and chip production is consistent then the machine can be considered to be operating efficiently and within its capabilities.

Reasons for increasing the depth of cut are linked either to productivity where fewer passes to remove an amount of material result in a reduction of machining time (and cost), or where material properties such as work hardening demand fewer but deeper cuts.

It is also worth mentioning that almost without exception manual centre lathes take material from the diameter, so while a cut of 2.5mm may be applied on the cross slide dial, the actual depth of cut is only 1.25mm. The reason for this is solely due to rotation of the workpiece and the lathe effectively removes material from the diameter. Where DRO displays are being used to determine depth of cut the operator should check the default settings and be aware if the display is showing absolute dimensions, incremental depth of cut or resultant depth of cut etc.

Example 2

Using a general purpose carbide insert tool set a 2.5mm depth of cut and restart the machine and checking that the workpiece is rotating down onto the top of the tool. It is then passed along the periphery of the workpiece along the z-axis towards the chuck. This is achieved by using either the hand wheel controlling the saddle, or by engaging the automatic feed. The material is removed from the surface reducing the overall diameter, or forming a parallel cylinder from an irregular shaped object.

To reduce the diameter of a piece of 30mm mild steel bar to a finished diameter of 24mm of no set length a typical method would be as follows:

1 0.2mm clean up cut and measure to establish datum size (or use DRO).
2 Set cross slide scale to 0mm as shown in Figure 3.11 (Figure 3.12 where DRO used) and return saddle to just beyond end of workpiece.
3 Apply cut of 2.5mm as shown in Figure 3.13.

4 Apply second cut of 2.5mm then measure to determine actual size (or use DRO).

5 Apply finishing cut of 0.8mm (or exact depth to achieve 24mm).

Figure 3.11 Cross slide hand wheel scale set to zero

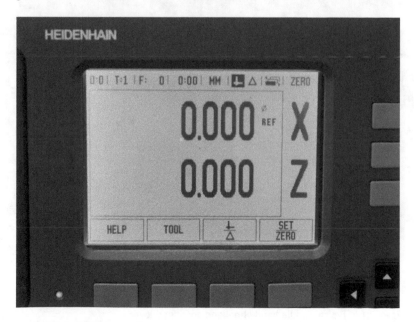

Figure 3.12 DRO set to zero on *x*-axis

Figure 3.13 Tool taking 2.5mm cut

Figure 3.14 Typical saddle handwheel and auto feed lever

Where automatic feed is utilised this is engaged at the start of each cut, and disengaged as tool reaches the end of the cut by engaging or disengaging the feed control lever. A typical example of a feed control lever is shown in Figure 3.14, however some machines may have a control switch located away from the saddle apron or may have no power feed mechanism fitted.

Where a parallel diameter has already been established but a zero or datum reading has been lost then it needs to be re-established. With a DRO this is fairly easy where an accurate measurement can be taken of the workpiece and an absolute value set on the DRO. However, where machines do not have a DRO alternative approaches have to be taken. One of the oldest of these is to use a wet cigarette paper applied to the surface of the workpiece. Many cigarette papers are .001″ thick and the cutter is then advanced until the paper is swept away by the tool. The tool point is then considered to be less than 0.001″ above the surface and the scales re-zeroed.

The surface finish that is achieved when turning can cover a wide range and in average applications values covering a range of 6.3–0.4µm are achieved, and while lower or higher values can be achieved those stated tend to be the average values. Quite fine turning can be comparable with surface finishes associated with grinding. However, where a surface finish is observed to be rough it can be due to one or a combination of several causes which include material properties, cutting speed, lubrication or coolant type, tool wear, depth of cut, machine stability, feed rate, or tool geometry.

Methods of improving the surface finish generated include one or several of the following actions:

- Check the tool tip for damage and/or excessive wear.
- Check that the cutting speed is in accordance with the recommended surface metres/min. Generally increasing speed assists with improvement of surface finish to a point and awareness needs to be maintained that as a diameter reduces the machine speed should increase given a reduced peripheral distance.
- Reduce the feed rate.
- Where low cutting speeds have to be used utilise higher quality cutting fluids or oils.
- Increase the nose radius of the tool or tool insert being used.
- Use tools with an increased relief angle.

If general cutting carbide tips are being used the depth of cut can have an impact on the surface finish produced. Generally carbide tipped tools like to be 'worked' and the appearance of a workpiece when excessively small finishing cuts have been applied can be less aesthetically pleasing. This occurs especially where a tip with a small nose radius is utilised, and resultant surface finishes can appear less bright despite the workpiece being of the correct size. A method to alleviate this is by ensuring that roughing material removal pecks are arranged such that the finishing cut is within 0.8–1mm in depth and this often produces a fine surface finish but has not been such a heavy cut that any spring back or error in diameter is produced, so the final size is within the desired tolerance.

It is not unusual for long lengths of continuous chips to be produced especially when there has been a confluence of free machining material properties, speed, feed and coolant type. These continuous chips can be of significant length if they are pushed down onto the cross side, onto the bed of the machine and proceed along the workshop floor. Lengths in excess of 10m have been seen from a basic workshop lathe, and while amusing and interesting, the continuous chip is swarf and as such is often razor sharp. This presents a safety hazard within the workshop.

While long lengths are not unknown it is more common for the continuous chip to form and then break when coming into contact with parts of the machine or other chips. This in itself is not a hazard but when a ball of swarf forms it gets pushed into the chuck it becomes a large rotating mass which damages the surface finish of the workpiece, tends to scratch an vision panels in the chuck guard and also smashes off sections which can be ejected from the machine towards the operator.

Techniques to reduce of remove these issues, can be as mundane as removing swarf between cuts to prevent a build-up, changing the feed to one that is less likely to produce a continuous chip, or utilise a tool tip with a more pronounced chip breaker groove. High Speed Steel tools do not have this feature and in some conditions continuous chips are difficult to avoid.

3.6 TURNING TO A KNOWN LENGTH

While it is common for turning to only involve reducing a diameter of material to a known diameter, the next stage of turning requires reducing the diameter of material to a known length to create a shoulder. The precision required to achieve a known length directly affects both the approach taken to setting a length on the machine, the type of tooling, and machine slide combination employed.

The simplest approach is to mark a line on the workpiece and then reduce the diameter up to this line. The line is either marked off the machine using a vee block and vernier, or digital, height gauge; on the machine using odd leg callipers, or simply utilising a rule and marker pen. The approach used links to the precision required for the length. While use of height gauges and vee blocks can be extremely precise, sometimes using a marker pen and rule can be equally adequate especially if the workpiece can be subsequently faced off a parallel surface in a later operation, such as that to remove a morse centre drill cavity.

The general approach to turning to a known length is exactly the same as that for parallel turning with the exception that the automatic feed or hand traverse of the saddle is stopped at a known point. It is normal for this operation to be undertaken in a series of cuts, either because the depth of material to be removed is greater than the capacity of the tool or machine, or because good practice suggests that a roughing cut, followed by a smaller finishing cut to size would be more prudent. A common effect of undertaking a series of cuts (or peck's), are a series of concentric witness marks on the plane face of the shoulder the diameter of which generally equate to each peck taken as shown in Figure 3.15.

The tooling used also directly affects the geometry of the shoulder produced. Tooling with significant side cutting edge and lead angles will produce an angular shoulder. Tooling with a negative side cutting edge angle has the capacity to create an undercut which is not always desirable. This occurs when the visible rear section of the tool tip is being observed as the material is cut and the negative cutting edge angle is such that the tool point, which is generally invisible when roughing, digs further into the material. Figure 3.16 illustrates aspects of lathe tool geometry.

The solution to both the build-up of concentric rings, the effect of positive cutting edge angle, positive lead angle, and negative cutting edge lead angle is to turn down to a length but provide an allowance for finishing to depth. For tooling that has a negative cutting edge lead angle this is quite simple and merely involves undertaking what is essentially a facing operation. The amount of material to be removed is rarely significant as it is relatively uncommon to find tooling with a negative cutting edge lead angle of greater than 5°. More importantly, the majority of material to be removed is located at the outer diameter not the inner. While it is tempting to push a tool into the smallest diameter and corner of the workpiece then withdraw the tool to cut a perpendicular shoulder, this is bad practise especially when using indexible carbide inserts as it relies on the clamping mechanism of the tool holder rather than pushing the insert into its seat. Where a tool with positive edge cutting and lead angles have been used there is a greater build-up of material towards the centre of the workpiece. The only method to removing this is to change the tool to on with a negative or zero angle. This may be either by using another standard turning tool with suitable cutting edge angles, a parting tool, assuming the material to be removed is within the width of the tool, or by using a bespoke form tool. However, it is likely that if this approach is taken it is probably going to be combined with an additional operation such as forming an undercut in compliance with a standard.

When undertaking a turning operation that involves machining to a diameter and length that results in the generation of a shoulder, consideration should be given to the type of tooling used and whether the use of particular tooling will require additional machining operations. While there are good reasons for using tooling with positive edge cutting angles and

Figure 3.15 Witness marks on workpiece shoulder

lead angles, particularly those relating to reduction of cutting forces, these tend to be more applicable in a production environment where material removal rates are a significant factor. In manual machining this is generally less significant and there is a trade-off between material removal rates and

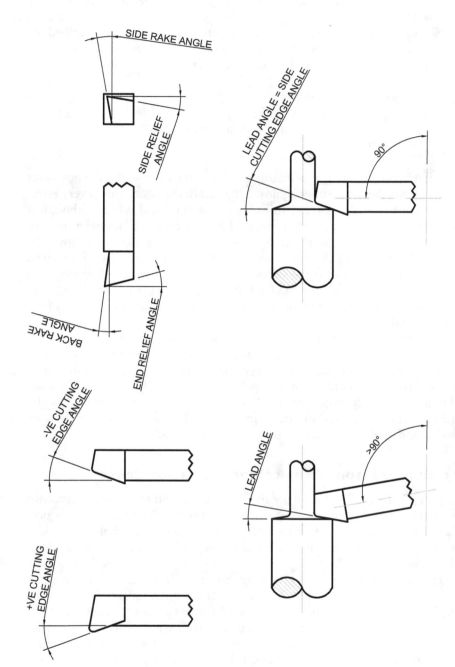

Figure 3.16 Diagram showing positive and negative lead and edge angles

the addition of operations to remove an angular shoulder resulting from the tool geometry used.

While turning material down to a length and diameter to a scribed, or otherwise marked line, is surprisingly accurate; often a greater level of precision is required. The approaches to achieving this are fourfold:

- bed stop set using callipers;
- bed stop set with gauge blocks (slips);
- gauge blocks (slips) and dial test indicator; and
- use of DRO.

The intention of setting a Bed Stop is to provide a physical reference point and stop for the saddle moving down the bed of the lathe. This is very effective when removing material utilising the saddle hand wheel, although if the machine automatic feed is engaged almost all machines are powerful enough to overcome the clamping force and push the bed stop down the bed of the lathe and lose any datum set. Where automatic feed is engaged close attention is paid to the reducing distance between the bed stop and the approaching saddle and the feed disengaged just before the mating surfaces touch, with the remaining small distance undertaken by rotation of the saddle hand wheel.

Bed stops are normally clamped to the bed of the lathe between the machine headstock and the saddle using one or more clamping screws. The bed stops can be of a simple plain type or incorporate a vernier adjustment screw. Irrespective of the type the operation is essentially the same. They are clamped at a set distance from a reference point on the workpiece (typically the end, or an existing shoulder). The methodology for setting this distance varies, and again, is dependent on the precision required.

Setting bed stops using callipers or depth gauge

One approach can be placing the tool in close proximity of the workpiece to be machined and then traversing the saddle towards the machine headstock. Callipers placed against the end of the workpiece or workpiece shoulder is pre-set and the tool is moved until the depth and tool point coincide as shown in Figure 3.17.

This approach is simple and easy, however, is limited in accuracy to the resolution of the calliper used, which for vernier scales, will at best be 0.02mm or 0.05mm. It is more likely that operator competence and their ability to hold a measuring instrument very still while observing a precise alignment between the end of the calliper and the tool point will significantly degrade this accuracy further, although accuracies of between 0.1mm and 0.2mm can easily be achieved with a skilled operator. Another factor degrading the accuracy of this approach will be the tool point radius. Any

Figure 3.17 Depth gauge of calliper used to set length of cut

tool point radius greater than 1.5mm introduces an additional problem as the end left-hand edge of the tool point moves away from the surface of the workpiece creating greater handling difficulties.

Irrespective of this following positioning of the tool point in the desired longitudinal position the previously loosened Bed Stop is slid up the bed until it is placed against the saddle, and then locked in place by tightening of the Bed Stop clamping screw(s) as seen in Figure 3.18. The depth is then set.

Setting bed stops using slip gauges

A more accurate approach involves the use of gauge blocks (slips). In this case the tool point is placed accurately on the end of the workpiece to be turned, or a selected shoulder. The loosened bed stop then has a pack of slip gauges placed adjacent to the saddle, or saddle register, and the bed stop is slid down the machine bed until the slip pack is trapped between the two and the bed stop clamping screw(s) are tightened as shown in Figure 3.19. The accuracy of this method in perfect conditions is limited to the accuracy

Figure 3.18 Bed stop secured in place

Figure 3.19 Gauge block pack used to set bed stop on known depth

of the gauge blocks used, which in almost all circumstances are sub-micron. However, handling errors, and more likely, minute movement of the bed stop as the clamping screw(s) are tightened will reduce this accuracy as will the machine parameters e.g the amount of machine backlash present. That said, this method provides extremely accurate setting of the bed stop such that diameters can be turned to lengths with an accuracy in the region of 0.1mm or better.

Setting bed stops using slip gauges and utilising vernier adjustment

Building on setting of bed stops using a slip pack, an alternative approach is to set the bed stop slightly short of the length required and where fitted utilising a vernier adjustment where fitted to the bed stop. The approach taken is to set a depth approximately0.1mm less than the depth required, or one matched to the machine dynamics and take a cut. The depth of this cut is then measured accurately and the vernier adjustment rotated an exact amount to increase the length of cut on the next pass. Figure 3.20 shows an example of a bed stop fitted with a vernier adjustment screw.

Figure 3.20 Bed stop with vernier adjustment screw

Setting bed stops using slip gauges and utilising compound slide adjustment

Another approach to producing a turned diameter to a known length with accuracy is to utilise a bed stop set to a known depth, again slightly short of the length required, and for machines that do not have a bed stop with vernier adjustment fitted, use the compound slide to provide an accurate cut to length. In this case the length required is set slightly short of that required and a cut, or cuts, taken until a shoulder is generated. The depth of this shoulder is then measured and the final length set by rotating the compound slide adjustment handle by an amount indicated on the handle scale.

An alternative to this, or for an extremely precise setting, is achieved by attaching a dial test Indicator to the bed of the machine and placing the indicator plunger against the tool post as shown in Figure 3.21. This is then used to measure the lateral displacement of the tool post as the compound slide handle is rotated until the set length to a desired value is achieved.

Digital read outs

The installation of digital read outs (DRO) has removed much of the tedium of setting up machines to turn lengths and diameters to levels of accuracy up to those of the machine, however, there is much to be said for setting lengths using a DRO but still installing a bed stop to a correct length. Especially when higher feed rates are being utilised, reading a rapidly changing DRO display can be difficult, and stopping a machine short of any perceived distance will in most likelihood leave a ring on the workpiece. The fitting of the bed stop provides a fixed visual reference marker for the operator to assess when they are close to a required length and take timely action to prevent the machine from powering beyond a desired length.

Issues and undercuts when turning to a known length

Rough turning is often focussed on significant material removal and generally utilises higher feeds and speeds than those used in following finish cuts. The danger is that momentary inattention leads to the machine pushing through a bed stop, and any effort to precisely set a length is made worthless, and focus retained. Setting the machine to cut slightly short and then undertaking a finishing plunge cut to the length required often negates this danger.

Of additional consideration is the root corner of the shoulder. The shoulder corner radius will match that of the tool used to perform the cutting operation. For tools that have a significant radius the mating shoulder radius may require removal to allow fitment of a mating part. Where small tool radii have been employed this may not be a problem or may have been ameliorated by the mating part having an internal chamfer of the tool nose radius (or larger). However, if a square corner, or thread undercut, is a requirement then cognisance

Figure 3.21 Fine adjustment of tool position using compound slide and dial test indicator

needs to be taken of this when determining what method is employed to set a length, as any subsequent precise facing or undercutting operation used to generate the finished length to the required precision, may remove any requirement for precision when setting the length of roughing and finishing cuts.

The selection of cutting tool material and tool geometry is complex and often driven by the material type, rate of production, depth of cut and part configuration. Awareness of the resultant shoulder corner and machining operations to provide a suitable geometry needs thought to provide the most efficient selection.

3.7 USE OF REVOLVING AND DEAD CENTRES

Two types of centre are used to support work in turning, and the two places these can mounted are the machine headstock or the tailstock. By far the most common location is a centre fitted to the tailstock, and as shown in Figure 3.22 these are used to support work such that it does not deflect away from the tool when cutting operations are being undertaken. This support ensures an accurate parallel cut, removes the potential for chatter marks or poor surface finishes, and protects the tool from damage. The length that extends from the face of the chuck can vary with the diameter and stiffness of the workpiece material, but a rough estimate of a length to diameter ratio of about 6 can be used as a guide. That said, very small diameters would require a very small centre, and the machine size may be an issue before being able to support a workpiece with a centre becomes a major issue.

Figure 3.22 Revolving (live) centre mounted in tailstock

Generally only one centre is used although turning between centres, is somewhat self explanatory and requires a second centre fitted to the machine spindle in the headstock. As shown in Figure 3.23 there are two types of centre used; known as a 'dead' centre' or a 'live' centre.

Live, or revolving, centres most common, and most useful, as they can cope with high loads and high rotational speeds. Live centres comprise a rotating tapered shaft fitted within a barrel which is fitted with a thrust bearing. They are generally sealed for life and require no lubrication. The centre has a tapered mounting shaft which allows it to be inserted into the machine Tailstock using a morse taper matched to the machine or matching sleeve fitted to smaller sizes.

A variety of sizes can be purchased which can fit a wide range of machine sizes. Selection of live centres is generally quite simple with only two considerations being of major importance: Morse taper size, and size of the centre nose radius. All centres be they live or dead, have some degree of

Figure 3.23 Fixed (dead) centre mounted in machine spindle

nose radius, as a perfect sharp point cannot be maintained, given handling considerations and the risk of stab wounds to the user. Therefore, a radius, or truncation is provided. The size of this radius or truncation is matched to the size of the centre with small centres designed for model making machines being significantly less than those provided for a large production lathe. The size of morse centre that needs to be drilled into the workpiece to ensure a secure support is matched to the centre being used.

A large centre being inserted into a small workpiece that has been drilled with a small centre drill does not provide a secure support. Other than ensuring full engagement of the centre in the workpiece, and that the rear edge of the cutting tool is not so large that it impacts the rotating portion of the centre preventing engagement with the workpiece, there is little wrong with using a large centre on a small diameter workpiece. However, this is not strictly true for the opposite. Small centres designed for smaller machines can be sleeved to fit a significantly larger tailstock. Small centres may not be designed for the loads imparted on them when fitted on larger machines.

While rotational speed is unlikely to be an issue, as smaller diameters, mainly drive higher rotational speeds, longitudinal and transverse loads can easily outstrip the design characteristics of the centre. A 'rule of thumb' is to use a centre that either fits into a machine Tailstock directly or only requires a sleeve of 1 or 2 morse taper increments. This approach ensures that all work can be accommodated, although with care and minimal loads applied to a centre, quite small centres could be used in large machines, but significant care must be taken.

Centres that are classified as 'dead' are solid and have no moving parts. The advantage of this type of centre is that they are extremely robust and provide the highest level of precision as there are no moving parts for manufacturing tolerance to be an issue. The disadvantage of this type of centre is that the interface between the point of the centre and the centre cavity drilled into the workpiece requires lubrication. Lubrication is normally achieved by using a suitable grease, although in previous decades, tallow or similar soap like compounds were used.

When using dead centres the pressure applied to the centre from the Tailstock needs to be managed such that the force does not displace all of the lubricant from the cavity leading to a metal to metal contact. However, using too little force either lets the lubricant leak out, or allows for movement. Use of dead centres needs attention to be paid to periodically refilling the centre cavity with lubricant.

Failure to effectively lubricate the cavity normally leads to the point friction welding into the workpiece and then shearing off the centre. This happens at an earlier point if high rotational speeds, coupled with significant pressure from the Tailstock is imparted on the centre. Use of dead centres has happened as the cost of revolving, or live, centres has fallen. The migration from use of high speed steel tooling to carbide tooling substantially increased cutting speeds and improved productivity, but effectively made dead centres fitted to a tailstock obsolescent. The almost universal adoption of this type of tooling has resulted in a significant increase in the rotational speeds the machines are run at, which quickly leads to failure of a dead centre.

It is not normal to use a live centre in the headstock as this has no relative rotation with the workpiece and as such there would be no reason to have a revolving centre. That said if it is the only option available then, if it fits and does not interfere with the workpiece drive mechanism, there is no real reason not to use one.

The advantage of dead centres is the simplicity and lack of moving parts, which removes potential for inaccuracy, however, they all require lubrication with grease or tallow etc. Failure to do this, and applying excessive loads or rotational speed generally results in the friction welding of the centre into the workpiece or the failure of the centre and consequent destruction of the workpiece.

3.8 PARTING OFF

Parting off is the operation used to make a transverse cut through the diameter of the workpiece such that it is separated from a section of bar stock being held in the chuck. Many machine operators have feared parting off given a number of parting failures but focus on tool condition and confidence make it a simple process.

Modern carbide tooling typically fitted as an insert into a blade type holder has transformed parting operations not least because a narrow tool tip can reach to a significant depth. Figure 3.24 shows examples of both

Figure 3.24 Blade type and carbide insert type parting tools

carbide and HSS blade type parting tools. The traditional high speed steel tool had to be either of a very substantial size to avoid weakness at the tool point root, or limited in the depth it could reach into the workpiece. While parting off workpieces into an internal cavity was generally not a problem, parting off solid workpieces over 20mm diameter required a much bigger tool to reach to the centre of the workpiece and often people resorted to sawing off the workpiece manually, then facing it off to length.

Irrespective of the tool type used correct set-up is essential. The tool must be on centre height, and be set at 90° to the machine x-axis. This ensures that as the tool plunges deeper into the workpiece it is not bent by the shank coming into contact with the workpiece and allows for efficient swarf clearance. Speed settings are less critical for carbide inserts than those for HSS tooling given their tolerance. However, consideration must be given to speed settings. The peripheral surface distance changes markedly as the tool plunges into the workpiece, which when the centre of the work is reached becomes zero.

Machining materials, especially steels and complex alloys, requires a little thought. For machines where the control system manages speed automatically one a surface speed rate has been set into its control panel, rotational speed rates can be ignored as the machine automatically speeds up as the tool point advances towards the central axis. For manual machines, one approach is to set the rotational speed to the correct setting for a diameter approximately one third of the way into the workpiece, which means that some overspeed is experienced at the surface, and less than optimum as the tool moves into the material, however, it is a pragmatic approach. Alternatively, stopping the machine mid-cut and increasing the speed corrects any problems that may be experienced.

Correct speeds and feeds are crucial for many machining operations, but parting off a significant range of engineering steels and non-ferrous materials has little problem given the tolerance of modern tooling. However, some materials such as stainless steels or titanium alloys can be sensitive to quickly work hardening and as the tool plunges deeper into a workpiece effective reduction of surface machining rates increases the load on the tool and work hardening may quickly develop to a point where further progress is impossible.

The use of significant amounts of coolant not only ensures the tool and workpiece remain cool but also assist with swarf removal from what is a confined space. Tool life is also extended by advancing the cutter into the workpiece at a steady rate such that swarf is generated at a steady rate and work hardening from periodic rubbing is eliminated.

Excessive tool overhangs are often tempting, as the close proximity of tool posts to the workpiece restricts the view of the cutting edge, however, this can leave the blade of the tool holder unsupported leading to chatter, or tool failure.

When parting it is also important not to support the workpiece on a tailstock mounted centre. The tailstock provides an axial loading, but more importantly, prevents the workpiece from moving away from the cutter and workpiece stub held in the chuck, and almost without exception then gets damaged as it tries to fall away from the driven rotating section. For larger workpieces with a central cavity, it is not unusual to mount a suitable substantial bar in a tailstock chuck or morse taper. This bar has significant clearance from the workpiece cavity an is intended to prevent the parted off workpiece dropping onto the bed of the machine, damaging the slide way or workpiece.

3.9 UNDERCUTTING

Undercutting can be a similar operation from parting off but differs from in that it is undertaken to a known depth, leaving the workpiece as a single entity, and is most commonly used to remove any corner radius left after parallel turning which has created a shoulder. This can also be a more complex operation creating a specific profile, in particular to provide a runout for screw threads which approach a shoulder. The specific form for this type of undercut is normally specified in national standards, with BS1936–2:1991 'Undercuts and Runouts for Screw Threads: Specification for ISO Metric Threads' being a common example.

A simple undercut is one that is generated using a parting tool of suitable width. Common carbide insert parting tools are in the region of 3mm wide and this is generally acceptable for anything other than finer or small diameter work. When the undercut is being used merely to remove a corner radius then undercutting to a depth of less than 1mm is generally acceptable. Depths in excess of the width of the cutter being used would be unusual as this involves removal of a significant amount of material and will weaken the component.

Example – Simple undercut

Manufacture of a simple undercut is undertaken in exactly the same way that an operator would approach parting off. The correct speed setting is selected and the tool tip placed adjacent to the corner being undercut. The machine is started and the cross slide used to advance the tool into the workpiece until the desired depth is reached as shown in Figure 3.25. It is worth mentioning that when using the cross slide scale the increments relate to removal of material from the diameter of the workpiece and not the depth of plunge. Therefore, to create a 1mm deep undercut the cross slide must be advanced 2mm on the cross slide scale.

If using a carbide insert mounted within a blade type tool holder there are little clearance issues to consider. However, if using a HSS tool then a check

Figure 3.25 Undercut created by plunging parting tool into workpiece

must be made to ensure the tool tip will be able to plunge to the desired depth without any offset of the tool from grinding resulting in the shank of the tool holder impacting on the external surface of the workpiece. Generally, trying to generate an undercut in a shoulder with a step depth of greater than 10mm risks this, and careful checking of tool geometry is prudent.

Example – Generating undercut for thread runout

Manufacture of an undercut to act as a screw thread runout can be undertaken in two ways. Firstly is by use of a form tool. This is where a tool, normally HSS has been ground to the exact profile required. This can be a time consuming, or if contracted out expensive process, as small undercuts for thread runout can have some very precise requirements with respect to corner radii. However, use of form tooling ensures that generation of the undercut is undertaken with a plunge cut undertaken using exactly the same method as a simple plunge cut. The only drawback to this method is tool life and occasionally, experiencing poor surface finish.

The second method is significantly more complex. Firstly either selecting a carbide insert with the required nose radius, or grinding a HSS tool to the correct corner radius is required. This is inserted in the tool post and the tool post and compound slide rotated such that the correct ramp on angle is achieved but the tool cutting edge is at 90° to the machine x-axis. Using

Figure 3.26 Complex generation of undercut

the compound slide wheel the tool is advanced at an oblique angle until the correct depth is achieved. The saddle is then traversed until the tool reaches the shoulder. The undercut has then been correctly generated. Figure 3.26 shows an example of this.

3.10 CHAMFERING (FORMED AND GENERATED APPROACHES)

Chamfers generally have two purposes. The first being to remove a sharp edge and improve handling and the second is to provide protection to a feature such as a thread runout or knurled surface. In keeping with this there are two approaches to producing a chamfer; one being formed, and the second being generated. Both of these approaches have their merits and also they both have limitations.

Forming a chamfer is relatively straightforward and has an advantage of speed and simplicity. Forming is where the cutting edge of the tool is ground or adjusted to the correct angle. These angles vary, however, the vast majority tend to be either 45° or 60° and it is not uncommon to have tools set at these angles. Generally HSS is used for this type of tooling as carbide inserts are rarely provided with a 45° or 60° positive cutting edge angles, and self-grinding HSS steels to the correct angle and size is a relatively simple task.

To form a chamfer the cutting tool is inserted on centre height in the tool post and then using the cross slide move the tool obliquely into the desired edge of the workpiece to form an angled corner. Chamfers are usually dimensioned on a drawing by showing a longitudinal, or trans axis, displacement and angle (i.e. 2.0 × 45°). The depth that the tool is plunged into the workpiece for a 45° chamfer will require twice the length of chamfer required. This is again because a lathe has a scale that removes material from a diameter, however, the normal method of dimensioning is either a displacement from the surface, or more typically, a longitudinal length and angle. Therefore, to generate a 2.0mm × 45° chamfer the tool would have to be advanced into the workpiece by rotating the cross slide handle a distance of 4mm. Figure 3.27 shows the generation of a 45° chamfer on the corner of a workpiece.

Figure 3.27 45° chamfer generation

Producing a chamfer by this method provides an easy way of establishing a known length to the feature, and is especially useful for 45° chamfers. For chamfers that are of a less typical requirement, or one that is 60° the principle of forming the chamfer is exactly the same, however, car must bc takcn to ensure the tool has been ground, or set to the correct angle such that the long length is correctly orientated. In the case of these angles it should be noted that the depth the tool should be advance into the workpiece will almost certainly be less than the dimensioned length of the feature. This requires a simple calculation to be undertaken to determine the plunge depth, and an example is shown in Figure 3.28.

While forming a chamfer using a tool set at the correct angle and utilising a plunge cut to form is a quick method for the feature formation, and provides the easiest technique for creating a chamfer of the desired length, it has some drawbacks. Forming a chamfer requires the cutting edge of the tool to be sharp and free of any defects. If the tool has been rough ground then grinding marks will be replicated in the surface of the formed chamfer. If the chamfer is to be of any appreciable size then it is not unusual to get chatter marks evident in the surface of the formed edge. This is because the loads being imparted onto the tool and workpiece are high, and substantial tooling and machine rigidity become a factor. Where larger chamfers are required, a generated approach can be a better methodology in order to provide a well-formed feature with a good surface finish.

Chamfers that are produced using a generated approach require set-up that is a little more complex that using a formed approach. However, size

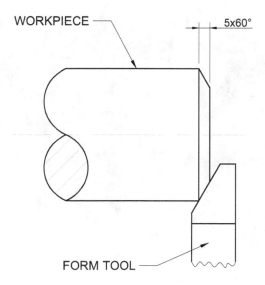

Figure 3.28 Calculation of plunge depth for formed unequal chamfers

of the chamfer is less of a problem and it is also the best method for producing chamfers on a workpiece that is not round such as hexagonal or square bar.

To produce a generated chamfer the compound slide is brought into use. To create a chamfer with a dimension of 2.0mm × 45° the compound slide is rotated counter clockwise until the 45° graduation on the base of the compound slide is aligned with the datum mark on the cross slide. With this rotation the tool post remains in alignment with the tool post and it is not uncommon for the travel of the cross slide to such that the tool cannot reach the work piece. Moreover, even if the cross slide does facilitate reach then it is also not unusual for the handle of the cross slide to become entangled with the machine coolant pipe, preventing rotation and providing a cramped area for the operator to turn the handle. The solution to this is to rotate the tool post through approximately 45° such that the cutting edge is presented to the workpiece in a normal orientation, which while used to generate an oblique cut, is rarely an issue. Rotating the toolpost ensures that the cross slide has enough reach to allow the generation of a chamfer on extremely small diameter workpieces and ensures there is enough clearance between any coolant pipe and the compound slide handle as shown in Figure 3.29.

Figure 3.29 Generating chamfers allowing for cross slide travel and coolant pipe clearance

To generate the chamfer, a cut depth is applied using the cross slide handle, and the compound slide handle is used to move the tool point obliquely across the corner of the workpiece.

While generating a chamfer has many advantages over the forming approach, identifying the correct length of chamfer by direct displacement of the cross slide is much harder. When forming a chamfer the tool has a straight cutting edge and it is easy to position this exactly on the corner of the workpiece. General cutting tooling is used to generate chamfers and as such, invariably has some degree of nose radius on the tool, even if it is quite small. Setting a radius nosed cutting tool adjacent to the corner of the workpiece at the correct tangent is difficult and takes time. The alternative is to set the machine up, take a small cut and measure the distance generated along the workpiece. Repeated small cuts are then used until the correct length has been achieved. Measuring these distances is often not easy, especially when the workpiece chamfer is small, and consideration of the tolerance required may affect the method used to machine the chamfer.

The set up described above deliberately uses a rotation of the compound slide counter-clockwise. While it may be possible to rotate the compound slide clockwise and successfully generate a chamfer it requires a right-handed tool to be fitted into the tool post, or a standard left-handed tool used to undertake a facing operation. For larger chamfers it is good practice to use tooling to cut on its designed cutting edge, and setting up another tool is an avoidable cutting operation.

More seriously, and the principal reason for avoiding a clockwise configuration, is that it puts the operators hands close to a rotating chuck. It also potentially prevents the chuck guard from being fully down, and has the risk of tools fitted to a multi-tool tool post coming into contact with the rotating chuck jaws. While a counter-clockwise configuration requires reaching across the machine, it is at a point away from the workpiece and chuck. As such this is a significantly safer approach for the operator to take.

For chamfers other than 45° such as the 30° chamfer found on bolt heads the principle remains the same for either approach. However, have awareness of where the angle is dimensioned from on the drawing as generating a 30° chamfer for a bolt head will require rotation of the compound slide to 60° to create the correct tool path angle.

The creation of chamfers on non-cylindrical workpieces such as hexagonal or square workpieces often only has an angle and no chamfer length stated. They are often best created using a generated approach and normally the chamfer is at the correct size as soon as a complete circle is witnessed at the end of the bar. In the case of a bolt head the corners of the hexagon are removed and as soon as the chamfer elements meet up in the centre of the flats to form a complete circle the chamfer is complete as shown in Figure 3.30. This operation is best undertaken in a series of small cuts as a small rotation of the cross slide handle will provide a significant length of feature on the workpiece.

Figure 3.30 30° chamfer for bolt head

3.11 DRILLING, INCLUDING TAILSTOCK SET-UP AND DRILLING LIMITATIONS

The tailstock of the lathe has two principal functions. Supporting long workpieces using a Revolving, or Dead centre, and providing a mount to facilitate drilling operations into the end of the workpiece.

The method of mounting drills can be achieved in two ways. Firstly directly inserting a morse tapered drill into the centre of the tailstock quill. Smaller drills will require a small taper sleeve(s) until the taper of the drill matches that of the machine. However, it is normal to use parallel shank drills, generally up to 12.7mm diameter within a three-jaw Jacobs type chuck fitted with a morse taper shank. While larger chucks are available, larger diameter parallel shanked drills are rarer, and the mix of larger morse tapered drills and smaller parallel shanks is very common.

Most tailstock quills have a scale etched or engraved into their upper surface. In order to insert a morse taper drill or chuck, the quill must be advanced using the tailstock handle until the zero mark is visible. This is because to eject a tool from the tailstock it is wound back until a pin engages with the tang of the tool and breaks the taper free. To insert a tool or chuck the tool is held such that the tang is horizontal and firmly rammed into the tailstock quill. If it does not rotate or extract when lightly pulled it is properly engaged and is ready for use.

Drilling is more often than not undertaken using HSS tooling. This means that cutting speed is more important than for carbide tooling as its less forgiving of abuse. Many manufacturers of twist drills provide small handbooks which contain, among much useful information, tables of cutting speeds for common diameters and surface cutting rates. It is important for a drill to cut well and clear swarf that it is used at its designed speed. Excessive speed and lack of cooling rapidly destroys the drill cutting edge, and often work hardens the bottom of the hole. In tool steels this can often result in significant hardening at the bottom of the hole cavity which cannot be drilled through.

Example 3

To calculate the cutting speed for a Ø10mm HSS twist drill used on mild steel:

Material cutting rate for mild steel = 27 surface m/min

Firstly convert Ø10mm to m.

Ø10mm ÷ 1000 = 0.01m

Using π × D calculate the peripheral distance:

= 3.142 × 0.01m
= 0.03142m

Then to find the number of revolutions/min, divide the material surface rate by the drill peripheral distance:

27 sfm/min ÷ 0.03142 = 859.33 rpm

It is unlikely that an exact RPM can be set, especially where gear selection gearboxes are fitted. In this case the nearest speed is utilised. Where the calculated speed is close to the middle of two gear ranges, then with larger sizes it can be advantageous to select the lower speed to assist with coolant entry and swarf removal. For, more modern or complex machines it may well be possible to set an exact speed.

To drill a hole in the end of the workpiece it needs first to have an initial hole generated using a morse centre drill. The very centre of the workpiece has no effective rotation and a twist drill with its 118° cutting angle will fail to bite into the material and in smaller sizes, can skate about on the surface until breakage. Therefore, a morse centre drill is used. This is specifically designed for creating a hole in a plane surface and its nose geometry and 60° conical shoulder provide the perfect starting hole for a drill.

The selection of drill length (i.e. stub, jobber, or long series) depends on the hole to be drilled. While rigidity is key to drilling holes at a precise diameter, stub drills are not often seen on machines.

Once the drill has been firmly inserted into the chuck or quill, the tailstock is slid up the bed until the drill is adjacent to the end of the workpiece and firmly clamped in place using the tailstock integral clamp lever. Ensure that any quill clamp fitted to the tailstock is released to allow free movement. Using plenty of coolant the tailstock quill handle is rotated to advance the drill into the workpiece. At sizes below approximately⌀14mm it is not unusual if smoothly rotating the tailstock quill handle to generate a continuous chip. These can be of significant length and are often razor sharp. The length can be controlled by momentarily pausing the advancement of the drill into the workpiece. The depth a hole can be drilled to is limited to the length of the drill. While jobber drills can be changed to a long series to increase the depth of drilling care must be taken not to attempt to drill beyond the length of the flutes as swarf has no egress route and coolant cannot enter.

Drilling to depths beyond 5:1 diameters can cause problems if a high degree of accuracy is required. For deep holes and ones requiring precision then utilising gun drills can provide a hole up to 50:1 diameters. While gun drills are capable of significantly greater depths getting coolant to the tool tip requires pumping, and if not delivered at the correct rate results in it turning to steam. Drilling to an appropriate length using a standard drill rarely presents a problem and if high surface finishes, or accurate sizes are required then reaming can be a consideration. For larger sizes post drilling boring is often a successful approach.

3.12 BORING

Boring is a cutting process on a lathe that requires the enlargement of an existing hole to provide a round diameter either completely through the workpiece or to a set depth. It is undertaken using a left-handed tool with a cutting tip set at 90° to the machine axis. They can be HSS or a carbide insert mounted into the tip of a boring bar. It is difficult to plunge cut into the end of the workpiece and can cause raised edges and difficulties with the tool holder rubbing the end face of the workpiece, therefore, it is usual to drill a hole which is enlarged by boring.

Drilling is often a faster method of creating a cavity than boring, but can only create diameters equivalent to the drill size and the bottom of the cavity will always have a drill point profile. However, drilling a single hole with a drill that approaches the required finished diameter, or series of stepped holes in a tapered cavity can often be an effective method of fast material removal that is finished by boring.

The process for boring a cavity is exactly the same as that for parallel turning, or parallel turning to a depth with one key difference. When applying a cut the distance is a subtraction rather than addition. While the cross slide handle is used to set the cut, rather than rotating it clockwise to advance the tool into the workpiece it is rotated counter clockwise. This in itself is not particularly significant, however, requires concentration by the operator to get into the mindset of advancing the tool away from the material centre, and making sequential cuts.

To control the depth that a cavity is bored to, is undertaken in exactly the same way as for parallel turning to a depth (i.e. use of bed stops, DTI or DRO). Tool geometry and tool overhang is the significant factor to successful boring of a cavity. The diameter of the starting hole, size of the boring bar diameter of the cavity to be produced and its length are the controlling factors.

The principal limiting factor is the diameter of the hole to be produced, and is closely linked to its depth. Boring by its very nature requires a tool to overhang the tool post often by a significant margin. This leaves it unsupported and as such subject to displacement or bending under load and chatter. Generally, the larger the size of boring bar the stiffer, and the deeper a hole can be bored out. There is a close relationship between tool size, diameter and depth of the cavity to be produced. As material removal is being undertaken with a cavity consideration needs to be made of the initial diameter being bored into as the bottom of the tool can interfere with the workpiece and the side opposite the tool point can impact onto the wall of the cavity. Figure 3.31 shows constraining aspects of the cavity and tool.

The length of hole being bored is the second limiting factor as the length of tool overhang must be at least the depth being bored. This is less of a problem where large diameter holes are being machined as substantial boring bars, often oil damped to reduce chatter and vibration, can be used. The problem is more acute where deep holes of a relatively small diameter have to be machined. This problem is less acute if pre-drilled holes can remove the bulk of the material leaving only a limited number of finishing cuts to be taken. However, tool constraints will determine that there is a minimum diameter and a maximum depth that can be machined for each size of boring bar.

The advantage of using drills to remove the substantial majority of material prior to boring is that it allows the use of larger boring bars. While it seems obvious that larger boring bars are more rigid than smaller ones,

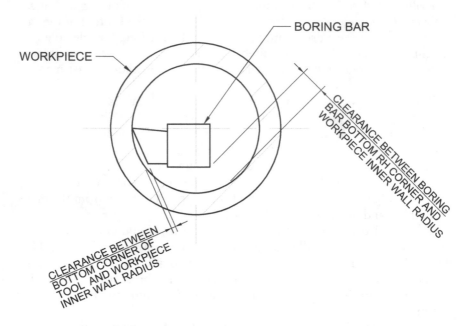

Figure 3.31 Boring bar cavity constraints

therefore, capable of boring out deeper cavities, Figure 3.31 identifies the problems of tool bar collision in smaller diameters which restricts the choice of tool. It is often a tedious, but necessary, task to commence boring with a small tool enlarging a small pre-drilled hole and then change to a more substantial tool to allow heavier and deeper cuts without associated chatter and swarf egress issues.

Boring a plain parallel cavity has few issues and is in itself no different to parallel turning, or parallel turning to a depth. However, the lack of visibility of the tool point requires careful understanding of any DRO data or close observation of bed stops. Following, use of the cross slide to establish contact with the cavity it is usual to take a light cut to provide a reference diameter and allow the cross slide scale, or DRO to be zeroed. Given the tool overhang substantial cuts are difficult to achieve without a large and very stiff boring bar, therefore, the size of cut that can be applied is normally substantially less than external turning and multiple passes need to be made. For machining an open ended cavity with a plain diameter, boring is merely a case of taking a series of cuts until the diameter is achieved. The combination of a series of boring bars in increasing size (if possible) coupled with a series or finish cuts of reducing depth is often a good way of generating an acceptable surface finish. The use of coolant can not only cool the tool and workpiece, but assist with washing away swarf. It is also normal to set a bed stop when machining a plain diameter such that the tool doesn't run into the chuck or work holding device.

Boring a closed cavity requires a higher degree of skill. In this case the tool is run down the internal diameter to a known depth set by use of a bed stop or DRO readout on reaching the desired depth the tool is then moved across the bottom of the cavity towards the central axis of the lathe to at least the centre to provide a plain face. The tool point on a boring bar is designed to take a cut on its front edge and nose radius. The point of maximum force moves slightly when going through the transition between parallel turning and facing at the bottom of the cavity. As the tool passes beyond the centre point the direction of force goes from being downwards to underneath the tool. This is of little consequence near the centre but can cause damage if the tool isn't moved away from the face as it passes the centre point. It is normal to apply a cut and run the tool into the workpiece and then transition across the bottom of the cavity, but where any central hole passes deeper into the work piece the facing operation can be undertaken first and then the tool extracted. This places the tool in tension rather than compression, but with a cut depth matched to the boring bar capacity this rarely makes much difference.

3.13 BORING ANGULAR FEATURES

Boring a tapered cavity has a little more complexity as consideration of the shank of the boring bar needs to be made, and brings into use the compound slide. When machining a plain tapered cavity the compound slide is rotated clockwise to half the included angle of the tapered cavity desired. The cut is applied by rotating the cross slide handle as for parallel turning, but the cut is taken by rotating the cross slide handle to machine an angle diameter. Where this tapered feature commences from with a recessed plain diameter then some rotation of the tool post may be required to prevent the shank of the boring bar hitting the end of the workpiece as cuts advance into the cavity. In addition care needs to be taken to check that the side of the tool opposite the cutting point does not interfere with the wall of the cavity. This method of boring can never produce a conical point and will always either require a small flat bottom, proportional to the size of tool bar used, or a clearance hole to machine into. An example of the set-up for boring a recessed cavity is shown in Figure 3.32.

Chatter marks and poor surface finishes are often a product of long tool overhangs, undersized tools, incorrect cutting speeds, poor choice of depths of cut for the material and cavity configuration. Effective swarf removal can also improve surface finish as it prevents scratching. Sometimes there is little that can be done for this and a subsequent internal cylindrical grinding operation is required, however, use of more advanced tooling such as oil filled boring bars can reduce this. With experience and correct set-up,

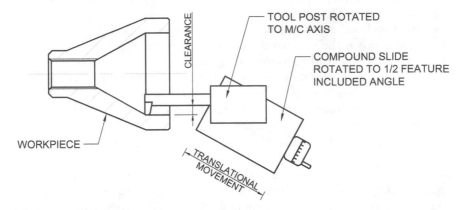

Figure 3.32 Tool set-up for boring recessed tapered cavity

surface finishes equivalent to external finishes are achievable, albeit generally at a slower rate.

3.14 TURNING ANGULAR FEATURES (EXTERNAL)

While most work undertaken on a lathe involves turning parallel features,
there can be a requirement for turning tapered features. There are three
approaches to this, one is by using a taper turning attachment, which will
be discussed in a later section. The second, and most common approach, is
by using rotation of the compound slide. The third, and very rare, approach
being offsetting of the tailstock.

While turning chamfers is probably the most common angular feature
machined on a lathe these are rarely of any significant length, however the
approach to setting up the machine for longer angled features is exactly the
same as for a generated chamfer.

Examination of the component drawing will identify if the angular
dimension is an included angle (i.e. one that is dimensioned either side of a
centre line or an angle from a face). Where included angles dimensioned the
compound slide should be rotated to half the dimensional angle required,
whereas an angular dimension from a feature should be set as dimensioned.

After loosening the securing nuts, rotate the compound slide until the graduations around the base of the compound slide line up with the zero line on
the cross slide. Once the slide has been rotated to the correct position these
are retightened and the compound slide is ready for generation of the angled
feature. When turning angular feature there needs to be some understanding
of where to tool and tool post will end up. If machining to a shoulder of any
significant size it is highly likely that the side of the tool or tool post itself will
impact the shoulder before the end of the cut is complete. Therefore, the tool

post may need to be rotated such that the tool remains perpendicular to the axis of the workpiece as shown in Figure 3.33. This will have a marginal effect on the tool geometry but in general practice this is rarely an issue. Successive cuts are taken by advancing the cross slide and then rotating the compound slide hand wheel to move the tool along the face of the workpiece.

Figure 3.33 Arrangement of compound slide for machining angular feature to a shoulder

If the angular feature requires significant material removal and the feature to be generated gets deeper towards the headstock then consideration needs to be made of the starting point. Cuts of this nature take the tool deeper into the workpiece as it advances and quickly gets to the point where the depth of cut is unmanageable. In contrast angles that open up do not have this problem as the tool advances out, rather than in, to the material. Therefore, consideration of the start point needs to be undertaken by the operator which may require the tool to be some distance down the axis of the workpiece, and work back towards the tailstock with successive cuts of a known depth. Figure 3.34 illustrates the problem of depth of cut when advancing into the workpiece, and the requirement to identify the start point to ensure the tool does not get overloaded.

There are some limitations to producing angular features through use of the compound slide. The most significant is accuracy, as even on the largest machines the angular scale is only marked in whole degrees, which effectively limits the accuracy of the angle produced to 1°. Smaller machines may only have increments of 5° if the machine compound slide base is of a small diameter. Another problem area can be surface finish. While the tooling is capable of providing a good surface finish most, if not all, compound slides have to be wound by hand and lack power feed. Unless rotation of the hand wheel is very smooth, it is likely that machining marks will be seen on the finished surface. Longer workpieces require the support of a tailstock centre, and this often prevents the rotation of the compound slide given an interference between the slide and either the workpiece, or the tailstock. While one option is to rotate the compound slide to the reciprocal angle and fit a right-handed tool into the tool post, care needs to be taken to ensure the operators hands do not come anywhere near the rotating chuck and that any guard can still be closed correctly. Reviewing the operation and

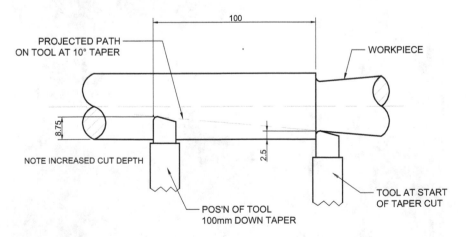

Figure 3.34 Tool start point and cutting depth issues

sequencing of manufacture can often negate this problem and the operator merely reverses the way a component is machined. Turning between centres allows the component to be easily reversed and can also negate this problem.

Offsetting the tailstock is a method of providing a taper. It is rarely used as it is only capable of generating a slow taper, and more significantly, requires the operator to spend some considerable time getting the tailstock back on centre when taper turning is complete. The method for setting the machine involves calculation of a right-angled triangle to determine the offset of the tailstock. This is quite limited on many machines which is accounts for its lack of use. Moreover when calculating an offset the length has to include the distance from the edge of the taper closest to the chuck to the centre hole of the workpiece. The tailstock alignment adjusting screws are then rotated to provide an offset. The offset is measured by using a dial test indicator placed perpendicular to the tailstock quill or body.

3.15 USE OF TAPER TURNING ATTACHMENTS

Some machines are fitted with an attachment to the cross slide that provides a mechanism for accurately turning internal and external tapers. Generally these attachments allow turning to an included angle of 20°. Those fitted with a vernier scale will allow angles to be machined to sub-degree accuracy, typically to within 15′ of arc as shown in Figure 3.35.

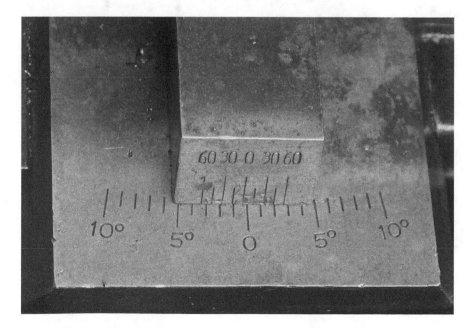

Figure 3.35 Taper turning attachment scale

The attachment works by anchoring the taper slide to the bed of the lathe as shown in Figure 3.36. This is generally achieved by inserting a clamp into a groove on the bed and tightening a clamp in a similar manner to the bed stop used in parallel turning. This anchor prevents the taper turning

Figure 3.36 Taper turning attachment general arrangement

attachment moving along the axis of the machine with the carriage forcing the cross slide to advance or retreat across the x-axis as the carriage moves along the bed of the machine. The degree of movement relates to the angle set, and the orientation of the angle i.e. increasing or decreasing, again depends on the angle set.

Once clamped in place, the cross slide has a follower which forces the cross slide into, or away from the workpiece generating a taper. The length of taper that can be turned is equal to the length of the guide bar and workpieces that require a long taper require careful setting up.

The method for setting up starts with setting the angle either as a known value in degrees and minutes or in a taper/unit. Some attachments are calibrated in both with degrees at one end and a proportion at the other. The method for setting the angle varies between type, but normally involves unscrewing a clamp, then offsetting the guide before re-clamping. The second operation is to move the tool to the start point of the taper and assess if the full length of the taper can be achieved without having to adjust the start position of the taper attachment. Where the start position has to be adjusted to allow the length of taper to be turned, the bed clamp is fastened tight and carriage moved towards the tailstock until the end of the taper turning attachment is reached. The bed clamp is then loosed and the carriage moved to the start point of the taper, at which time the bed clamp is tightened. This means that the machine has been set up and is now ready for use.

As the carriage is moved down the bed of the machine either by hand or using feed, the tool will advance into, or out of the workpiece. Successive cuts are applied in the same way as parallel turning by advancing the cross slide using the cross slide hand wheel. The same precautions for depth of cut need to be made for taper turning attachments as for tapers generated through rotation of the compound slide, in that reverse tapers (i.e. those that require the tool to advance into the workpiece) can rapidly become to greater depth of cut for the machine, or tooling resulting in catastrophic failure. Therefore, an allowance for depth of cut needs to be made such that the tool is not overwhelmed close to the end of each pass.

Taper turning attachments are used in exactly the same way for turning cavities, and all of the same constraints that apply for boring cavities using a rotated compound slide apply.

3.16 MEASUREMENT OF TAPERS

Measurement of tapers can be slightly difficult in practice, however, the basic theory that sits behind it is simple trigonometry. Tapers are either defined as 'taper per inch' or more commonly by an angle in degrees. While taper per inch dimensions are always represent the total taper, operators need to check that any angle is an included angle, i.e. dimensioned either side of a centreline axis or dimensioned from one surface only. There are two parts to defining a taper; the first being the angle of the taper, and the second being a reference diameter, or discrete toleranced inspection point. There are also two approaches to measuring a taper. One approach is measurement on the machine, which is normally conducted while roughing and finishing the taper, and the second approach is measuring off the machine. Workpiece taper measurements undertaken off the machine are often more easily achieved, and often more accurate than those taken on the machine given the slightly different approaches taken, however, there does not need to be a substantial difference and the tolerance band and precision required is a factor in the approach used. When turning between centres removal of the workpiece and re-insertion is a simple operation. Where a taper is being turned with the workpiece held in a three-jaw chuck, it is almost impossible to return the workpiece to the position it was previously in and as a result should not be removed from the machine to measure.

There are two aspects to measuring a taper. The first is establishing that the angle of the taper being machined, or the amount of taper/inch is correct. The second aspect is identifying that the feature has been turned down to the correct diameter. While when using a taper turning attachment precise angles can be set, they require checking to establish that the correct angle is being achieved. Having set the correct angle on the taper turning attachment, including using the vernier scale where needed, a series of roughing cuts are undertaken. until a tapered portion of the workpiece is established. Given that taper turning attachments are used to machine slow tapers i.e. ones up to 10° or 4″/ft, the longer the section to measure is, the more accurately the angle of taper will be determined. It is also common for the initial cuts of the taper to be away from any reference diameter or nominated inspection location. This is not a problem as establishing the angle is unrelated to the diameter, however, care needs to be taken not to cut too deeply such that any diametral limit is exceeded or that any error in the taper angle cannot be corrected.

To calculate the taper angle achieved the tapered section generated needs to be measured in two places at a known distance apart. Figure 3.37 shows the basic principle employed.

The approach taken by the operator to establish the measurements accurately can depend on the equipment available, however use of callipers with a chisel edge to measure the diameters make the task simpler, as the use of a micrometer creates problems given that the flat surfaces of the micrometer spindle faces do not lend themselves to measuring a tapered diameter. Establishing the length apart is a little more difficult. Lightly marking the taper in two places a known distance apart using the tool point is perhaps the simplest approach. The tool is touched onto the surface of the workpiece at a point close to the lower end of the taper, and the bed stop set. The workpiece is rotated by hand lightly marking the workpiece around its periphery. The tool is then extracted using the cross slide control wheel and the carriage

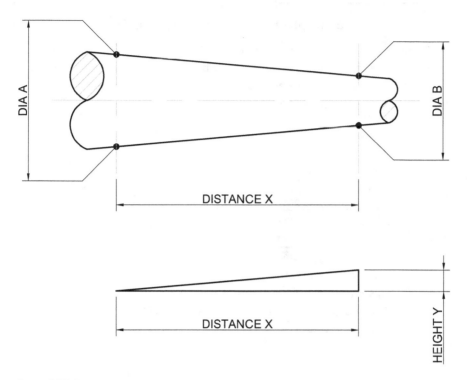

DIA A

DIA B

DISTANCE X

DISTANCE X

HEIGHT Y

Figure 3.37 Determining taper angle

moved along the bed towards the tailstock. A gauge block (slip gauge) of a simple size is inserted adjacent to the bed stop and the carriage brought back down the bed until the gauge block is trapped between the bed stop and the carriage. The workpiece is similarly lightly marked and the workpiece has two concentric rings marked a known distance apart. The workpiece is then accurately measured using the chisel points of the calliper and the angle calculated, as shown in the example below.

The difficulty in determining accurate values for tapers on machine is accurately establishing the distance between the points measured. Using fixtures fitted with two callipers or dial test indicators is one approach; another is using a slip gauge mount in a clamp and located against the workpiece where standard callipers are used to take a reading when pressed up against each end, or getting a colleague to assist. However, this is either expensive, or leads to errors and lightly marking the workpiece while roughing means that the workpiece does not need to be removed from the machine, and a degree of precision can be established, and where not accurate, the taper adjusted and the process repeated on a subsequent cut.

Example 1 – Using included angle

To determine if the taper attachment for a 10° included angle taper has been correctly set, two light marks have been scribed into the workpiece using a tool point 100mm apart. To calculate the angle achieved the measurements need to be resolved into a right-angled triangle.

$\emptyset$40.90mm – $\emptyset$23.40mm = 17.5mm

17.5mm / 2 = 8.75mm

Using tan angle $= \dfrac{\text{opposite}}{\text{adjacent}}$

$\qquad\qquad\qquad = \dfrac{8.75\text{mm}}{100\text{mm}}$

Tan of angle   = 0.0875
0.0875 × tan^{-1} = 5.0006°
5.0006° × 2 (to identify the included angle) = 10.001°

Example 2 – Using taper/foot

To determine if the taper attachment for a taper of 2″/ft has been correctly set, two light marks have been scribed into the workpiece using a tool point 4″ apart. To calculate the taper the following calculation is undertaken.

Measurement at position 1 = 2.100″ dia
Measurement at position 2 = 1.430″ dia

2.100″ – 1.430″ = 0.670″ taper over a 4″ length

To resolve taper to an amount over a 1 foot length, value multiplied × 3 (4″ × 3 = 1 ft)

0.670″ × 3 = 2.01″

Therefore taper generated = 2.01″/ foot

Once the taper has been correctly established the diameter can be reduced to the size required. This does not require measurement in two places, but often requires the accurate measurement of a diameter at a position identified on the component drawing. Identification of the position is normally achieved by placing the tool at the set distance, undertaken in the manner used for machining to a known shoulder depth. Once the point is identified the callipers are placed against the tool and the diameter measured. As the taper has already been established the operator only needs to reduce the workpiece to the diameter required and measure in one place.

Where the workpiece is being turned between centres and can be removed from the machine, the workpiece can be measured on a surface plate. By mounting the workpiece vertically it can be located on a datum edge and slip packs used to provide a linear position to measure both the taper and the taper diameter at a set point. It is a longer process than undertaking measurement on machine, however, removes any additional opportunity for error, however care must be taken to ensure centre holes and centres remain clean.

3.17 MACHINE BACKLASH ISSUES

Wear in the lead screw, or gearing of the lathe, or the precision that the machine was made to can lead to backlash varying amounts of back lash. There will always be some movement as gears and lead screws need to have some clearance to allow for movement and a film of oil to come between mating surfaces. Each component in a drive chain will have its own small clearance, however, when a number of these components are used a cumulative mount of backlash can be observed. Generally this backlash is not a problem as drive commences prior to the tool cutting into the workpiece and the load is only applied in one direction until the end of the cut. The tool is withdrawn or the workpiece stopped and the tool returned to the starting position ready for the next cut. However if machining in a pocket, where left- and right-handed tools are being used, this is again of little consequence if the cutting direction is aligned with the machine z-axis. It is only when taper turning inside a pocket, or thread cutting where backlash can be a problem.

Taper turning backlash

When taper turning within a pocket it can be quite difficult to ensure a consistent taper that does not have a step. Turning a taper between shoulders, using a taper turning attachment normally requires the use of both a left-handed tool and a right-handed tool where cuts are applied on feed in opposing directions. The build-up of backlash in the gear train and in the taper turning attachment, as well as the slow angle of taper makes it easy for a flat spot, or parallel section to become apparent where the changeover from le left-handed tool to a right-handed tool is undertaken. This can be seen in the component shown in Figure 3.38.

Figure 3.38 Parallel section on workpiece due to machine backlash

Taper turning attachments use a follower that slides down an angled guide moving the cross slide in or out as the carriage is fed down the bed of the machine. Clearances in the cross slide and the taper turning attachment mean that the workpiece will rotate and the feed engage before the tool advances when changing direction. This leaves a small parallel section on the workpiece that is broadly proportional to the amount of backlash.

This problem is not that simple to cure. As the taper is being cut between shoulders, the tool cannot be taken off the end of the workpiece and two tools have to be used. The simplest solution is for the operator to position the tool as close to one shoulder as possible and work down the workpiece i.e. go from the widest section to the narrowest. When the correct size of taper is achieved, change to the opposite handed tool. The operator will need to set the tool to meet the surface of the workpiece, well up the machined section, and then work up the taper. Figure 3.39 illustrates the adjustment that is required.

The machine is started and the feed engaged. There will be a lag in any movement on the z-axis until the backlash has been taken up as the tool moves down the body the tool can be advanced to ensure it takes a cut when the unmachined section of the workpiece is encountered. This technique progresses until the finish cut is required. The operator needs to use great care to ensure that the tool point is advanced just enough that it forms one tapered section and does not dig in too far.

An alternative approach is to establish the difference in the depth of cut caused by the backlash, and ensure that the tool working up the workpiece is positioned in the tool post. This only works when the length of workpiece taper being turned is at least twice the distance between the tools mounted in the tool post to allow the distance to be cut and the opposite

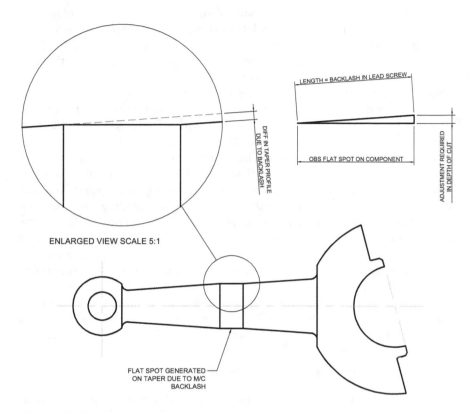

ENLARGED VIEW SCALE 5:1

LENGTH = BACKLASH IN LEAD SCREW

OBS FLAT SPOT ON COMPONENT

DIFF IN TAPER PROFILE
DUE TO BACKLASH

ADJUSTMENT REQUIRED
IN DEPTH OF CUT

FLAT SPOT GENERATED
ON TAPER DUE TO M/C
BACKLASH

Figure 3.39 Cut depth adjustment on workpiece due to machine backlash

tool introduced. Identifying the distance the tool needs to be advanced is shown in the example below.

Thread cutting backlash

Backlash in thread cutting only becomes an issue when returning the carriage back to the tailstock end of the workpiece when cutting threads. Because there is a build-up of clearances in the gear chain and lead screw, the cumulative effect of this is backlash. Backlash when screw cutting is embodied in difference between when motion of the carriage starts in relation to rotation of the workpiece. The greater the difference between when movement starts, compared to rotation of the workpiece is the amount of backlash. When screw cutting the end of each cut requires to tool to be returned to a precise starting position to commence the next cut. If the machine was merely placed into reverse the backlash would mean the workpiece would rotate before the carriage starts to move along the bed of the machine and the tool will come into contact with the flank of the thread being cut damaging it throughout the length the tool is left in contact. In extremis this would be all, the way to the start position.

To counter this problem, the operator only needs to withdraw the tool point from the thread form being machined. If the compound slide has been rotated and is being used to apply each cut, then the operator only needs to zero the scale on the cross slide to provide datum mark to return to. The operator can then fully withdraw the tool, and with the screw cutting feed remaining engaged, switch the machine into reverse, an allow the carriage to proceed towards the tailstock. The operator needs to ensure that the tool is moved beyond the end of the workpiece for a short distance to allow any backlash to remove itself, when forward drive is re-engaged. The tool is returned to its zero datum position on the cross slide, the next cut then applied on the compound slide, and the cutting process repeated. Where the compound slide has not been rotated to set each cut and the cross slide is used the operator only needs to withdraw the tool however, the cross slide scale will either need to be zeroed at the end of each cut or the setting noted to ensure the operator can return the tool point to the correct depth of cut.

The issue with backlash when thread cutting is only apparent when leaving the lead screw feed engaged. Where a thread dial is fitted to the carriage this problem is not apparent as the feed screw is released between each cut.

3.18 TURNING BETWEEN CENTRES

Turning between centres is a technique that is much less common than it used to be. Prior to the invention of three- and four-jaw chucks, this was the principal method for mounting a workpiece. Turning between centres has some advantages as well as disadvantages, with the principal advantage being that the workpiece can be removed from the machine to be inspected, or reversed in the machine to allow machining from the opposite end. When the workpiece is returned to the centres it remains on centre, and further operations will be concentric. While three-jaw chucks are immensely convenient, due to the scroll within the chuck, and the way jaws operate, it is unlikely the workpiece can be returned to exactly on the centre it had prior to removal and as such concentricity errors are introduced. Therefore, if concentricity tolerances are extremely small, or the workpiece is to be subsequently ground, turning between centres becomes the set-up of choice. However, this method does have some disadvantages. Firstly it is not possible to part off a component when turning between centres as the workpiece will 'grab' the tool as the cut comes to the central point. This is because it is retained longitudinally by the tailstock centre and has nowhere to go. The workpiece requires a centre hole in each end, which if the workpiece has already be machined to the required length, will be retained. While this can be avoided by machining a centre mounting stud to fit the centre into which can be removed later it can be an issue. There is less tolerance for the tailstock being slightly off-centre and any measurable error will show a slow taper on the workpiece.

The workpiece is mounted between two centres. One is a dead centre fitted into the machine spindle nose and has no relative movement. The other end is fitted into the tailstock, and while may be a fixed or rotating centre, will be considered live as there is relative rotation between the workpiece and centre. Rotational movement is imparted to the workpiece by fitting a driving 'dog' to the workpiece. This is generally a simple yoke with a clamping screw, but some have an angled leg which engages with a faceplate. More commonly the simple yoke is pushed around by a peg mounted on a catch plate. Figure 3.40 shows a typical headstock set-up.

Prior to turning between centres it is normal to face off the workpiece at both ends, machine it to the required length and drill the centre holes. General practise tends to achieve this involves using a three-jaw chuck prior to its removal, however, this is not always required. Sawn bars can have centre holes drilled into them, off machine using a pillar drill, or in extremis a hand drill. However, it is essential that the workpiece must have a centre hole drilled with a Morse centre drill in each end as close to the centre as possible to remove potential vibration. These holes need to be as deep as possible to ensure a firm engagement with the centres on the machine. This

Figure 3.40 Headstock set-up showing dead centre, dog and catch plate

allows for stability, reduces the chance of the workpiece being ejected from the machine, and allows for any centre nose radius reducing engagement and increasing the likelihood of workpiece ejection when under load.

Following centre drilling the chuck is removed from the machine. Generally there are two types of chuck mounting. One is simply a screwed thread, and while this can be tight, merely requires unscrewing. The second, and most common type are those that sit on a taper and are locked up using a cam lock arrangement. To remove a chuck fitted with these; using a square key rotate each cam lock to the unlocked position. It is prudent to place a piece of wood to protect the machine bed in case the chuck falls, and then the outer periphery of the chuck is given a firm tap with a soft faced mallet to break the grip of the taper. Normally the chuck hangs on the cam lock studs, but it can fall and care needs to be exercised to prevent damage to the chuck, machine and operator.

Fitting a dead centre into the spindle is easy. It is important to ensure the mating surfaces are clean and free from swarf, and the centre is just firmly inserted into the spindle and is retained. Ejection of the centre requires a bar to be inserted down the spindle from the rear of the headstock and sharply tapped with a hammer or mallet. It is rare for this centre to be anything except 'dead centre', however, it is possible to purchase blunt-ended centres which are inserted and then machined to a conical point.

Fitting the tailstock centre is undertaken in exactly the same manner as for parallel turning long bars. The choice of centre depends on the precision required, the rotational speed to be used and the axial and transverse loads to be applied. This centre is a live centre i.e. there is relative motion between the centre and the workpiece. If a fixed centre is to be used then lubrication is required in the centre hole. Even when lubricated it is not unknown for the centre to be friction welded into the centre cavity, if axial loads and rotational speeds are relatively high. There is a risk when this happens of workpiece ejection, and the low cost and precision of rotating centres suggests that this is a better option.

The catch plate and driving dog (if being used) is fitted in exactly the same way as a chuck. For machines that use a taper and cam lock arrangement, it is simply a case of cleaning the spindle nose taper, and mating catch plate taper, then inserting the cam studs into their mating holes in the machine spindle and tightening up the locking screws. For machines with a threaded spindle, it is a case of cleaning the threads and tightening until it locks.

The driving dog is a device fitted to the workpiece that transmits rotational movement from the spindle via the catch plate to the workpiece. They come in a range of sizes that equate to the diameter of bar being used which ensures that the load is balanced and vibration or transverse loads are not transmitted to the machine. The use of a large driving dog on a small bar would have a substantial off-centre load. However, the clamping screw fitted to the driving dog is sized for the expected workpiece diameters and

Figure 3.41 Driving dog assembled onto workpiece

should not be changed. The workpiece is inserted into the centre of the driving dog yoke and the clamping screw tightened. The tightness of this clamp is important as it is the way that rotational movement is transmitted to the workpiece and if the friction between the end of the screw, and the two opposing tangential faces is less than the force imparted by cutting, then it will slip. However, significant tightening will mark the surface of the material being turned. If the area is unimportant or going to be subsequently machined then this is not significant. Alternatively use of a sacrificial strip of soft metal between the screw and workpiece is an alternative. It is also not unusual for a flat to be machined to prevent rotational slippage. Figure 3.41 shows a driving dog fastened onto the workpiece and ready to be mounted between centres.

The workpiece and driving dog assembly is the inserted between the centres on the machine. For larger workpieces this can be awkward as the workpiece needs to be supported by hand, engaged in the headstock dead centre, and the tailstock simultaneously slid along the bed to engage the live centre in the workpiece. Where the weight of the workpiece is significant or requires an awkward lift that risks injury to the operator use of some mechanical handling equipment or wooden chocks is prudent.

Figure 3.42 Workpiece mounted and ready for machining

Once the workpiece has been mounted in the machine as shown in Figure 3.42 some basic checks need to be undertaken which can be summarised as:

1 Is the peg on the catch plate engaging the driving dog or its clamping screw? If engaging the clamping screw, the workpiece needs to be removed and rotated such that the driving dog leg is engaged.
2 Is the driving dog longitudinal position such that it is fully seated on the workpiece and fully engaged with the catch plate drive? If not loosen and adjust until correctly seated.
3 Is it possible to machine to a length required without turning into the driving dog? Where this is an issue, the best approach is to determine if a smaller dog can be utilised, whether it can be moved closer to the end of the workpiece while retaining sufficient engagement, or whether a longer workpiece can be used. Where none of these options are available it is not unusual to just machine into the dog and treat it a sacrificial tooling.
4 Is the catch plate drive of a length such that it will impact with the tool or tool post before the length of cut has been completed? Where there is a significant protrusion it is not unusual to cut the

length of it down. Most, if not all drives are merely a threaded peg, secured by a locknut and are easily replaced. The length of drive after the section engaged on the drive dog is doing nothing, and therefore, not required. Where a bent drive dog is being used and inserts into a faceplate this would not be an issue.

Following all of these checks machining operations can commence, using all of the approaches, feeds and speeds that would be used for parallel turning, turning to a depth and taper turning. What cannot be achieved is boring and drilling as an axial load provided by the tailstock centre is required at all times.

3.19 OFF-CENTRE TURNING

Off-centre turning is a technique used to produce a turned diameter parallel to the central axis of a workpiece. The classic example of this being turning the journal for a crankshaft. The set-up for this, is in many respects the same as for turning between centres, and in fact has to be undertaken between centres as an offset cannot be achieved at the headstock using a three-jaw, or four chucks unless complex tooling is manufactured for the purpose.

When a feature offset from the main axis is to be machined a second set of centres have to be drilled at each end of the workpiece offset from the principal axis. The offset distance is that required to form the centre of the offset feature. When marking the new centre positions it is important that they are on identical axis at each end of the workpiece or a skewed feature will be generated. The most efficient method for marking these centre positions is to clamp the workpiece into a pair of matched vee blocks and then scribe the horizontal and vertical marks using a vernier scriber or equivalent. Once marked the new centre holes are drilled using a Pillar drill or other tooling that produces an accurately drilled centre at each end.

Use of vee blocks is particularly useful where multiple offset features are to be generated. It is common for offset features to be 180° apart, and as such all should be marked at the same time to reduce alignment errors. Similarly those that are required to be generated 90° apart can be accommodated by this method. Angles other than these can be produced by a number of methods, but the most precise would be using a dividing head relocated to a surface plate. While merely marking up the workpiece ends and using co-ordinates is an option, the potential for introducing alignment error is significant, and this approach should only be used when all other methods have been exhausted.

Off-centre turning will always involve the creation of a pocket and requires careful thought about which tooling is to be selected and how the cutting point will be presented to the material. The ends of the workpiece may have to have a circular flange into which the offset centres are drilled. It is common for these to be machined away to leave a desired plain diameter following completion of the offset features. This often means that all machining linked to the offset features needs to be completed prior to their removal, which may include operations on a different machine, such as a cylindrical grinder prior to the completion of the general turning. Therefore the sequencing of operations is an important factor for the operator to consider.

One method of creating the pocket is to plunge cut using tooling such as a parting tool to create an initial slot in the central part to be removed. A pair of left- and right-handed turning tools are then used to widen the slot to its required width. It is normal to allow for a finishing facing cut, given that off-centre turning can occasionally be a little brutal with vibration transmitted from tool impact and sub-optimal cutting speeds used.

Depth of cut is not different to normal parallel turning, however, initial cuts do not remove much material and create an offset arc on the surface of the workpiece. As the depth of cut increases the area of surface affected increases and the radius generated reduces. This process continues until a feature that has a completely round profile is generated. The process of plunge cutting and then widening using left- and right-handed tooling continues until the offset feature desired diameter is achieved. The respective ends of the pocket are then faced off using each tool until the required width of gap and surface finish is achieved. This process becomes simpler if careful attention is paid to setting the tool offset lengths from the tool holder to ensure the right-handed and left-handed tools protrude the same amount such that the cross slide setting does not have to be changed as the tool post is rotated, or tool holder inserted between each tool.

While off-centre turning is relatively straightforward there are some issues that become apparent. It is not always possible to use an optimum cutting speed if the workpiece is of significant mass and offset. A $\varnothing100$mm workpiece 300mm long has a mass of approximately 18kg. If a $\varnothing20$mm journal is to be machined offset by 25mm then a substantial mass is going to be off-centre. If machining at 90 surface m/min then the machine RPM will initially be set at approximately 280rpm, however, as material is removed speed should be increased to approximately 1420rpm. While 280rpm will not trouble machines capable of mounting a $\varnothing100$mm × 300mm long workpiece set-up for turning a $\varnothing20$mm journal offset by 25mm, however, the vibration and loads on the centres introduced by having 18kg rotated off-centre at 1400rpm will be significant. Therefore, caution needs to be exercised when setting rotational speeds for workpieces of significant size. The operator should judge the effect of the workpiece on the machine and adjust accordingly. Best practice approach involves starting slow and bringing the speed up as required. Close inspection of the surface finish will be a clear indicator of tool performance and while cutting material at the optimum rate is good, the vibration may degrade tool performance in proportion and the operator will be able to judge what the best combination of variables are.

Another issue that is prevalent in off-centre turning is tool impact. As the tool is advanced into the workpiece there is a section where the tool is dug into the material and cutting and a period where the tool is in free air. As the tool advances the period the tool is in free air decreases until it continuously cuts. While this is generally not of major concern, the selection of tool needs to be such that the tool can accept and irregular loading and the impact of the workpiece swinging into its cutting sector. Narrow pockets generally demand smaller tools, deeper pockets demand longer overhangs and consideration needs to me made of the robustness of tooling and the depth of cut to be applied for each pass.

3.20 FOUR-JAW CHUCK USE

Four-jaw chucks are a good solution to clamping workpieces that are not round or hexagonal etc. While they are at their best when clamping square or rectangular workpieces, they are flexible enough to be used to firmly hold a variety of irregularly shaped objects. The key function being good engagement with the chuck jaws to ensure stability while undertaking machining operations.

The principal difference between a four-jaw chuck and a three-jaw is that each jaw is independent of the others, rather than being mounted on a scroll as a three-jaw chuck is. While independent jaws allow for clamping both circular, as well as irregular and multi-sided objects, clamping items such as hexagonal bar, are often better held in a three-jaw chuck to avoid damage to the apex of the workpiece sides in two places. However, the principle of the operation of a four-jaw chuck is no different than that for a three-jaw although the set up for a jaw is more time consuming.

Setting up a four-jaw chuck can be frustrating if the operator is not well practiced, and it takes time and experience to be able to mount a workpiece rapidly, however, it should not be feared or avoided when it is the best work holding solution. The correct positional mounting is dependent on what feature is to be generated on the workpiece. For instance if the operation is merely to face off a square of rectangular bar, it doesn't matter if the axis of the workpiece is off-centre as long as it is square to the axis. However, if the operation is the turn a parallel circular feature that must be located equally around a central axis then workpiece setting is critical. It is also possible to set up a four-jaw chuck to generate off-centre features, by deliberately mounting a workpiece a known distance off-centre.

There are a number of techniques which can be used to successfully set up a four-jaw chuck which go from the relatively agricultural, through the highly sophisticated; however, below are descriptions of two basic methods:

- use of a centre; and
- free set-up.

Use of a centre

For regular shaped workpieces such as square or rectangular bar that have a plane cut face, mark two diagonal lines across the opposing corners on the end of the bar to form a 'X'. Where they cross in the centre of the bar drill a hole with a morse centre drill. Open up the jaws on the four-jaw chuck, place the workpiece up against the front face of the chuck (assuming its large enough to be supported across the chuck centre aperture) and support in place with a revolving or dead fixed centre mounted in the tailstock and engaged with enough force to prevent the workpiece slipping. Then use the chuck key to wind in the jaws until they lightly engage the work piece. Once four faces of the workpiece are engaged each is lightly tightened and then fully tightened after which the supporting tailstock mounted centre is removed. The workpiece can be checked for equal alignment by hand rotating the chuck to check for any visible 'wandering' or oscillation of the workpiece, or by using a dial test indicator (DTI) to check that opposing faces provide the same indicator deflection. When using a DTI, it is useful to have the indicator probe as close the machine centre line axis as possible as this reduces any errors from the workpiece not being absolutely vertical, or horizontal. Where there is a positive or negative offset a jaw is loosened and the workpiece pushed across using the opposing jaw by half the offset. The process is then repeated. Once one place is centred to process is repeated by rotating the chuck through 90° and going through a similar exercise. Once the operator is happy with the adjustment, a check to ensure all jaws are tight and machining commences.

Free set-up

A second approach is to wind in the jaws without the workpiece inserted using the until they meet in the centre. In turn, each pair of opposing jaws are opened up by the same number of turns of the chuck key, with this progressing until the gap between them is approximately symmetrical about the axis and the gap big enough to accept the workpiece. This process is repeated for the second pair of jaws. Following this the workpiece is inserted and while being supported by hand or using a suitable mechanical support the jaws are tightened. Again this is done in pairs with each jaw being brought in by rotating the chuck key a similar number of turns on each side until the workpiece is securely retained. For precise alignment a DTI is then used to get the workpiece as central as required, and the jaws then fully tightened.

To create an offset the above process(s) are undertaken to find the central point and then by loosening one jaw pushing the workpiece across a known amount either by use of a DTI to measure displacement as shown in Figure 3.43, or by use of a slip gauge inserted between the DTI and workpiece. a similar process can be undertaken to offset in both an x and Y direction.

Figure 3.43 Four-jaw chuck with workpiece securely mounted on centre

Any workpiece that is fastened into a four-jaw chuck, where off-centre turning is taking place generally does not require the use of counter-weights, as the mass of the chuck is normally significantly greater than the workpiece and it would be highly unusual for a four-jaw chuck to require any type of balancing.

3.21 USE OF COLLET CHUCKS

Collet chucks are an additional type of work holding device. The have a number of advantages over conventional three-jaw and four-jaw chucks in that they provide more precise alignment of the workpiece. A key advantage of collet chucks is that they allow machining of extremely small diameters of workpiece on a machine that would not normally accommodate such fine workpieces. Traditionally very small diameter workpieces had to be machined on small lathes such as watchmakers, however, the use of collet chucks makes machining workpiece diameters at approximately ∅2mm relatively straightforward on common workshop lathes.

This type of chuck allows for removal and reinsertion of the workpiece with introducing any significant loss of concentricity. The ability to remove and reinsert a workpiece is particularly useful where difficult or complex measurement of the workpiece has to be undertaken between operations.

The method of workpiece holding methodology used in collet chucks is such that there is significant engagement between the collet and the workpiece. This is often in excess of 50% of the surface area, depending on the type of collet. This gives a clear advantage over common three and four-jaw chucks as there is a reduced chance of slippage where cutting loads exceed the frictional resistance of the clamping area, and the consequent generation of circumferential scoring. The design of collet and the required clamping loads also tend to prevent indentation marks that are commonly found on workpieces that have been securely clamped in a three or four-jaw chucks. While these indentations are not an issue if the workpiece is to be parted off, where a workpiece is repositioned to machine another feature, such as the opposite end of the workpiece, either reduced clamping pressure or use of soft sacrificial pads such as copper sheet need to be employed. The use of a collet chuck removes this requirement, and has better positional concentricity than common chucks tightened using a scroll.

Collet chucks are generally found in two discrete types. A one piece collet that is essentially identical to R8 type collets use in some turret milling machines, or a multi-component type. The R8 type offers more complete engagement with the workpiece periphery, and the multi-component type have a series of small plain parallel jaws which grip a substantial amount of the workpiece periphery. Figures 3.44 and 3.45 show R8 and multi-component collets respectively.

Both types require the insertion of a collet into a larger chuck which is connected to the spindle nose via cam lock, or a screw thread depending on the design of lathe. The collet is tightened using a chuck key which differs from common three and four-jaw chuck designs in that the locating section of the key is normally of a bevel gear type which is inserted into a peripheral gear ring on the chuck, rather than a square socket. However, tightening is achieved in essentially the same manner i.e. rotation of the chuck key until the workpiece is tightly retained.

There are significant advantages of using collet chucks, however, there are a number of disadvantages. The foremost disadvantage being that it is not possible to hold multi-sided or irregularly shaped workpieces. Moreover, to hold a wide variety of diameters requires a wide selection of collets as each has a range it will be able to grip between. Multi-toothed collets allow for a greater span of sizes for each collet used and a range up to 6mm is not uncommon. R8 style collets are less tolerant and are normally sized to take a standard size, typically in whole millimetres. This restricts the bar stock that can be used and also is somewhat restrictive in the diameters that have been previously machined to a requires size and fall between sized R8 collets. R8

Figure 3.44 R8 style collet

style collets are also limited to diameters of ∅25mm given the dimensional constraints of the R8 system.

Collet chucks are particularly useful for holding small diameters of bar stock and remove the requirement for smaller lathes. However, the requirement to hold a wide variety of collets to accommodate a range of workpiece diameters, the time taken to insert the collet prior to insertion of the workpiece, and the inability to hold irregularly, or multi-sided workpieces significantly reduces their utility. At times they are indispensable, however, the should be seen as an alternative, or enhancement to work holding options on the lathe, rather than a true replacement for three- or four-jaw chucks.

Figure 3.45 Multi-toothed style collets

3.22 FACE PLATE WORK INCLUDING CONSTRAINTS AND USE OF WEIGHTS

To machine a workpiece that cannot be held in a three-jaw or four-jaw chuck due to its size of irregular shape it is not uncommon to mount the workpiece of a face plate. Faceplates are generally a large circular disc with a flat front face central hole and a series of radial slots and an example is shown in Figure 3.46.

The faceplate mounts onto the lathe spindle nose in exactly the same way that a three-jaw or four-jaw chuck does (i.e. utilising a cam lock system to pull a taper onto the machine spindle nose, or a threaded mounting system). The slots in the face plate are intended to be used for workpiece clamping.

The method utilised to clamp the workpiece into place is a significant factor in the successful use of a faceplate, in that an unbalanced load rotated, and any speed above those considered relatively slow, will cause significant vibration. This vibration almost certainly leads to poor surface finishes, tooling damage, and in extremis, damage to the machine. To address this issue two approaches can be taken. The first is a relatively straightforward approach of using diametrically opposed clamps across the narrowest part

Figure 3.46 Common faceplate

of the workpiece. Unfortunately this is often not possible and this requires the fitment of balancing weights.

The type, position, and mass of the balancing weights are all factors to be considered, as is the method of attaching them to the faceplate. The workpiece mass and centroid are the features that drive the positioning of balance weights, and what the operator is seeking to do is reduce any significant eccentric loads when the workpiece is rotating. Lightweight workpieces retained using simple direct bolts to clamp it to the faceplate may require little, if any, counterbalancing especially if being machined at relatively low rotational speeds. However, larger workpieces such as steel castings or those using heavy steel clamping systems which are offset will almost certainly require substantial counterweights.

There are three main approaches which can be taken to determine the size and position of counter weights. The first is simply placing the lathe gear selector to neutral such that the faceplate runs freely and the faceplate will rotate until the resultant mass is centred at the bottom. Weights are then applied at a diametrically opposed position and the faceplate is spun until it settles. Weights are then either added or removed, moved in towards the centre or out until the entire assembly seems reasonably balanced. While this method is often pragmatic and does not require any specific mathematical ability, it assumes that the machine headstock bearings are free running and allows unimpeded rotation of the faceplate.

The second approach is to weigh the workpiece and weigh any significantly sized clamps used to mount the workpiece. The location of the approximate centre of each component (i.e. workpiece and clamp) is established, and either a single large weight, or multiple weights are fastened in opposing positions to balance the entire assembly.

A more scientific approach, however, one that requires a degree of mathematical ability is to calculate a resultant force through the addition of vectors. The mass of each component (workpiece, or clamping device) is established and its relative angular position. The resultant force is then calculated, or established using a graphical method, and a balance weight applied at the resultant distance and angle required. One advantage of using a modern 3D parametric computer-aided design (CAD) system to produce the component designs is that the mass and centroid of the component is easily accessible.

Methods use to retain a workpiece to a faceplate vary widely and depend on the type, size and complexity of the workpiece. By far the simplest and probably most secure is by inserting bolts or screws into tapped holes in the workpiece from the rear of the faceplate. The difficulty with this approach is that it requires the workpiece to have tapped holes in a fortuitous location, or the design permitting specific holes to be drilled and tapped for tooling purposes, which remain in the finished component. It is also not unusual for the workpiece to have to be mounted onto the faceplate off the machine to give access to slots in the faceplate close to the centre. The combined weight of the workpiece and faceplate can be substantial and subsequent mounting of the assembly on the machine spindle nose may require lifting equipment. An alternative approach is often the use of standard clamps. These take a number of forms with either 'swan neck' or parallel clamp with a packing piece or wedge, however, almost all will have a slot allowing for rotational and longitudinal adjustment. The advantage of using these type of clamps is that initial insertion is easy with the faceplate mounted on the machine and the workpiece can be positioned and adjusted with relative ease. However, it is important to ensure that the engagement between the clamp and any packing piece or wedge used to support the rear end of the clamp, is fully secure. Failure to fully tighten the clamp retaining bolts, ensure proper and robust engagement with clamp wedges or packing pieces, can result in them being ejected from the faceplate assembly. When this occurs the clamp immediately loosens and if any other retaining clamps are not strong enough to hold the workpiece until the machine can be stopped the workpiece will also be ejected. Where the workpiece is of any substantial size it is likely that the machine slide ways will be damaged potentially causing writing off the machine.

An additional benefit of utilising rear fixed bolts or screws to fasten a workpiece to a faceplate is that it provides unimpeded access for turning operations. Where clamps, and to a degree balance weights are utilised the tool path must be considered. While boring operations may seem straightforward issues such as protruding clamping bolts impacting with the tool post as the boring tool reaches the bottom of a cavity, or facing tools being brought out from the centre is not uncommon.

3.23 SCREW CUTTING

Screw cutting on a lathe is a relatively straightforward process once some basic principles have been understood. The formation of a screw thread is similar to ordinary turning, however, there is a precise ratio between the rotation of the workpiece and the feed rate of the tool along the axis of the workpiece. The speed of workpiece rotation selected is also significantly slower than that which would ordinarily be selected for material removal, and the depth of cut is also smaller than that which would commonly be used. It is also worth noting that multi-start threads can also be formed, although this does introduce an additional level of complexity.

There are a number of approaches which can be taken, and the method selected is often driven by the type of tooling being used. There are two principal types of tool used for thread cutting, HSS and a carbide insert. Both types of tool have the thread form ground into tool, however, HSS tools tend to have a modified 'vee' end and carbide inserts have a thread form and depth protruding from a plain edge as shown in Figure 3.47.

The shape of the tool must exactly match the thread form of the screw thread to be cut. While this is relatively straightforward for common thread forms such as Metric, Unified, Whitworth, BA, Acme etc; threads which are non-symmetrical such as a buttress thread require care to ensure that the form ground into the tool is the correct hand for cutting. Larger square threads can also present a challenge, especially when being used in multi-start threads as in extreme cases the thread helix angle can foul the tool sides and a relief angle greater than the helix of the thread needs to be ground into the tool flank.

Tools that are self-ground need to be checked to ensure that the angles are symmetrical, the nose radius (where required) has been ground to the correct size for the thread being cut, and the length of tool exceeds that of the depth of thread to be cut. Figure 3.48 shows a simple gauge that is used to check angles ground are of the correct profile.

Once tooling has been procured and inserted on centre height in the tool post the lathe needs to have a series of gear settings made. The single most obvious difference between ordinary turning on feed and screw cutting is that an alternative lever is used to engage the feed and that the drive is delivered by the machine lead screw rather than the feed rod. The lead screw

Figure 3.47 Carbide insert tool for screw cutting

Figure 3.48 Tool grinding gauge

directly controls the feed or rate of advancement along the workpiece and the consequent pitch of the thread being generated.

To set the required lead screw feed rate two adjustments need to be made. The correct gear cog ratios need to be set internally within the machine headstock, and the series of gear levers on the front of the headstock need to be configured. Most of the lathes commonly found in workshops are capable of turning threads with pitches from 0.1mm to approximately 14mm. Given this wide range of thread pitches the gear cogs need to be set in the headstock to cover the range of pitches required. The information to set the correct gear configuration, as well as gear lever settings is normally displayed on a panel on the front of the machine headstock. Figure 3.49 shows a typical information panel.

Typically there are four cogs in the headstock which control the lead screw speed. By using the information panel on the front of the headstock the correct gear configuration for the thread pitch being machined is selected. Changing the gear cogs is relatively easy and the information panel shows which cog with their respective number of teeth needs to be mounted on which shaft. As the cogs differ in size it is normal for at least one of the shafts to be mounted on an moveable arc that allows for correct engagement as shown in Figure 3.50. The arc and the gear cogs are normally held in place by nuts with a friction device such as 'Nyloc' or stiff nuts, although some have thin locking nuts as well, however, adjustment only requires the use of a spanner and is not a complex activity to complete.

Once the correct gear configuration has been set the gear levers need to be moved to the correct position for the thread being cut. The change of gearing allows for a range of thread pitches to be generated and machines will also allow for the machining of both imperial threads (designated in threads per inch) and metric threads (designated in mm of pitch). Therefore, the gear lever setting ensures that one specific pitch is achieved, or alternatively a feed rate of an exact amount for one rotation of the machine lead screw. Figure 3.51 shows the gear lever settings for a metric 12mm thread (M12).

Following setting of the machine gearing and mounting of the tool in the tool post the operator needs to determine if any rotation of the compound slide and tool post needs to be undertaken. It is common to rotate the compound slide to an angle equal to half that of the thread included angle, and adjust the tool post such that the tool is positioned at the normal to the workpiece. The reason for doing this is two-fold. It ensures that as the tool is advanced it only cuts on one side of the tool thereby generating a thread, rather than plunging in and cutting on both sides of the tool simultaneously which would be forming the thread. The benefit of this is that it reduces the load on the tool and prolongs its life, and ensures that the thread form is an equal angle. The second benefit of this approach is that it allows for the withdrawal of the tool at the end of a cut using the cross slide and if zeroed at the commencement of machining will allow the tool to be returned to the

Figure 3.49 Thread pitch setting table on machine headstock

same depth at each repeat cut, as the subsequent advance of the tool is being undertaken by the compound slide.

The tool has to be withdrawn at the end of each cut to return the tool to the start point for subsequent cuts either by releasing the lead screw and using a

Figure 3.50 Gear configuration in machine headstock

dial attached to the machine apron, or by leaving the lead screw engaged and reversing the machine. Where the machine is being run in reverse to return the tool to the start for a subsequent cut the machine lead screw and gearing system backlash has to be accommodated. If the tool was left engaged in the workpiece the machine backlash would damage the workpiece due to the underside of the tool impacting with flank of the thread being cut. Whatever method being employed by the operator the tool has to be withdrawn at the end of each cut to allow it to be returned to the start, and having the machine set such that the cross slide handle is returned to a '0' at each cut is an efficient way to proceed and one that is less subject to error.

It is possible to use a carbide tipped tool and 'plunge cut' to a required depth, however, this relies on the operator correctly starting from a known

Figure 3.51 Gear levers set for M12 thread generation

point. It also requires the cut to be applied using the Cross Slide and advanced for each pass until the correct depth is achieved. This approach is generally less controlled that utilising the combination of compound and cross slides when screw cutting.

The configuration of tool and cross slide can be seen in Figure 3.52. This has been set for a metric thread with a thread angle of 60°.

The workpiece needs to have been prepared for screw cutting in that the portion of the workpiece to have a thread machined into it needs to have a good surface finish and be of a diameter that is within the stated tolerance for major or minor diameter (if turning a female thread form). The start of the thread should have a chamfer applied that is slightly deeper that the depth of thread to protect the completed thread start. The end of the thread should also have an undercut or runout. This is a normal design feature and there are specific standards that denote the form of this feature. It is of key importance in screw cutting as it provides the runout point for the tool. If there was no undercut or runout on the work piece the tool would have to be stopped in exactly the same place as each pass was completed, or what is a fragile tool would impact a significant shoulder of workpiece material.

Figure 3.52 Compound slide and tool post set for machining of metric thread

Screw cutting on a manual lathe is usually undertaken at slow speeds with small incremental depths of cut. It is often stated that the cutting speed should be in the region of 25% of the normal material cutting speed. However, the primary reason for a slow speed is to remain in control and stop the machine before the tool impacts into any shoulder at the end of a pass. The length of any undercut provided to give a thread runout is often quite short and often within two thread pitches. This means that the operator needs to allow the tool to exit the workpiece into an undercut void within two revolutions of the workpiece less any overhang of the tool point. Typically this requires the operator to stop the machine by using the machine break within one revolution. A slow rotational speed provides the reaction time to do this. Thread cutting at speeds that were correct for the tool/material combination would not allow for the reaction of the operator, and this is why threads being produced on a CNC seem to be undertaken at a speed that is an order of magnitude faster.

Having set the gearing, cutting speed and oriented the tool post and compound slide, the tool point is brought very close to the workpiece and the cross slide and compound slide scales zeroed. A depth of cut is applied on the compound slide. This depth is generally quite small as heavier cuts will damage a fragile tool point and also prevent any corrective activity. Typically depths of cut would be in the 0.2–0.5mm range.

To begin to machine the thread, the machine is started in forward and the lead screw engaged using the handle on the machine apron. The tool will immediately advance along the workpiece and concentration must be maintained as generally the pitch of the thread is such that it quickly completes its pass.

As the tool exits the cut into the undercut or runout area, the machine is stopped by the operator stepping on the brake. The tool is withdrawn from the workpiece using the cross slide and withdrawn to a distance where the tool point is outside of the workpiece. If using a thread dial, the lead screw handle is released and the carriage wound back to the start. If the machine is not fitted with a screw cutting dial the lead screw is left engaged and the machine run in reverse to take the tool back to the start ready for the second cut. It is important to leave the lead screw engaged, or to utilise the screw cutting dial as each pass must commence in exactly the same position or a series of parallel grooves will be machined rather than a continuous thread. Successive passes are undertaken, with a small cut being applied on the Compound slide scale and the cross slide scale returned to its zero position. This process is continued until the correct depth of cut has been achieved.

The cutting of internal threads is essentially no different to external threads except that there are sometimes issues with the compound slide colliding with coolant pipes when rotated. There is also a restriction to the length of thread that can be machined, and the minimum diameter of workpiece as the shank of the tool impacts the opposite side of the workpiece because the angle it is rotated to is fixed to half the thread angle, hence restricting the length it can cut across the axis and the diameter it can enter. This means that for the majority of internal threads cut a simpler approach is taken. This often involves the use of taps to cut internal threads, which is easily, economically, and accurately undertaken up to quite large diameters before single point cutting is utilised.

It is therefore not uncommon to use a 'form tool' approach and not rotate the compound slide. In this case the tool is set up in a manner similar to that of a boring bar (i.e. clamped into the tool post along the axis of the lathe and brought into the work piece as for boring). The end of each cut needs to be accurately identified and the use of a bed stop can be an advantage when a clear view of a cavity bottom is unavailable. At the end of the cut the brake is used as for an external thread and the cross slide is rotated to move the tool out of the workpiece before engaging reverse.

3.24 USE OF THREAD DIAL

An alternative to leaving the lead screw lever engaged and running the machine in reverse to return the carriage to the start point for the next cut, is to use a thread dial where one is fitted to the lathe. The thread dial has a gear that is engaged in the lead screw and allows for the lead screw feed handle to be disengaged and the saddle returned to the start point by the operator rotating the saddle handwheel. The thread dial has a rotating disc, and register mark, and the handwheel is rotated until the number the disc was initially set to is aligned with the register. As the thread dial gear wheel has remained engaged in the lead screw thread the tool will be returned to a position such that it remains in sync with the pitch of the thread being cut. Once the tool is positioned off the end of the workpiece to be cut and the correct number on the thread dial lined up, the lead screw feed lever is re-engaged. Figure 3.53 shows a typical thread dial.

Figure 3.53 Thread dial

Thread dials typically have a number of divisions, and these are used when cutting multi-start threads. If a two start thread was to be machined a double pitch setting is selected in the gearbox, the thread dial set and aligned with a number or mark, and the thread cut. To ensure that the second thread form start point is exactly 180° from the start point of the first thread the number or mark on the thread dial opposite the initial mark is used, and the second thread form cut in exactly the same way as the fist but using the new number, or mark. Where thread dials do not have numbers operators are advised to mark the dial position with a removable mark to avoid confusion.

Failure to correctly return the carriage to the correct start point or fail to properly align the number or mark with the datum, will result in the pitches in a multi-start thread being out of phase, or for both single or multi-start threads, where cutting is approaching full depth will result in a heavy cut being taken down one flank of the thread form. This generally breaks the tool and also severely damages the thread form being machined. Where this is suspected, the operator should position the tool point just above the major diameter of the thread, switch the machine on and observe the position of the tool point following the thread. If it is ahead or behind the expected point, the thread position has been lost and needs to be re-established.

Thread dials can become disengaged. The dials are not always left engaged and can be swung out of engagement when not required. To engage the dial a screw or lever is loosened the dial assembly swung into position and the screw tightened. If there has been some wear or swarf has entered the mechanism it is not unknown for them to become disengaged. To pick up a thread when the position has been lost, with the machine switched off the operator needs to position the tool such that it has entered the thread groove and re-engage the lead screw feed using the handle. The tool should not be entered to full depth at this point to allow for any backlash in the feed mechanism, and once the operator is happy that this has been allowed for the machine is switched on. As the tool progresses down the workpiece the operator, uses a combination of cross slide and compound slide adjustment until the tool point is fully re-engaged. At this point the machine is switched off with the tool remaining engaged and the scales set once again to zero. The toll can then be withdrawn using the cross slide adjustment control wheel, the machine run backwards until the tool point is beyond the end of the workpiece and the machine made ready for the next cut. At this point the thread dial can be re-engaged, and the thread cutting completed in the normal manner.

3.25 MEASURING THREADS

As threads are cut it is important to determine the depth of thread cut an ensure it complies with the thread form sizes mandated within the relevant standard. There are three approaches taken to measuring threads, which are as follows:

- using a nut;
- three-wire method; and
- wire gauges.

Using a nut

The simplest, although least accurate method for measuring a thread is to assess the depth of thread being cut by using the scale on the compound slide and comparing it with the depth identified in the thread form standard. Note, this requires an allowance for the angle of the compound slide, as the tool is cutting along the hypotenuse of a triangle. This approach is backed up by observing the width of land remaining and comparing it with a standard thread, such as a bolt. As a complete thread form is being approached the machine operator tries to screw a nut of the correct size on the thread. When it freely screws on, the thread form must be complete. However, what this doesn't confirm is whether the thread is cut too deeply, or is within tolerance, it only confirms that the thread can have a nut screwed onto it. However, if it is cut too deep then the only indication using this method would be radial or trans-axial movement in the nut. Threads that are over cut run the risk of stripping when any significant load or torque is applied to the nut.

Three-wire method

Where a set of wire gauges are not available threads can be checked by using a micrometer to measure across three wires that have been inserted into the thread grooves. This measurement, as with most thread measurement is a check of the pitch diameter. The angle of the thread is controlled by the shape of the tool as is the root radius. For external threads the thread diameter will have been generated by turning to a diameter and tolerance, mandated by the thread standard. The pitch will be determined by the feed of the lead screw, which only leaves the most important dimension the thread pitch diameter to be measured. The pitch diameter is a point roughly half way down the thread face, and is the nominal point where two thread meet and load is transferred.

To measure the pitch diameter accurately inserting two wires into the thread groove on one side of a thread, and a single wire into the groove opposite and between the two, provides a surface for a micrometer to measure across. This technique is extremely accurate, however, it is also requires the operator to be adept at holding the wires in place while using a micrometer one-handed. Where regular thread measurement is undertaken it is common for workshops to have a floating micrometer, which can be fitted to the cross slide and hooks to suspend measuring wires from. However, where thread cutting is less regularly undertaken it is unusual to find specialist tooling. Where specialist sets of measuring wires are not available it is not unusual for drill shanks to be used. However, it is important to establish the correct size, if using tables to derive a correct distance over the wires.

Each thread form, and each thread size will require a different sized wire. This is to ensure that they are able to have a tangential engagement at the thread pitch diameter and preventing form entering the groove due to an overly large diameter, or too small and seat below the pitch diameter, and potentially the crest of the thread. Good-quality handbooks and internet resources all have formula for calculation of the correct wire diameter for each thread form, and size. However a rough approximation to identify a wire size for metric threads is 0.577 × thread pitch.

The value for the pitch diameter, will either be stated on the component drawing or available in many simple workshop pocket books which give thread details. Once the correct size wire has been selected the thread is measured over the top of the wires, and the value written down. Workshop hand books have formula for calculation of the pitch diameter from the diameter over the wires, and the difference between the value obtained and the value required is the depth of cut remaining to be take. However, to save the operator from the mathematical computations good workshop handbooks contain data tables for each thread form and thread diameter, giving wire sizes to be used and distances across wires for the correct pitch diameter.

If the operator has access to a workshop handbook, or internet resource they will be able establish the correct wire diameter to be used in advance, and be aware of the correct size, or toleranced limits to be achieved, prior to the commencement of thread cutting. This provides the operator with limits to be achieved and the depths of cut adjusted as the limit is approached.

Where sets of wires are not available, and correctly sized alternatives such a shard wire or drill shanks are not available, non-standard wires can be used. However, to identify the over wire size for the correct pitch diameter, trigonometrical calculations can be undertaken, or more simply a 2D CAD system used to accurately draw thread, identify the pitch diameter and then insert the wire diameter available. As long as the upper surface of the wire is shown to protrude above the thread crest a diameter across the wires can be established for the correct thread profile. This approach only works where wires, or drill shanks are available close to the optimum wire diameter. However, it does provide the operator with an alternative when no other options are available.

Wire gauges

Wire gauges are probably the easiest to use and as accurate as the three wire method. Wire gauges look not unlike a torsion spring, with a pair of opposing metal tabs attached. The tabs are compressed towards each other and the gauge screwed onto the workpiece. The operator releases the tabs and the gauge is measured over the wires. The upper and lower limits to be achieved in order for a thread to be manufactured to the correct sizes are printed on one of the gauge tabs. These limits do not relate to the effective diameter of the thread but of the tolerance band to be achieved when measuring over the wires. Achievement of a thread that sits between the upper and lower limit sizes will comply with the standard for the thread being cut.

3.26 CUTTING MULTI-START THREADS

Multi-start threads are where there are a number of thread 'starts' for a thread form cut into a workpiece. They are generally used to increase the speed of tightening a nut while retaining the thread form strength of finer thread forms. What this means to the machine operator is that the cutting of the thread is undertaken in the normal chosen method for single point cutting, but has the pitch increased by the multiple of the number of thread starts. A typical multi-start thread would be a two start type.

Two start threads are where there are two thread forms machined at 180° to each other. They combine to have the appearance of a single thread with the pitch between the two thread grooves being exactly that for a single start thread, and the only visible difference is the thread helix angle which is double that for a single start thread.

When machining multi-start threads the gearbox settings have to be adjusted such that the pitch the tool moves for each revolution is multiplied by the number of thread starts. An example of this is that the gearbox setting for an ISO metric coarse M20 thread (2.5mm pitch) would have to be set for a 5mm pitch if a two start thread was to be cut. This limits the number of starts a machine can cut, or the largest diameter that can have a multi-start thread. With the ISO metric coarse series of threads pitches increase until M64 after which they are 6mm. However, given that the majority of machines found in a general workshop have a maximum pitch of 14–16mm then there is a limit to two starts after M36. That said this is generally no great problem given the rarity of machining multi-start threads with greater than two starts.

The use of a thread dial is of greater importance when cutting multi-start threads, although is not essential. It is key however, that the commencement of each thread groove is the correct angular distance from the preceding one. In the case of a two start thread each groove should commence 180° from the first groove. Failure to achieve this will result in a pitch error between the two thread grooves identified by the land at the top of the crest of the first groove being eroded and the trailing thread having a wider land. Generally this error of pitch will prevent any correct nut being screwed on.

Where a thread dial is being used the second thread commences when the opposite number on the thread dial lines up with dial register. However an alternative or the approach that needs to be taken where no thread dial is fitted to the machine, is to cut the first thread to full depth. Place a dial test indicator on the bed of the machine with the DTI plunger placed against the tool, advance the cross slide one standard pitch distance, and then commence cutting the second thread in the normal manner chosen. Having offset the compound slide yet retaining the lead screw pitch set for double the standard thread pitch the second groove appears in the mid-point between the first thread groove machined.

3.27 TOOLING CONSTRAINTS IN MULTI-START THREADS

The number of thread starts increases the helix angle of the thread significantly especially as threads tend to be cut in even numbers. A two start thread has twice the helix angle of a single start thread, and a four start thread would have 4 times the angle etc. What is significant about this is that there is a minimum effective diameter that threads with multi-starts can be machined to. The principal driver is the ratio between the thread effective diameter and the effective thread angle thread and while a ⌀20mm metric thread with four starts would only have a helix angle of 9.04° there needs to be cognisance of whether the effective thread angle goes beyond a point that the thread was able to lock when mated with a nut. However, if the thread form for an M20 thread was used on a workpiece of ⌀100mm then the thread angle for a four start thread would only be 1.6°. From this it can be determined that there is a relationship between the number of thread starts, the thread helix angle, and the minimum diameter of the workpiece. The specific values for this are going to relate closely with the thread form being machined and the maximum angle that permits thread locking on full engagement.

A further constraint is the maximum pitch that the machine is capable of cutting a thread to. For many general workshop machines it is not unusual for this maximum to be 14–16mm and as such a limit for many machines would be an M30 four start thread, however, this would give a thread helix angle of approximately16° therefore this may not be a practical constraint unless holding power was not a requirement for the feature.

3.28 ISSUES WITH NON-STANDARD THREAD CUTTING

While ACME, Buttress and Square threads are recognised and well used thread forms, they are less common to cut than metric, unified or BA thread forms, although perhaps more common than Whitworth thread forms. Taps and dies are readily available, albeit at a cost for Acme threads, but tooling for square and buttress threads are harder to source, and while taps are relatively easily sourced split dies are much less so, or entirely unavailable and require single point cutting. The approach to single cutting these threads is not different than for any other thread form, in that a HSS tool is either ground and checked against a thread gauge of the type shown in Figure 3.48 above, or purchased as a carbide insert.

Thread forms such as buttress threads have one semi-vertical face, normally angled at 7° which is the load bearing contacting flank of the thread form, and a trailing angle of 45°. Square threads and acme threads are often used as biaxial drive threads rather than a unidirectional drive thread or locking thread such as a metric or unified thread form. Being a symmetrical thread form it can be used to drive on either flank, and this type of thread does not benefit from being advanced down a flank, which is entirely feasible for an Acme thread form, but not for a Square thread form, which must be plunge cut with the tool used in a similar manner to a form tool rather than one that is being used to generate a thread form. This increases the load on the tool as it is cutting on both sides at once and operators should take care to ensure that the depth of cut does not overload the tool leading to breakage, or tearing of the workpiece materials, leaving a poor surface finish. Cutting these thread forms may therefore take longer than common threads that can be easily and accurately generated.

Where more geometrically constrained threads are to be cut, such as square or buttress types, the depth of the thread and the thread helix angle becomes an issue. Square threads have a groove cut that is half the pitch of the thread form and the depth of the groove is equal to its width. Where smaller diameters are being cut with multi-start threads, care must be taken to ensure that the trailing edge the groove does not hit the top of the tool due to an excessive thread helix angle. Where this is a risk, it is normal to grind away any part of the trailing edge to form a relief angle similar to that found on the cutting edge. This ensures that the crest of the thread does not become eroded by fouling the tool training edge. That said this is unlikely to be an issue for any thread form other than a square thread, as buttress and acme should have enough thread trailing flank angle to avoid this problem.

3.29 USE OF WORKPIECE SUPPORTS

Occasionally, the standard lathe set-up of chuck and tailstock cannot accommodate, or facilitate effective turning of long think workpieces, or allow the facing or drilling of workpieces with a long overhang. To allow these limitations to be overcome, two types of steady can be fitted to either the lathe carriage, or the bed to provide additional support to the work piece and hold it steady while a machining operation is performed. The steadies are of two distinct types. A travelling steady is fixed to the lathe carriage and moves down the workpiece. A fixed steady is secured to the machine bed and does not move. Both fixed and travelling steadies can only be used with round workpiece materials. Therefore square or hexagonal sections cannot be used with these steadies, although if a workpiece has had a circular section milled on the end of a polygonal stock bar of sufficient width, it would be able to be mounted in a fixed steady.

Travelling steadies

When long thin workpieces are to be turned there is a risk that the workpiece will bend away from the tool as it naturally resists cutting. This deflection is not consistent throughout its length as the workpiece is well supported at the tailstock end and is progressively less well supported as the tool passes down the workpiece until it reaches the centre of the workpiece length, Support the progressively increases as the effect of the support given by the chuck or spindle centre, becomes increasingly apparent. The effect of this is for the cutting process to generate a slightly convex workpiece profile with the workpiece being the correct diameter at each end and wider in the centre where the tool has pushed the workpiece away and cut less material. The surface finish also degrades where the tool has not been cutting effectively. The amount of deflection, and the degradation of the surface finish depends on the workpiece material type, the relationship between its length and diameter, and its stiffness. Softer more ductile materials will tend to deflect more than alloy steels, for the same length diameter ratio. However, tubular workpieces, will tend to be stiffer. Irrespective of the length/diameter ratios, physical profile or material properties, this deflection can be overcome by fitting a travelling steady to the front edge of the saddle to support the workpiece opposite, and just ahead, of the tool point to prevent deflection occurring.

Travelling steadies tend to be a bespoke attachment for each design of lathe, and are either supplied with the machine or bought as an optional piece of tooling. A standard travelling steady has a curved or angular frame that supports two adjustable pins, mounted at 90° to each other. One pin is normally oriented horizontally to directly oppose deflection away from the tool point and one is mounted vertically to resist the vertical force component generated by the workpiece trying to lift away from the top of the workpiece. The pins are normally manufactured from a soft metal such as copper and can be advanced or retracted using a screw mechanism. They come into contact with the rotating workpiece, and will both lightly mark the workpiece and wear. Some designs incorporate small roller bearings at the end of the pins to remove frictional issues, and reduce workpiece marking, however, this introduces a potential accuracy issue, but allows easier support for engineering plastic workpieces. The majority of steadies have soft metal pins with metal to metal contact. Figure 3.54 shows a typical travelling steady fastened with bolts to the front of a lathe carriage.

Figure 3.54 Typical example of travelling steady fitted to lathe carriage

The size of the travelling steady is matched to the size of the lathe, and are only used on smaller diameter workpieces. That said some substantially sized travelling steadies do exist, such as those used on the very long bed lathes that turn large diameter but long components such as a ships propellor shaft. However, the effective length to diameter ratio for a given material which requires the use of a steady remains very similar.

When using a travelling steady the operator has to take a number of issues into consideration. Firstly when fitting the steady to the carriage, the pins of the steady should be retracted to allow clearance around the workpiece, and assist with bolting it in place. It is not uncommon for the pins to be very stiff, given that the pins are often dissimilar materials to the frame, and can corrode into the bore of the frame. Irrespective of the materials used, travelling steadies tend to be used irregularly and unless cleaned and lubricated when stored away, it's not uncommon for the threads and bores requiring some penetrating oil to be applied before they freely move.

The travelling steady is fixed to the front of the carriage to ensure that the contact point of the steady pins are in contact with the workpiece ahead of the cutting point. The tool point should be as closely opposed to the support such that the steady pins remain fully engaged on un-machined material. The diagram shown in Figure 3.55 shows an ideal arrangement.

While the longitudinal position of the travelling steady pins will always be fixed, given they are not provided with any adjustment for movement

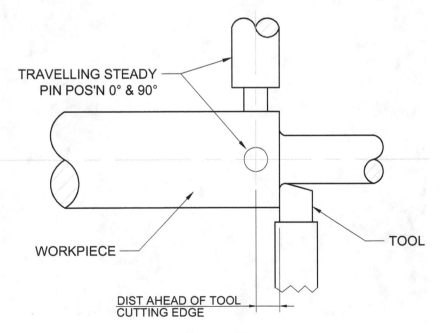

Figure 3.55 Travelling steady supports with respect to tool point

along the z-axis, the tool can be moved to an ideal position using the compound slide adjustment, and the handwheel is rotated until the tool point is aligned with the steady pin trailing edge.

With the workpiece held in a chuck and mounted on the tailstock centre, or held between two centres, the support pins are advanced by rotating the pin thumbwheel or other mechanism. This continues until the pin comes into contact with the workpiece. The operator should be careful of forcing the pin into the workpiece causing it to be deflected towards the operator or down towards the lathe bed as this introduces the potential for a deeper cut to be made than required. The steady is there to support, rather than provide a point loading of the workpiece. and both pins are adjusted until in contact with the workpiece.

Lubrication of the workpiece to reduce friction is much debated, with some guidance suggesting use of grease, tallow, oils or nothing at all. In general, at relatively high speeds oils and greases to do not stay in contact and get wiped off the workpiece. For slow turning speeds they become an option. Correctly set steady pins that also get the benefit of a coolant flow rarely suffer from any significant heating. The positioning of the pins slightly ahead of the tool means that any light surface marking is removed by the trailing cutting action.

As the travelling steady consists of a relatively substantial casting, or fabrication that is mounted to the front of the carriage, the operator needs to ensure that there is enough workpiece material remaining at the end of cut to prevent the steady from being driven into the lathe chuck. It is not unusual for in excess of 50mm of material to be left between the chuck and the end of cut to allow for the width of the steady. When turning between centres it is important to check that any driving pin arc swing is not ignored, and if required should be cut back to a minimum length to prevent a clash. Operators should check for any potential interference by moving the tool point to a point where it is at a position that represents the end of the cut and manually swinging the chuck to ensure that nothing will impact on the steady when cutting at speed. Where a specific length does not have to be machined to the depth stop should still be set to ensure the operator does not overrun, and accidentally drive the assembly into the chuck while concentrating on the tool point.

When turning using a travelling steady, material is removed using the same speeds and feeds as for any other material or diameter combination, and is only there to support the workpiece. Heavy cuts that generate significant heating at the support pin interface are probably best avoided. Between each pass the pins will need to be advanced until back in contact and each subsequent cut taken until the required diameter is achieved. When single point cutting long threads, the use of a steady should not wear the crest of the thread form as the loads are much reduced given the small depth of cut normally employed, however, the use of soft pins such as brass or copper should be used to ensure that there is no wear to the read form crest.

Fixed steadies

Fixed steadies have a completely separate function from travelling steadies. They are designed to support a workpiece that is of a diameter that is too large to pass through the central hole of the chuck, and is of a length that means the chuck jaws do not have enough engagement or resistance to the trans-axial cutting forces leading to the workpiece tearing out of the chuck. They cannot be used when the lathe is set up for turning between centres as there is no method of retaining the headstock end of the workpiece on the dead centre. Fixed steadies will retain a workpiece on axis but not along it, therefore, they are only used when a workpiece is held in a three- or four-jaw chuck.

Fixed steadies are directly bolted to the bed of the lathe and positioned at some convenient longitudinal position. They are used to allow the end of a workpiece to be faced off, drilled with a centre drill to allow subsequent workpiece tailstock centre support, or other machining operations. A fixed steady normally has three pins located 120° apart which contact with the workpiece in a similar manner to those used in a travelling steady. The steady normally has a hinged section to facilitate insertion of the workpiece, and is clamped down by a locking screw. Figure 3.56 shows a typical example of fixed steady with the hinged gate section open ready to accept a workpiece.

Prior to mounting a workpiece in the lathe when using a fixed steady support, pins need to be withdrawn to a point where the gate of the steady can be closed. The lower two pins should not be withdrawn so far the gap between the would allow the workpiece to slip through, but should also not be so advanced that the workpiece is canted upwards and cannot be properly fastened in the chuck.

It is common when using fixed steadies that the workpiece will require the chuck jaws to be widened beyond their normal maximum, however, most jaws on three-jaw chucks can be reversed, changed for a set of jaws designed for large diameters or a set of soft jaws, previously bored out to the correct diameter. The operator needs to carefully examine the type of jaws, as some standard jaws have a convex clamping face on the external steps of each jaw. These are primarily designed for clamping on the internal diameter of tubes, and while the jaws can be reversed, they provide a poor clamping holding, and workpieces have been known to be easily ejected from the chuck. However, some chuck jaws have a concave jaw profile and will provide clamping at each point, however, often mark the surface of the workpiece where they bite in. Figure 3.57 shows an example of convex faced jaws and Figure 3.58 shows concave faced jaws.

The ideal solution for the operator is the use of a second set of jaws designed for holding large diameters or the use of bored out soft jaws. That said these are not always available, and reversing the jaws is entirely feasible, however, the operator needs to consider how secure the workpiece holding mechanism is, given that large workpieces often have significant

Figure 3.56 Fixed steady ready to accept workpiece

mass, and release from the chuck while rotating is a significant safety hazard to the operator and has a high potential for damage to the machine.

The fixed steady is clamped to the bed of the lathe at a longitudinal position that leaves it as close to the end of the workpiece that is feasible, and the gate should be swung open to accept the workpiece when it is inserted into the chuck, and laid onto the lower steady pins. It is important

Figure 3.57 Convex faced chuck jaws (not reversed in chuck)

that the lower pins have been set to the same protrusion distance to avoid any *x*-axis error. This is normally achieved by simply measuring the protrusion of the pins from the steady frame using a vernier or digital calliper, or simply a steel rule.

Figure 3.58 Concave faced chuck jaws

Once the work holding arrangements have been made, the workpiece is inserted into the chuck, and where possible, supported by the operator while the chuck is tightened to ensure it is as aligned with the axis of the machine as possible. Where the workpiece is of such a size or mass that it cannot be supported by hand, the workpiece is placed on the lower jaws of the chuck, with the opening between the jaws has set to a diameter close to that of the workpiece, and rested on the lower pins of the fixed steady.

The next process is to align the workpiece with the axis of the lathe. Workpieces that self-support in the chuck are relatively straightforward to centre. The lower pins are advanced until they come into contact with the workpiece, The gate of the steady is closed and clamped in place with the locking screw. The upper pin can be advanced, however, it is sensible for the operator to not wind it down onto the workpiece until it is established that the workpiece is concentric with the machine axis, as the lower pins will be used to align the workpiece.

To align the workpiece, it is normal for a DTI to be used to identify any eccentricity. However, gross errors are normally removed by eye. If the end of the workpiece inserted in the chuck, as a square end then a quick check for squareness can be made by looking to see if the end face of the workpiece is butted up against all three, or four, chuck jaw faces, where this is the case the workpiece is then checked for concentricity using the DTI and hand rotating the chuck. However, it is common for the end of the workpiece to have been sawn off at a slight angle, or if the end of a bar to have a significant burr or distorted end. In this case having assessed by eye that the workpiece appears aligned with the machine axis, then the DTI is employed to 'clock in' the workpiece. Figure 3.59 shows a workpiece being 'clocked in' and it can be seen that the DTI has been placed as far along the workpiece as possible to ensure that the greatest deviation is measured.

It is usual for the workpiece to be marked at top dead centre on the periphery or end of the workpiece when commencing adjustments to ensure the operator has a datum mark. Once the DTI has been set with some displacement of its spindle and, for analogue instruments, the DTI bezel set to zero, the workpiece is rotated by hand through 180°. The bottom adjusting screws are then advanced or retracted to remove half the error and the DTI is zeroed and the workpiece rotated back up to top dead centre. This process is repeated until the workpiece is running true. The operator needs to recognise that stock bar itself is not always perfectly round, and irrespective of how well the workpiece has been clocked in a zero reading will never be achieved. It also makes sense to reposition the DTI to the 90° position, or use two DTI's to ensure that the workpiece does not have any significant lateral eccentricity. This is relatively unusual, if the lower pin distances have been evenly set, but checking is good practice. To remove any error on the x-axis a similar approach using a DTI is used, however, a final check with the DTI placed vertically should be taken. Once a concentric workpiece has been established the top pin is advanced until it is in contact with the workpiece. The pin should not be tightened such that significant pressure is applied to the workpiece as this causes the pin to be pressed into the workpiece, causing frictional heating and significant marking. While there will always be some marking of the periphery of the workpiece, it will be more significant where plain soft pins provide the supports. The workpiece will be resting on the pins, and the top pin merely has to be advanced until it touches the workpiece preventing movement. Figure 3.60 shows a workpiece having been clocked in, and ready to have its end face machined.

Lubrication of the workpiece is again subjective. Machining speed used should be set for the correct material removal rate for the workpiece material and the tooling being used. Where a relatively large diameter workpiece

Figure 3.59 DTI used to assess eccentricity during fixed steady adjustment

is being turned using HSS tooling the number of revolutions may be quite low allowing for greases or oil to be used as a lubricant. However, the contact of solid pins, rather than bearings, tends to scrape off greases and oils, and the use of coolant is often the best approach. A good flow of coolant directed onto the face being machined and onto the periphery of the workpiece will remove any significant frictional heating of the fixed steady pins.

Figure 3.60 Workpiece correctly mounted in fixed steady

The use of fixed steadies, especially those with soft fixed pins will mark the surface of the workpiece. However, a correctly set up fixed steady, with a workpiece that has been lubricated and cooled with a sufficient flow of coolant will only have light surface marking as seen in Figure 3.61.

The fixed steady is only there to support a workpiece, and turning operations undertaken on any part of the body or end of the workpiece are undertaken as for any normal turning operation. While there are rarely any issues with tuning undertaken on the workpiece at the tailstock end, operators may have some access issues when machining the body as it is entirely possible that the compound slide will clash with the frame of the steady. Where this may be a problem, operators should consider whether drilling of a centre hole in the end face and subsequently supporting the workpiece on a tailstock centre in a more mainstream fashion. Where this is not possible, the compound slide can be rotated through 90° degrees to reduce the length of clash, however, the width of the toolpost will ultimately control the minimum distance between the tool and the steady. Operators could consider placing the tool of the right hand side of a universal tool post, however, that introduces the potential for running the tool post into the chuck, and operators must carefully think through how the body of the workpiece will be machined, if a tailstock centre cannot be used.

Figure 3.61 Workpiece marking post machining using fixed steady

3.30 MANUFACTURE OF SPRINGS ON A LATHE

For obscure sizes, 'one-offs' or to obtain a spring not easily sourced from a spring manufacturer at an economic cost; it is perfectly feasible to manufacture springs on a lathe. The method employed is very similar to that used for screw cutting with a few additional and specific operations, linked to the design of spring required to be manufactured. However, it is key that the operator understands that the settings of the lathe are those required to manufacture the spring at its 'free length' and not for its length when compressed.

In order to manufacture a spring a mandrel has to be sourced. The diameter of this mandrel is exactly that of the spring internal diameter. Its length needs to be long enough such that the entire free length of the spring can be manufactured, plus an allowance for the wire insertion hole along with clearance at the headstock to ensure the Tool Post and wire drag device does not impact the machine chuck. Figure 3.62 shows an example of a mandrel for winding a spring with an internal size of 10mm. The mandrel needs to have a cross hole of a diameter slightly larger than the spring wire diameter. It is generally good practice to countersink this hole slightly to take away any sharp edge such that the wire does not fracture under tension.

Figure 3.62 ∅10mm mandrel for winding spring

It is also useful to mark the mandrel with the start point for winding the spring and the end point. These marks are a guide which aids manufacture, rather than something that is toleranced, however, some precision is always required especially where specific end conditions are a part of the spring design, such as 1.5 coil ends, or closed and ground ends etc. While multiple closed coils can be used up to the start point, as they are ground off post manufacture, a closed and ground end or requirement for 1.5 closed coils are a part of the free length and any additional coils would affect the spring rate of the spring. Therefore, once the mandrel start line has been crossed close attention must be paid to the number of closed, and pitched rotations. Similarly when approaching the second line indicating the end of the spring allowances must be made for the end closure. Once this line is passed additional closed coils are ground off post manufacture. Where no end conditions are required by the spring design, manufacture typically starts before any line, and finishes after with the final spring being cut to the exact length required post removal from the mandrel.

Manufacture of springs on a lathe requires some kind of wire drag device. While there are a number of commercially available devices many machinists manufacture their own. These drag devices add tension to the wire such that it is formed into a tight coil as it is wrapped around the mandrel, however, do not provide such tension that the ultimate tensile strength of the wire is exceeded and it snaps while being wound. This is particularly prevalent when winding stiff springs of smaller diameters using comparatively larger wire diameters to form the coils. Figure 3.63 shows a typical type of drag device used in spring manufacture.

Typical materials used in the manufacture of springs are stainless or carbon steels, such as piano wire, however, it is possible to manufacture them from almost any material, however, aluminium alloys and brasses make poor springs, but can be used for show. Some plastics can be utilised, however, their generally low levels of strength and ductility make them a poor choice for manufacture on a lathe, and are better purchased from a spring supplier.

The spring to be manufactured will be that identified by the design. It is not uncommon for a spring drawing to consist of a right-angled triangle with a set of loads identified at specific distance from the narrowest angle. These loads represent the pre-load and compressed load of the spring. Of interest to the operator manufacturing the spring is the free length represented by the bottom of the triangle, the number of coils, the wire diameter and material type, along with any specific end conditions. The free length may be dimensioned or stated in any notes or tables on the drawing. Care needs to be taken when examining spring drawings to ensure that spring diameter does not refer to the mean diameter i.e. the diameter representing the centreline of the wire, rather than the external diameter, as any error would affect calculation of the diameter the mandrel need to be manufactured to.

Figure 3.63 Spring wire drag device

To determine the machine pitch settings for a spring with no end conditions stated, it merely requires the machine operator to divide the free length by the number of coils, and the result of this calculation is the pitch that the machine needs to be set to, in exactly the same way that the machine would be set for cutting a screw thread. It is almost always that springs have larger pitches at smaller diameters than threads and correct gearing selection is crucial. It should also be noted that manufacture of larger springs will be restricted to those having a pitch no greater than the maximum screw cutting pitch such as 14mm or 16mm, and the wire diameter will be limited by the ability of the machine to withstand the forces involved in plastic deformation of the wire.

To determine the pitch for springs with a closed end type an allowance for this needs to be incorporated. For example a spring that had closed and ground ends would have one full turn of the wire at each end without any feed applied. Therefore when calculating the pitch, one wire diameter would be subtracted from the free length (as half of each end will subsequently be ground away) and the number of coils then divided into the remaining length to establish the pitch. If the ends were required to have double closed ends then four wire diameters would be subtracted from the free length, etc.

Once the correct screw cutting pitch has been selected the mandrel mounted and marked and the drag device fitted to the tool post manufacture can commence. A length of wire with sufficient length is cut from stock. This length includes the full amount required to form the spring spiral and any end closures an amount to cover the distance for the mandrel cross hole to the spring start point, and a remainder to cover the distance from the end of the spring to the drag device mounted in the tool post. The length is often a surprising amount.

The wire is led through the drag device and inserted through the mandrel cross hole until it protrudes out of the other side ensuring full engagement. The drag device is then tightened and the chuck rotated by hand to ensure a good lock of the wire through the cross hole is determined. Then with the machine set to a slow speed it is switched on and the screw cutting handle of the machine apron is then engaged. Figure 3.64 shows a plain spring in manufacture, and both the wire is clamped to the mandrel by a locking screw and wire feed from the drag device can be seen.

Figure 3.64 Spring in manufacture

If there are no end conditions required for the spring the wire will wrap around the mandrel at the set pitch until the second marked line is passed at which point the machine can be stopped, the remaining length of wire coming from the drag device snipped off with wire cutters and the section adjacent to the mandrel cross hole cut. The spring can then be slid off the mandrel and cut to the exact length required.

Where specified end conditions are required a similar approach is taken, however, as the spiral approaches the start point, the machine is stopped using the brake and the screw cutting feed disengaged. It is normal for the chuck to be rotated by hand, or with the assistance of the chuck key, to form the correct number of plain turns required (i.e. 1.5 etc.). The screw cutting feed is then re-engaged and the machine turned on. The spring is allowed to form until the opposite end is approached at which point the machine is stopped using the emergency brake, the screw cutting feed disengaged and the second end has its end close formed.

It is not unusual for springs to be nested in pairs. This is to increase the spring rate that can be delivered within a given pocket. Where springs are nested they are almost always manufactured with opposite spirals to ensure they do not bind. This means that one spring will have a right-handed wound spiral, and the second will have a left-handed wound spiral. To manufacture a left-handed wound spring the lathe is operated in reverse such that the chuck rotates away from the operator. The feed setting also needs to be reversed to ensure that the wire progresses towards the lathe chuck as for a right-handed would spring.

3.31 KNURLING

Knurling is a process by which a pattern can be formed in the surface of a workpiece. Generally there are two uses for knurling, which are to provide a surface which is easily gripped and can be held securely; or to raise a surface in an interrupted pattern such that it can be driven into a hole or cavity to retain a component. The second use is only used where it is impractical to utilise a standard interference fit.

Typically there are three types of knurl:

- diamond;
- straight; and
- diagonal.

There are other types, such as concave, however, these are rare and only require a plunge cut rather than continuous feed along the surface of a workpiece.

Common factors of knurling tools are the circular pitch of each tooth on the major diameter. The size of this pitch controls how coarse the pattern being produced will be and the depth of knurl. Typically knurling tools are classified as coarse, medium, or fine.

The finer the pitch the lower the depth of impression. Selection of the knurl to be used is normally stated on the component drawing, although it is worth noting that there is no significant material loss during knurling and the process deforms the workpiece surface. This means that as a pattern is indented peaks are raised above the workpiece nominal diameter and operators may often find that where sizes are stated on drawings for knurled features care needs to be taken to establish if this is the 'mean' knurled diameter, or an external 'nominal' diameter.

There are two generic types of knurling tool. Those that are used to 'push in', and those that clamp down. There is no better type, although push in types are perhaps more common due to greater simplicity and lower cost. Knurling tools consist of a pair of wheels and sometimes a third plain surface riding wheel. For diamond patter knurling the each wheel of the pair has a series of diagonal grooves machined into the wheel with the opposing wheel being mounted such that the grooves appear to be 90° to the first wheel. The advantage of the push in type of tool is that it allows for multiple sets of knurls to be fitted to one tool holder, often with coarse medium or fine knurls combined in one tool. Top knurling tools are essentially the same, however, have a clamp containing a pair of knurling wheels and a surface riding roller on the opposite side. The principal advantage of this type is it removes any axial bending moment, although excessive clamping force can lead to flattening of the knurl by the opposing roller in extreme conditions. Figure 3.65 shows a typical workshop 'push in' type knurling tool holding multiple sets of knurls.

Knurling is a comparatively forgiving process, however, for a high quality crisp knurled finish it is important that the feed of the machine is matched to the linear pitch of the knurl to prevent over knurling. The methods to achieve this can either follow those for screw cutting, or adjusting the machine gearing such that the normal feed per revolution using in parallel turning equates to that of the knurl. This may require adjustment of the gearing ratios to achieve, however, given the quality of knurl that can be achieved it is often work spending the time to adjust the machine, especially where longer lengths of knurl are required as changing patterns and over knurling detract from the component visual appeal.

It is normal for the knurling wheels to gather a small amount of swarf as a by-product of the knurl generation process. Prior to commencement of any knurling operation it is good practice to clean the wheels and remove any remaining swarf to ensure a clean knurl.

There is considerable debate about the correct speed and feeds for knurling and many operators have their own opinion. Literature shows that the surface machining rate for the workpiece material using HSS tooling is the correct approach with a feed rate in the region of 0.1mm per revolution.

Figure 3.65 Common workshop knurling tool

However, slower speeds at approximately half the surface machining rate can produce high quality knurls especially is significant amounts of coolant are being used to flush away any swarf generated that may degrade the knurled appearance.

Where 'push in' (side) knurls are used the amount of force being applied needs to be assessed. If thin bars of tool steels are to be knurled it is entirely possible that the force applied can cause the workpiece to bend away from the tool. Where this is apparent the use of a centre and/or travelling steady may be required. Where a knurled head is machined onto a long thin body, such as a thumb wheel on a threaded shaft an alternative approach is to knurl the section prior to machining the thin body to ensure there is enough rigidity. Similarly the operator could consider the sequence of operations and orientation of the workpiece such that any heavy loads applied by knurling take place adjacent to the chuck.

The actual process of knurling is straightforward. The knurling tool is inserted into the tool post in the same manner as any other tool. The speed and feed rate has been correctly set and the tool is then brought into contact with the workpiece at the location the knurling is to commence. The machine is started and the compound slide (for 'push in' type) rotated until a clear pattern is witnessed on the workpiece. For clamping type knurling tools the tool is applied over the top centre of the work piece and the clamp tightened until a suitable pattern generation is witnessed. The feed is engaged and coolant applied unto the entire required length of knurling has been formed at which point the feed is disengaged and the machine stopped. Figure 3.66 shows knurling in progress.

Figure 3.66 Knurling operation (worn rollers)

3.32 FEEDS AND SPEEDS

Setting of feeds and speeds on a lathe is relatively straightforward, and there is a vast amount of information available to assist the operator, especially that provided by tooling manufacturers. However, the machine operator only needs a few pieces of information to successfully configure his machine. The key information the operator needs is as follows:

- surface machining rate for the workpiece material;
- cutter material type (i.e. high speed steel or carbide); and
- recommended feed rate.

The surface machining rate varies for material types and is either quoted in surface meters/min (SMM), or surface feet/min (SFM). The value given is related to the way the material performs when placed under load by a cutting tool. This information relates only to the material, suppliers or stockholders rarely have this information, and unless the operator can access manufacturers data, the internet and good quality machining handbooks are often a good source of values for differing materials. It is also common to find different surface machining rates for drilling as opposed to turning or milling. As a broad guide, typical values for common engineering materials are shown in Table 3.1; however, experience or sourced data can provide an optimum value for each material type.

These values are a guide and the achievable rate varies with cutter design, material and the behaviour of the material. As a general rule, softer materials can be machined more quickly than harder materials, but some higher feed and speed rates on softer materials can cause tearing of the material and lead to poor surface finishes. Some experience and experimentation to balance the materials, cutters being used and machine dynamics assists.

Table 3.1 Typical values for surface machining rates using high-speed steel and carbide cutters.

	High-speed steel (SMM)	*Carbide (SMM)*
Mild steel	27	60
Carbon steel	20	40
Stainless steel	18	60
Cast iron (soft)	24	75
Cast iron (hard)	15	60
Aluminium	76	150
Copper	46	70
Magnesium	91	170
Bronze	46	70
Brass	46	70
Acetal co-polymer	400	800

As can be seen in the table, the cutter material (i.e. HSS or carbide) has a fundamental effect of the machining rate. While the table suggests that it is always going to be more efficient to use carbide tooling, that is not always the case as some HSS tools, can lead to better surface finishes on some materials. Of crucial importance is the tool material Machining rates are calculated for material removal at 'X' surface meters/min. This provides a standard factor accommodating cutters of differing diameters. A simple formula for calculating can be seen in Table 3.2.

Once the spindle speed has been calculated the operator changes the machine gear settings to the closest speed setting.

Feed rates can be slightly more complex to set, however are normally expressed or calculated as feed/revolution. Feed rates, as with machining rates can vary considerably depending on material type, and also cutter type. Recommended feed/revolution rates for carbide insert tools will vary from 0.2–0.9mm/rev (0.008–0.036") depending on the material type, however, HSS tooling may require a lower feed rate, and materials that are known to quickly work harden, such as some stainless steels, may require a slightly higher feed rate than any ideal calculated, which could be compensated for by slightly reducing the rotational speed.

Most lathes have variable feed rates which are changed using a gearbox setting. Where this is experienced the closest setting to that calculated is utilised. The plates fastened to the front of most machines identify not only pitch settings for screw cutting or worm gear cutting, but gear settings for the feed screw. Figure 3.67 shows a typical example of the gear settings for screw cutting and feed rates. The upper part of the plate relates to thread pitch settings, but the lower section relates to the feed rates.

It is worth noting that the feed rate for facing is often half that for turning, due to the way that the machine gearing works. It is rare for the operator to change the gears settings to ensure that a workpiece is faced at an optimum rate as it is unlikely that any time would be saved. The feed rate set is expressed as the amount of movement along the z-axis for each revolution of the workpiece. Coarser feed rates remove material more quickly and is more productive, but will provide a poorer surface finish than a fine

Table 3.2 Calculation of feeds and speeds

	Calculation of spindle speed
Data	S = spindle speed D = tool diameter (m) Vc = surface machining rate (SMM)
Formula	$S = \dfrac{Vc}{\pi D}$

Figure 3.67 Gear setting information for screw cutting and feed rate

feed. This can be partially offset by using a tool with a larger nose radius, however, this may be a problem when machining to a shoulder. In extremis this may mean that a series of roughing cuts are undertaken using a coarse feed, and the gearing changed to a fine feed for the finishing cut, or cuts.

3.33 USE OF COOLANTS

Most workshop machines have a facility for, or actively use coolants supplied through a pipe system to the cutting tool and work piece. The primary role of this flow of coolant is to reduce the temperature of the cutting tool and workpiece. The intention of this is to extend tool life, increase productivity and help prevent any heat treatment of the workpiece as a result of heat inadvertently introduced as a bye-product of the cutting process.

Most machining operations can be undertaken without using coolants, however, the use of them allows for increased feed rates, greater depth of cut and significantly increases the tool life. The application of coolant to the cutting area of tooling has a positive effect, although the degree of tool life extension is more marked for HSS tooling than carbide inserts.

The cutting action of a tool involves a shearing process where the material is deformed and converted to a chip, either continuous or as a series of small swarf chips. The majority of energy involved in this process is converted into heat which is focussed at the tool point and along the surface of the tool where swarf rubs over the surface of the tool. Some tools are more resistant to heat than others, and carbide tool inserts can withstand significant amounts of use without coolant, however, in time will develop thermal cracks which lead to cutting edge failure. High-speed steel tooling degrades quickly under significant load, and it is not unusual to see HSS tool cutting edges glow bright orange due to heat. Tooling under these conditions has a short life, and creates a poor surface finish. The addition of relatively small amounts of coolant rapidly quenches this heat build-up and substantially extends tool life.

The selection of coolant can be quite involved as they vary from specific cutting oils, solids and dry powders to gas cooling. However, most machines found in a general machine shop use water-miscible fluids. These fluids are an emulsion of oils, water and an emulsifying agent. The proportion of mixture varies, but a mix of 1 part oil/emulsifying agent to 40 parts water is quite common. The advantage of this mix is that water has excellent cooling properties, and the oil content provides good lubrication, with a secondary advantage of providing some corrosion protection to the machine. While there are a wide variety of oils which can be tailored to differing materials it is normal for a machine just to have one type which is used across a wide variety of workpiece materials. The only exception to this would be where a machine was being used in a repetitive production environment, and machining one type of material.

Delivery of coolant to the workpiece and tool cutting edge is normally via a sectional/flexible pipe, or a rigid pipe consisting of an elbow joint and extending delivery section to accommodate differing orientations. Both designs have their advantages and disadvantages, in that rigid pipes are stronger, but can be more difficult to keep oriented in the right direction, while sectional pipes can be more easily oriented but are more vulnerable, when swarf wraps or knocks the pipe away. Irrespective of this the machine operator needs to ensure that coolant continues to be delivered to the tool point as the cut progresses. When considering what orientation to deliver coolant to the tool and workpiece it is worth assessing where the coolant delivery pipe, and coolant stream will be at the end of the cut to ensure that coolant does not impact the chuck in any appreciable volume. While this does not cause any harm to the machine it is normal for the coolant to be sprayed out onto the floor below the guard, and out of the back of the machine creating a safety hazard.

While coolant manufacturers and literature suggests that coolants should be delivered at 10lt/min – 20lt/min, this represents a best case. In general it is not a requirement to deliver large quantities of coolant to the tool point which can often be a rather wet process. However, the delivery of a stream to the tool such that there is little or no steam witnessed rising from the tool/cutting area is often a good indicator. Workpieces that are more than hand hot when checked at the end of each cut are an indicator of insufficient cooling. Poor surface finish and 'blueing' of swarf chips is also an indicator of insufficient cooling.

As stated earlier, it is not essential to use coolant, however, its use can have a marked difference. When machining materials such as mild steel peck depths of 2.5mm are not unusual for an average machine. If tool life is to be maintained for any appreciable period, and to prevent the workpiece from becoming excessively hot, this depth may need to be reduced to approximately 0.5mm.

When machining more exotic materials such as magnesium or metals that may react with water no coolant would be used. In general this should not be a problem for the operator, as the cutting speed and feed rate should not be so high that heat does not dissipate fast enough, and the operator needs to have an awareness of these marginal conditions.

3.34 BEHAVIOUR OF DIFFERING MATERIALS

When machining materials on a lathe it is often seen that differing surface finishes and a wide range of chip formation is produced for the same machine settings. Some of this is directly related to speed and feed settings for differing types of material and the machine operator needs to have correctly set the machine speed for the material removal rate and tool type being used.

While the surface removal rate will stay the same as diameters are reduced the speed should be increased to retain optimum material removal conditions. While this is not reasonably achievable for facing operations unless the machine is fitted with a variable speed controller rather than gearbox, setting the machine speed for turning an external surface, and later boring out a cavity at a much smaller diameter, should involve the operator remembering to change the spindle speed.

Differing workpiece materials and differing types of tool will produce differing types of chip. Drilling into workpieces especially when using larger HSS drills will produce long continuous chip as long as the operator winds the tailstock quill feed continuously. These continuous chips can be both hot and sharp, and the operator only needs to cease advancing the quill for a single revolution to manage the length and remove a safety issue.

Similarly heavier cuts of aluminium and some steels or plastics on higher feed rates will produce long continuous chips. This in itself is not any significant hazard, however, the swarf often tends to wrap itself around the workpiece and as it gets pushed into the lathe chuck begins to fragment and be ejected from the machine, and scratches the workpiece creating a poor surface finish. The selection of carbide inserts for the material being machined having chip breakers moulded into the insert upper surface is one way of avoiding this.

Machining some plastics creates significant heat, and significant tool wear. The operator needs to ensure that the heat generated when machining plastics does not start to burn the workpiece. This is controlled by reducing the depth of cut and feed rate of the machine.

Cast irons, and phenolic resin based workpiece materials can be a joy to machine. The metal swarf produced tends to be in the form of a fine chip, and presents little problem for the operator. However, when machining materials such as phenolic resin based workpiece materials such as Tufnol, a fine dust is produced and it is unusual to use any coolant when machining this type of material. Coolant is not generally used as the heat is removed in dust production, but also it clogs the coolant recirculation system, leading to reduced flow or failure. Any fine dust is easily inhaled by the machine operator, and consideration should be taken to either providing dust extraction directly above the workpiece or the operator wearing a dust mask of an approved type.

Materials such as cast irons, brasses, bronzes, most aluminium alloys, mild steel and alloy steels cause little problem to machine operators, when the correct speeds feeds and tool tip inserts. However, some materials such as some stainless steels, titanium and titanium alloys can demonstrate significant work hardening. When performing some operations such as drilling these materials, especially when using HSS drills it is important that the operator continues to cut until the depth required, or amount of material is removed. Failure to do this will result in excessive work hardening of the face being machined due to frictional heating. It is not unusual for this surface to become harder than the cutting tool such that a tool cutting edge cannot be maintained leading to tool failure and an inability to complete the cut. Preparation, focus and a confident operator is the methodology to avoid this problem.

Other materials such as extremely soft or tough alloys, leads and some plastics require skill to machine if a good surface finish is to be achieved. Key to achieving this is correct tool selection especially with respect to the tool point radius, as is speed and feed selection. Where tool selection is not optimum tearing of the workpiece can be observed creating a rough surface finish which is both unsightly but may also be dimensionally unacceptable. It is recognised that some designs require sharp corners at the root of shoulders, which prevents the use of tooling with large nose radii, and the operator through judicious use of roughing cuts, changes to the speed and feed settings, and some practice can achieve a reasonable finish.

Some of the softer engineering polymers, and harder materials such as glass filled nylon can cause significant tool wear. It is not unusual for tools to wear out more quickly when machining these types of materials that alloy steels, irrespective of optimised machine settings. Operators need to examine the tooling used when machining engineering polymers to check that the cutting remains in good condition prior to any degraded workpiece surface finish becoming apparent.

Chapter 4

Milling

4.1 THE MILLING MACHINE AND DEGREES OF FREEDOM

Milling machines differ from lathes in that the workpiece does not constantly rotate, but is passed obliquely under, or against a rotating cutter. A standard workshop lathe is only used for the generation of round features, whereas a milling machine has the ability to produce flat surfaces, slots, squared features, elliptical and helical features, as well as round surfaces and bored cavities.

There are three basic types of milling machine, which are the horizontal mill, end mill, and turret mill. Figures 4.1, 4.2, and 4.3 identify the three types respectively. Horizontal mills and end mills have almost entirely been replaced by either computer numeric controlled (CNC) machines or turret mills, which is the most common type of milling machine found in today's workshop environment. There are many similar features across all three types of milling machine. All milling machines have a strong base from which a column rises at the rear of the machine. This column has a knee attached to it which can rise up and down on the z-axis. Attached to this knee is the machine table, which can travel left and right on the x-axis. The knee has handles or hand wheels for raising and lowering the machine knee and operating the cross slide (y-axis), and all machines have the handle or control wheel for moving the table on the x-axis located at one or both ends of the machine table.

Fitted to the top of the column on a Turret mill is a dovetailed ram which mounts the motor and spindle assembly which is located on the z-axis. This ram can move forward and backwards on the y-axis, and is commonly fitted with the ability to traverse on the z-axis. It is also common for the head assembly to be able to rotate on the y-axis, and have some capability of rotating on the x-axis as well.

Vertical milling machines tend to have less flexibility in their design, as the motor tends to be located within the machine column, and as a result the head is fixed, so there is generally little or no option for a sliding ram, or

DOI: 10.1201/9780429298196-4

rotational features that can be commonly seen on turret mills. While this can be a significant limitation, is does provide a more rigid machine.

A horizontal milling machine has many similar components with an overarm being mounted on top of the column from which an arbor support is hung. The spindle is mounted on the y-axis below the overarm and the

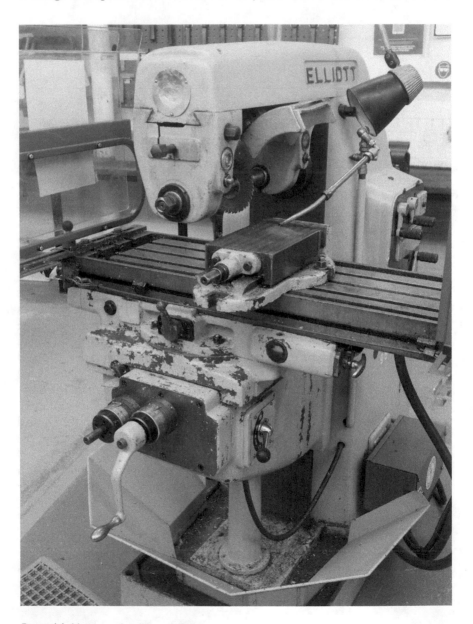

Figure 4.1 Horizontal milling machine

Figure 4.2 End milling machine

arbor support is used to support the arbor cutter assembly inserted into the spindle.

The turret milling machine differs from and end milling machine by having its motor mounted on the top of the milling head adjacent to its spindle.

Figure 4.3 Turret milling machine

This allows for the head of the machine to be rotated in at least one axis, providing the ability to easily create angled features without having to reposition the workpiece or to create elliptical features.

The number of degrees of freedom a milling machine has is key to its flexibility to generate shapes in materials. All milling machines are capable of translational movements in three axes, which, when the operator is standing at the front of the machine facing it, are as follows:

- x-axis = machine table move left (+ve) to right (–ve).
- y-axis = machine table moving towards operator (+ve) or away from operator (–ve).
- z-axis = machine table moving vertically up (+ve) or vertically down (–ve).

However, milling machines also have rotational movements along these axes. In turret and end milling machines there is always a rotational movement on the z-axis which is used for generating motion of the cutter. In horizontal milling machines this rotational movement is provided on the y-axis where a cutter is mounted onto a horizontal arbor. Therefore, as a minimum milling machines have four degrees of freedom, three translational and one rotary. This can be increased by the utilisation of a dividing head mounted onto the machine table. The dividing head, either powered or moved using the manual winding mechanism provides a fifth degree of freedom as it introduces a rotational movement on the x-axis. The combination of translational movements, coupled with rotary movements and the ability to manipulate the angle of the turret mill head allows for extremely complex shapes to be produced and a milling machine that has a dividing head option can be considered a five-axis mill. Figure 4.4 provides identification of the translational and rotary movements that create five degrees of freedom.

The method by which a milling machine removes material is by clamping a workpiece either directly to the machine table, or utilising a vice or other fixture and then feeding it into the rotating cutter in a controlled manner such that the cutter removes material.

4.2 MACHINE CONSTRAINTS

There are a number of constraints to the size and complexity of the shapes that can be generated on any machine, and the milling machine is no exception. To determine the basic lengths widths and heights that can be accommodated on the machine, assessment of the length of traverse on the x-, y- and z-axis is a good starting point. However, when assessing whether a machine is correctly sized the type of operation and size of the cutter also need to be factored in.

An example of this is where a workpiece is 350mm high and requires a 75mm deep slot to be machined into its upper surface. On a machine that has a maximum height traverse of 400mm, this would require the tool to have a 25mm deep cut on its first pass. While this could be countered by utilising a stub end mill to take the first few cuts and then changing to a longer tool; assessment of whether a workpiece will fit into the machine envelope must include assessment of tool lengths and depths of cut. Figure 4.5 shows the maximum workpiece height that would fit under a cutter, and

Figure 4.4 Turret milling machine axis notation

illustrates the effect that cutter length has in reducing maximum workpiece height.

Similarly the length of traverse must also allow for the diameter of the tool such that it clears the workpiece at each end of the cut. Some face

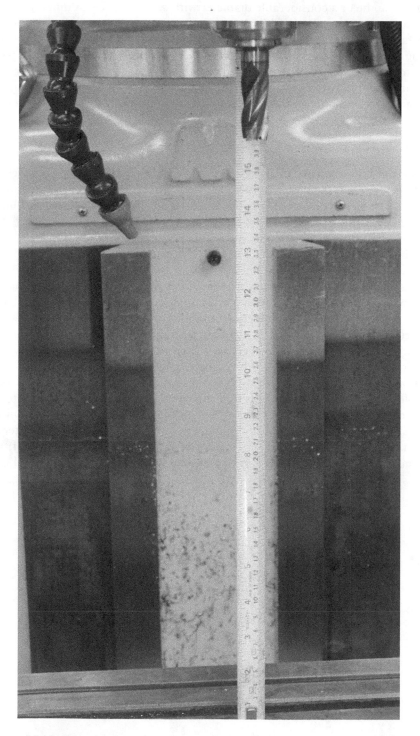

Figure 4.5 Maximum height under cutter on turret milling machine

mills can be of a considerable diameter with cutters $\varnothing$75–100mm not being in any way uncommon. While machining a face to a size would only require a face mill to have passed over the workpiece for half its diameter (i.e. so the full width of cut is achieved). However, failing to allow the tool to pass completely off the workpiece when cutting has finished often leaves surface marks due to swarf scratching, movement in the spindle when not under load or backlash when the direction is reversed.

4.3 DIRECTION OF CUTTER ROTATION

There are two approaches to material removal when milling and these relate to the direction the cutter is rotating in relation to the machine direction of feed. The two types are normally classified as 'downcut' milling or 'climb milling', and the second type is 'upcut' or 'conventional' milling. Figure 4.6 shows the difference between conventional and climb milling.

In upcut milling the cutter rotates against the direction of feed and the swarf chips are rotated up and away from the cutter. This type of milling uses significantly less power than down cut milling, however, often has a poorer surface finish than down cut. In some conditions the cutter can have a tendency to dig into the material and this tendency while providing poorer surface finishes, can make holding tight tolerances a challenge for the operator. The chip starts being formed with no width and increases to full width as the cut progresses. This produces increasing heating and reduces tool life. Workpieces being conventionally milled can often require more robust work holding devices as the cut tends to lift the workpiece and the action of lifting up chips can produce re-cutting of chips, further degrading the surface finish.

Downcut milling, or 'climb' milling, is where the rotation of the cutter is in the direction of the feed. The load is reduced on the cutting edge of the machine and the swarf chip is rotated round with the cutter and ejected out

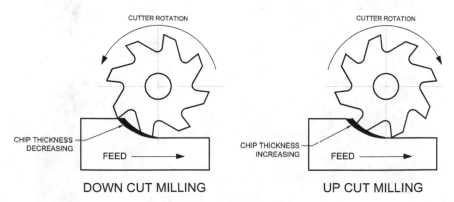

Figure 4.6 Conventional and climb milling

behind it. This approach requires more machine power. However, it provides a better surface finish and due to the lack of tendency to dig in, can hold required tolerances more easily, and generally enjoys slightly better tool life. The chip formation starts at full width and reduces through the shearing action and as such is cooling. It is this mechanism that assists with longer tool life. As the chips are deposited behind the tool it is unlikely that there will be any chip re-cutting that might impact on the surface finish.

4.4 MACHINE BACKLASH ISSUES WHEN CLIMB MILLING

One limitation of down cut milling is that it requires a machine that is either fitted with a backlash adjuster, or is a solid, well-constructed machine without any wear or significant play in its lead screws.

Where the machine has significant play it is not unusual for the machine to 'shudder'. This motion is a result of the workpiece initially resisting the cutters shearing action and the table is pushed away from the cutter due to clearance, or back lash, in the lead screw. However, as the table continues to advance the cutter begins to shear the metal and the table is dragged through until it comes up against the flank of the next thread on the lead screw. This shuddering can sound quite alarming to the machine operator, and while not disastrous, will provide a poor surface finish and is indicative of too greater cut depth and an aggressive feed rate being employed by the operator for the condition or rigidity of the machine.

In extremis where machines with significant wear are being used it is possible for down cut milling to pull the workpiece through. The cutter drives the workpiece through biting into it overcoming the friction of the thread helix at which point it begins to rapidly rotate until either the cutter impacts with the end of a pocket, or the table stops are reached. This is not a desirable condition as it is normal for this to cause significant damage to the workpiece and generally snaps the cutter.

Down cut, or climb, milling is the generally preferred method given its significant advantages. However, if the machine has significant wear or no back lash adjuster fitted then the operator needs to consider whether the depth of cut be reduced from optimum, and/or the feed rate similarly slowed to reduce the load on the machine. It is common for machine operators to utilise the table clamps to assist with reducing backlash. However, this is abuse of the machine, increases the power required to deliver the cut, and as such is not recommended. A more measured approach is to reduce cut depths and feed rates is recommended.

4.5 WORKPIECE SET-UP

There are a number of ways that a workpiece can be set up on a milling machine but key is ensuring that the maximum number of operations can be completed without having to move the workpiece. The reasons for this are twofold. Firstly, by undertaking multiple operations to generate features, any errors that could be attributable to moving the workpiece are removed. An example is where parallel slots are being machined, any movement of the workpiece between machining each slot may lead to a lack of parallelism. Similarly, if a square aperture is to be machined, mounting the workpiece such that slots or a pocket can be machined without movement of the work piece then any error in squareness achieved will be that of the machine, rather than any attributable to movement of the workpiece between operations. The second advantage is that setting up, clamping and ensuring squareness is one of the more time-consuming tasks, and greater operator productivity is achieved when the number of workpiece mounting operations are reduced.

There are a number of ways that a workpiece may be set up and clamped to the machine table ready for machining. The simplest and quickest method is to hold the workpiece in a vice. While these are a quick and reliable method the size of workpiece is limited to the width the jaws can widen to, and the longitudinal support is limited to the width of the vice jaw. However, larger items are often either directly bolted down to the table using standard bolts and tee nuts, or using slotted, or swan neck clamps. While this is often a more complex set-up task, it allows for multiple clamps to be used, allowing for removal, and reinsertion as a tool passes along the workpiece, greater stability. It allows for the mounting of irregular shaped workpieces, such as rough castings which could not be accommodated in a vice and may require the use of screw jacks to establish a flat plane to commence machining.

When setting up the machine to undertake a task the operator needs to consider the sequence of tasks to be undertaken and assess the order in which they should be undertaken. There is no one best method of commencing or undertaking tasks, and most operators will have a preferred style. What is important is giving consideration to the most efficient method of material removal, consideration of swarf egress channels, and the generation of datum surfaces for subsequent feature generation. Another key aspect being identification of the tool path such that impact with any clamping mechanism is avoided.

An example of this is where a casting needs to have a flat face machined, a longitudinal slot and a series of holes drilled through. The operator could choose to machine the flat surface first to provide a datum face to more easily establish the depth of the slot which would be the second operation. However, drilling of the holes and machining of the slot could be undertaken first, in order to provide an egress route for swarf and coolant, prevent overcutting and scoring of the flat machined surface. The correct sequence, is a combination of operator style and the types of operation to be performed and experience normally identifies the best sequence.

Another important workpiece set-up is establishing an initial datum. When milling it is important to understand where the centreline and end of the tool is in relation to the workpiece, in the x-axis, y-axis and z-axis. There are four basic methods of establishing these datums which are:

- use of a probe linked to a digital read out (DRO);
- use of a centre finder (wobbler);
- use of cigarette papers; and
- trial cut to establish a datum.

Probing is quite rare in manual machining given their expense, however, some more modern and advanced machines have spindle and tool probes that are linked to a machine DRO. These probes require the operator to bring a probe fitted to the machine spindle and brought into contact with the workpiece in three planes, and a second probe which is used to probe the tool end and side. This probing allows the machine DRO to understand where the workpiece is and establish a datum. This technique is widely used in the majority of CNC machines, however, while rare it does exist in some manual machines and provides a highly accurate method for the operator to establish a datum, prior to commencing machining.

The use of a centre finder (or wobbler, as they are colloquially known) is widespread. Figure 4.7 shows an example of a centre finder.

The tool is fitted into the spindle of the machine and the operator uses their fingers to guide the centre finder spindle until the ball on the end is running true. The height of the ball is adjusted by raising or lowering the table on the z-axis until the periphery of the ball is just below the top edge of the workpiece. The machine table is then traversed along the x-axis until the workpiece is very close to the ball. Any rapid traverse is then stopped and the operator slowly traverse the table by hand until the ball touches the workpiece. Touching the workpiece immediately moves the ball out of alignment and centrifugal force throws the centre finder spindle to its extremity. From that point the operator traverses the table half the diameter of the centre finder ball, which are typically $\varnothing$6mm or $\varnothing$4mm. At this point the centre of the machine spindle is directly of the datum edge on the x-axis. The same technique is used to identify a datum edge on the y-axis. The use of a centre finder to identify a datum surface for the z-axis is possible, however, cannot be considered quite as accurate as the x and y-axis datums. The centre finder is stabilised and rotating and brought down until it touches the surface of the workpiece. The majority of the time the centre finder spindle will spin out, however. not always. What also needs to be established is the distance from the machine spindle nose to the end of the centre finder, which can be troublesome to achieve with any accuracy, and then an offset applied when the tool length of the tool being used to machine a feature is established. While centre finders are an excellent method of establishing a datum for the x-axis and y-axis, they lack utility for the z-axis. Machines that have fully enclosed guarding are problematic for the operator to access the rotating

Figure 4.7 Centre finder fitted to machine spindle and ready to probe

spindle to get it running true in order to use it, and use is limited to machines that have spindle guarding only.

The classic method, and one that is surprisingly accurate, is to identify datums using cigarette papers. This method provides a quick and easy method of identifying datums for the x-axis, y-axis and z-axis irrespective of the guarding system in place on the machine. Depending on the make of cigarette paper they tend to be in the range of 0.001–0.003″ of an inch thick (0.025–0.076mm). A cigarette paper is soaked in water or coolant and applied to the surfaces for which the operator is trying to identify a datum. An example of this is shown in Figure 4.8.

The revolving cutter is then gently brought into contact with the cigarette paper in a similar manner to that used for a centre finder. As the cutter makes contact with the paper it is eject from the face it was applied to. For the x-axis, and y-axis, datums, the table is lowered until the end of the cutter is above the workpiece, and the table then advance half the diameter of the cutter and the thickness of the cigarette paper. The DRO or scales are then set to zero as the centre line of the tool is directly over the datum edge. Using cigarette papers to establish the z-axis datum is undertaken in very much the

Figure 4.8 Cigarette papers applied ready for datum definition

same way. The table is carefully wound upwards until the end of the tool connects with the paper and sweeps it away. An adjustment is then made for the thickness of the paper and the scales or DRO, set to zero. The widespread use and true advantage of this approach is not only that it is quick and easy, but also it lends itself well to workpiece set-up of components that have already been partly machined and that accuracy is important.

The final method for establishing a datum is the simplest, and just involves taking a light first cut. This approach lends itself best to an initial roughing cut to create a flat face or datum surface for subsequent operations. The approach is simple the cutter is brought close to the surface of the workpiece and the machine rapid traversed along the x-axis and y-axis to check there are no significantly raised areas that may bring the cutter into an overly deep cut. The cutter is then traversed off the workpiece, a light cut applied across the workpiece. This approach applies for all faces of the workpiece and the operator merely has to choose which edge they wish to use as the axis datum, and the scales and/or DRO are set to zero.

4.6 SQUARENESS

Squareness of a workpiece when being set-up can be of significant importance. It is, however, easy to introduce errors if some simple checks are not made. The use of vices to hold a workpiece is quick, convenient and a widespread practice, however, they have two common problems. First the fixed jaw of the vice needs to be perfectly in line with the table axis or an angled feature will be machined as the workpiece is passed under the cutter. The second most common error when using vices is the workpiece being rotated between the jaws as wear in the sliding jaw, or limited precision in the vice manufacture results in rotation.

Clocking in a vice using a long parallel clamped in its jaws, is the method by which longitudinal squareness is established. However, machining a block that is essentially a square cube with all faces being 90° to each other requires more care.

Initial roughing operations to create a basic block that has multiple square faces requires some thought. Material that has a poor surface finish that may not be necessarily flat, such as hot rolled steel or metallic castings can be problematic. The material is placed in the vice and an initial cut is taken. The operator then has a choice to rotate the workpiece 180° and machine the opposing side. Using this approach allows the operator to use two machine parallels that 'lock up' and the upper surface is machined. If the parallels were 'locked up' the two surfaces will be parallel. The next operation requires the workpiece to be rotated by 90°. It is important that one of the previously machined surfaces remains in intimate contact with the fixed jaw face. If the vice is of poor quality or has significant wear, it will try and rotate the workpiece producing a face that is not 90° to the previously machined surfaces. The operator has a number of options to counter this. One method is to use a round bar between the workpiece and the sliding jaw. Figure 4.9 shows an example of this, and this approach ensures that angular accuracy is maintained as the flat face is pushed against the vertical fixed jaw.

A similar approach can be established by packing the vice jaw with shims or paper. Whatever approach is taken the intention is to ensure the fullest engagement possible with the vertical face to ensure that any error of squareness, is that of the vice, rather than induced error by the work holding device. Figure 4.10 shows paper packing, and while care needs to be taken

Figure 4.9 Clamping with round bar between sliding jaw and workpiece

Figure 4.10 Clamping with paper packing between sliding jaw and workpiece

to ensure full contact with the fixed vice jaw face it overcomes the problem where use of a round bar provides reduced surface contact, that could lead to the force of machining overcoming the friction of the work holding device and ejection of the workpiece.

Parallels are almost always employed when using a vice, but they can only be used to determine if a workpiece is seated correctly, trapped between the vice slide-way, or seat, and a machined face. Once a workpiece has been roughed into a square parallels become a good indicator of squareness.

Universal vices require additional checks. The rotational movement on the *z*-axis is removed through 'clocking' in the same manner for a fixed vice. The rotational movement on the *x*-axis or *y*-axis requires more care, and again is checked using dial test indicator 'clocking'. It is always worth the operator checking for any movement away from the vice 'zero' angles and either recording these. or marking them to ensure they can be returned to zero if moved.

4.7 CLOCKING' IN WORKPIECE/WORK-HOLDING DEVICES

When a vice or any other independent work holding device is fitted to a machine table a dial test indicator needs to be used to ensure that the fixed jaw is parallel to the axis of the machine. The process is often referred to as 'clocking in' due to the principal method using a dial test indicator, or 'clock' as it is often called. Failure to ensure that the work holding device is square to the machine axis will result in the subsequent machined slot or edge having 'runout' in relation to its datum edge.

While the process of 'clocking in' can be somewhat tiresome, it rarely takes long, and while it can be quite quick to get to a point where the squareness is well within any tolerance limit for runout, it is worth spending the time to ensure that the work holding device is set as parallel to the axis as is possible. Some vices and work holding devices such as dividing heads, are fitted with square registers into their base. Figure 4.11 shows the base of a dividing head which is fitted with registers to assist with alignment when attaching to the machine table. While these have been accurately fitted by manufacturers, there is always some clearance between the machine table slot and the work holding device register, and while generally a quicker process, these still need checking.

Simple types of vice rarely have registers fitted to their base and often the simplest way to get it aligned is for the operator to push the vice over up against the finger tight clamping bolts. Given that the clamping bolts are normally the same diameter, and the slots in the vice base are normally aligned with the fixed jaw, or at 90° to it, they can be a quick method of getting the vice roughly aligned and the mounting bolt nuts tightened slightly. This requires sufficient force that the vice cannot be moved by hand, but

Figure 4.11 Dividing head base registers

also that the vice can still be moved in small increments using light taps of a mallet.

The next process involves the clamping of a parallel bar of metal. Often a long machine parallel or piece of gauge plate that is known to be flat and straight is used. As a rough guide a section that is twice the width of the vice jaw is place into the vice so there is an even overhang either side and clamped. This item is not going to be machined so clamped by hand is sufficient.

A dial test indicator (DTI), normally using a magnetic base, or other device that prevents the indicator from moving in use is fastened to any rigid part of the machine that does not move with the machine table. The arm of the DTI and its head are then adjusted such that the plunger of the DTI is approximately at 50% of its displacement when pressed against the surface of the plate or parallel clamped in the vice jaw at approximately its centre point. Figure 4.12 shows the initial set up.

Once the DTI is in position it is normal to use rapid feed, or wind the table until the plunger of the DTI is close to one end. The bezel of the DTI is rotated until the it is 'zeroed' and rapid feed or hand traversing of the table is undertaken until the other end of the parallel or gauge plate is reached. The value for runout is then taken. The value will be either positive or negative and relates to whether the plunger has been depressed more during the traverse (+ve), or has extended (−ve). Irrespective of this, the vice needs to be rotated clockwise or counter clockwise, using light taps of a mallet, until half the positive or negative displacement is removed. The DTI is then zeroed again and the table once again traverse to the far end and a second reading taken. It is possible that the error has been completely removed and the pointer of the DTI stays at exactly zero, however, this is exceptionally rare. It is more normal for the DTI to show a runout, often indicating too much force has been applied by the mallet, however, the process of halving the errors while traversing the table left and right and adjusting the vice accordingly until zero runout is achieved.

If it is not possible to achieve a reducing runout, it often indicates that the vice clamping bolts are too slack, and require tightening slightly. Similar problems can arise from having the holding down bolts too tight, which requires heavy blows from the mallet to get movement. This often causes the vice to bounce rather than rotate, and little improvement to runout is observed by the operator and leads to frustration.

Once the runout has been removed the holding down bolts need tightening. Care must be taken to progressively tighten each side as it is possible to move the vice if heavy force, or hammering is used to tighten on side and then the other. Irrespective of the care taken it is good practice to check for runout after final tightening to check for movement. If this has occurred the nuts need to be slackened off and the process repeated.

For work holding devices such as dividing heads the process is exactly the same only that it would be normal to use a round bar held in the chuck,

Figure 4.12 Initial DTI set-up for 'clocking in'

however, it is also important to ensure that the head is not elevated and that the datum line can be seen to perfectly align with 0° on the dividing hear elevation scale. Failure to do this will result in runout being observed on the DTI scale, however, this will be compounded by the plunger of the DTI travelling up or down over a curved surface, and the observed error will partly, or in whole, be due to this tracking over a curved surface rather than a significant problem of alignment with the machine table.

This process is also to be used when workpieces are directly clamped to the machine table or mounting tooling such as angle plates are used. In these situations a straight workpiece datum edge would be utilised, or the face of the angle plate.

4.8 WORKPIECE CLAMPING

Clamping of a workpiece onto a milling machine requires consideration of a number of factors, which are primarily as follows:

- workpiece stability;
- tool access;
- potential for facilitation of multiple operations without re clamping; and
- damage to workpiece related to clamping method.

The simplest and often quickest method is to use a machine vice and these come in three generic types. A simple plain vice that has a plain fixed and moving jaw, or two moving jaws. The second type is similar but has the ability to rotate around the z-axis. Essentially the base of the vice is bolted to the machine in the same manner as a plain vice and has a second layer that can revolve, often marked with a scale calibrated in degrees, with a second set of clamping bolts to clamp the revolving section at a desired orientation with respect to the z-axis. The third type of vice is often referred to as a universal vice. This design has all of the features of a vice that can rotate around the z-axis, but also has the ability to rotate along a horizontal axis. Figures 4.13–4.15 show the three generic types.

Figure 4.13 Plain machine vice

Figure 4.14 Machine vice with rotating clamp

Figure 4.15 Universal vice

When using a plain machine vice consideration of the length of the jaws against the length of the workpiece needs to be assessed. Where significant overhangs exist it is common for vibration and distortion to become an issue when machining. While it is common, and often desirable to have overhangs the amount of overhang achievable will depend on the sectional size of the workpiece and the material it is made from. Wide and thin sections may experience vertical distortion while machining, and narrow and deep sections may experience horizontal distortion. Sections that while substantial, will experience movement if excessive unsupported lengths are used. There is no hard and fast rule to what is correct, and the operator needs to balance material type, feed rate, cutter type and depth of cut. Experience is generally the best indicator of what is correct, and while sometimes there are few alternatives than use of a vice with large overhangs, consideration of providing supports at the extremities is also advisable.

It is normal to support a workpiece with a vice on machine parallels. This requires assessment of two factors; which are height of the workpiece out of the vice jaws, and the area of the workpiece providing the clamping area. It is extremely bad practice to machine into the vice jaws, not only because they are normally hardened and damage a cutter, but primarily because it damages the vice reducing its useable life. However, it is also undesirable to move the workpiece for subsequent cuts, and the operator will need to consider the sequence of operations such that sufficient material remains clamped in the jaws to rigidly hold the workpiece while being able to undertake cuts to the full required depth, particularly on the outer edges of the workpiece.

The depth the workpiece is inserted into the vice directly relates to the force being imparted on to the workpiece by the forces acting on it by the cutter. In general the larger the vice, the stronger the holding force and if the vice jaws are wide the depth of engagement can be lower than where a smaller vice has been used. Where conditions are marginal or alternatives not available, small cuts on low feed rates can be an alternative.

The use of a machine vice provides a quick, easy, and generally accurate way of workpiece holding, however, for larger workpieces, uneven shapes and thin workpiece sections, clamping directly to the machine table is a secure and accurate method. If the machining operation involves piercing through the full thickness of the workpiece, sacrificial materials, parallels or other types of spacer must be used to protect the machine table. Where no perforation is required workpieces can be fastened directly to the table. Assessment needs to be made if the length of tool and tool holder is sufficient for the full depth of cut to be achieved when materials are bolted directly to the machine table. It is not uncommon for a milling machine spindle nose to be unable to reach the table, and while machines with a spindle that can be fed down on its quill feed can address this, consideration of the length of tool and tool holder needs to be made, and if the operator considers that the tool will not be able to reach its full depth then packing under the workpiece, or use of long series tooling needs to be considered. However, if the cut to be made requires a small diameter tooling then it is unlikely that long series tooling would be an option.

Methods of bolting directly to a machine table vary. The simplest is directly bolting through holes drilled in the workpiece with nuts and washers applied to the upper surface. The second method is by utilising slotted machine clamps. These come in two types, either a plain slotted plate, or a slotted plate with an angle rear section which has had a set of steps machined into it. These steps are used in conjunction with a mating stepped wedge as shown in Figure 4.16.

The advantage of using a slotted machine clamp is that there is significant latitude in the angle that they can be emplaced and the number that can be used. Where the workpiece is such that the clamping has to be used where machining needs to take place it is very common for additional clamps to be used such that one can be removed at a time to complete an operation and the replaced while another is removed.

Slotted clamps work best when parallel or slightly inclined towards the front. Some clamps have a slightly rounded front edge others have plain flat surfaces. When setting up these clamps it is important to consider if the face of the clamp in contact with the workpiece is going to damage it in any way. Clamps with a raised and rounded contact edge are more likely to do this than plain flat clamps, however both have the ability to do so, and the operator needs to use a sacrificial packing piece of some suitable soft material like brass/aluminium/plastic under the contact to prevent marking of the workpiece.

Figure 4.16 Slotted machine clamp with stepped wedge

The thickness of the workpiece and any sacrificial packing piece will drive the height of any resistance block or the size and setting of any stepped wedge at the rear of the clamp. The tee bolt used to provide the clamping force needs to be long enough to ensure that there is full engagement of nut and washer above the upper surface. While some tee bolts have removable threaded shafts many do not and a variety are needed to accommodate differing workpiece thicknesses.

When clamping down a workpiece the bolt wants to be as close to the front of the machine clam as possible to provide the greatest clamping force as the clamp is effectively pivot point about the resistance block at the rear. When the bolt is halfway between the point of clamping and the resistance block the force will be even on both. As the bolt moves towards the point of clamping the force increase and vice versa. It is unusual for the operator to tighten bolts to a known torque, unless the configuration of the workpiece requires it and use of a normal open-ended, or ring spanner tightened by hand is sufficient to provide rigid fixing over a number of clamps. Figure 4.17 shows a machine clamp in use; it is worth noting that machine clamps can be used for clamping workpieces of significant thicknesses or height, and it is only the length of the bolt and resistance piece that needs to extend, and this method of clamping can be extremely flexible.

When using machine clamps considerable force can be applied even when hand tightening with a spanner. It is important that when thinner sections of

Figure 4.17 Machine clamp in use

material are being used and the workpiece has been raised above the machine table that support is provided directly under the clamping area or bending will occur. This bending does not just occur at the edge being clamped but can cause the centre of the workpiece to bow upwards. The operator needs to insert a machine parallel or packing material that is capable of supporting the direct load under the clamp to ensure there is no plastic deformation of the workpiece and distortion or bowing elsewhere.

Where it is not practicable to use a vice, or machine clamps it is also quite common to use a face plate. The face plate is merely bolted square to the table in a similar manner to a vice, and then used to bolt or clamp, using machine clamps to retain a workpiece. This method is particularly useful where a workpiece may have been able to be fitted in a vice, however, the length of the workpiece would require machine at a distance substantially above the vice clamps leading to instability, or where machining is required to take place on the workpiece at an angle that would be difficult to set using a vice, such as where a sine bar needs to be employed. Figure 4.18 shows an example of this.

4.9 USE OF REGISTERS

Where a workpiece needs to be set up parallel to the slots in the milling machine table or a face plate/angle plate needs to be set square, a quick method

Figure 4.18 Workpiece clamped to angle plate

of ensuring alignment is to use registers. While 'clocking in' is an accurate method, if the workpiece does not have an edge that is of sufficient thickness to use a dial test indicator (DTI) on, or the item being mounted on the table is known to be square then using registers is perfectly acceptable.

Registers are rectangular blocks of steel that have been ground to a size that is a transition fit with the width of the slots in the machine table. These registers need to be long enough such that they have full engagement with the depth of the machine table slot vertical surfaces and protrude above the machine table sufficiently to provide an edge to push the workpiece or face plate against, after which it is securely clamped. Figure 4.19 shows a face plate being positioned using registers prior to bolting down securely.

4.10 WORKPIECE OVERHANGS

On occasion operators are required to machine a feature into a workpiece that would ordinarily not fit onto a machine. While the question could be asked, why not use a bigger machine, however, the driving force is often cost and availability. Where machines have an interlocked guard around the cutter, machining large objects is relatively straightforward. Where fully enclosed guarding is employed there may be some difficulty, or the operator may be prevented from completing the operation at all, and an alternative approach found.

Figure 4.19 Face plate located using registers (slip gauges utilised as register)

Many fully enclosed guard arrangements allow for overhanging work-pieces to protrude below the bottom of the guarding and also have non-interlocked opening end panels to facilitate larger work.

Irrespective, of the guarding issue the operator needs to be aware of the maximum and minimum distances from the rear edge of the cutter to the machine column. Where this is outside the size of workpiece all overhang will need to be off the front of the machine and the operator will need to establish that all of the traverse and elevation wheels remain accessible, and any feed controls that may be overhung by the workpiece. For longer work-pieces the operator needs to establish that the workpiece will not impact any solid object, or create an unsafe condition while feeding through the cutter.

Clamping of workpieces when undertaking this sort of machining rarely involves the use of a vice, and machine clamps are the normal approach as shown in Figure 4.20. Initial set up of the workpiece is also often linked to the use of registers, which can be utilised through 'stepping' a workpiece to

Figure 4.20 Overhanging workpiece

allow for consistent readjustment of the workpiece in known steps to allow successive machining operations.

4.11 USE OF MACHINE PARALLELS

Machine parallels are generally rectangular section hardened bars of metal which are ground on four sides to provide two sets of parallel faces. There are alternative types such as 'I'-shaped sections, however, all types have at least one set of ground parallel faces, and are supplied in matched pairs. Alternatives can be sourced, such as using ground tool steels, but it is key that the pair is of a matched size and the faces to be used are not bowed in any way. The normal orientation in use, especially for plain ground parallels, is to utilise the narrow face as any bowing, or distortion is likely to be along the length of a face rather than a thinner edge. Figure 4.21 shows a typical type of parallel.

Where there is some doubt as to whether a set of parallels are a matched pair, careful measurement using a micrometer with vernier scale, comparator, and surface plate, or simply re-grinding, to ensure both are the same height is appropriate. Given the multiplicity of uses for parallels it is normal for a variety of heights to be available to the machine operator to accommodate differing depths of cut and workpiece thicknesses. Therefore, an operator needs to ensure that parallels of a suitable height are available. Where they are not, it is not unknown for the operator to use parallels with a tool steel placed on top. While this achieves the lift required, it is often impossible to get them to 'lock up' and the mating faces often get swarf and debris getting between the faces,

Figure 4.21 Machine parallels

which can lead to dimensional errors. If the operator has access to a surface grinder, it provides an opportunity for additional sets to be quickly manufactured, and where facilities to case harden, or use through hardening steels, the use of workshop hardening techniques such as heating the machined parallels to bright orange using an oxy-acetylene torch and then turning off the oxygen supply while still applying a burning acetylene flame introduces carbon into the surface and as long as subsequent heavy grinding to the bearing surfaces is not undertaken, a hard ground surface is provided.

Parallels are used to lift a workpiece off the machine table, or lift it up in the jaws of a vice to allow machining. Where workpieces are being fastened directly to the machine table, parallels allow for tool penetration below the workpiece. Clamping down directly over parallels will ensure that the workpiece is mounted on level datum. Using pairs of parallels in a vice that are 'locked up' gives the operator confidence that the workpiece is level with the base of the vice, which in itself is level with the machine table.

4.12 USE OF DEPTH STOPS

Depth stops are utilised in two places on a milling machine. The most common are those fitted to the side of the machine table where it is normal to have both adjustable and fixed depth stops. The fixed depth stops are not adjustable and are fitted to disconnect the feed to ensure the machine table is not fed off the machine. However, some machines have moveable stops in addition to fixed to allow the feed to be disconnected where an operator wishes. Figure 4.22 shows an example of a microswitch-operated feed stop.

These stops are either of a simple cam type or electric micro switches and both types have a period where they begin to operate prior to stopping the machine. The operator needs to be aware of this period as setting the stops generally involves the loosening of a machine screw and sliding them along a groove to a desired distance or adjacent to a cam. When the cam hits the block, either a plunger on the depth stop is depressed or the block starts to operate the cam. Either method involves a traverse of several millimetres before feed ceases and the operator needs to establish this depending on its criticality.

Use of depth stops is useful when the operator is working in a busy environment, however, the accuracy and consistency of depth stop actuation is not always precise. It would not be advised to use depth stops as a method for stopping feed when machining pockets or blind slots, and the use of a physical fixed depth stop or value witnessed by the operator on a DRO, would be a more reliable approach. However, where a cutter passes over or through the workpiece they can be an efficient and effective device allowing the operator to concentrate on other aspects of the work as a cut on feed progresses.

A second type of depth stop is fitted to machines that have a quill. This device is where the spindle extends out of the machine turret in a similar

Figure 4.22 Microswitch-operated feed stop

manner to a drill which is what it is generally used for. When a hole is required to be drilled with a twist drill or slot drill to a known depth the depth stop fitted to the head of the machine is used.

Typically, without the cutter rotating the tool tip is brought into contact with the workpiece. This removes any backlash in the system as its pressed onto the workpiece in the manner it will experience when cutting. The most common type of depth stop is a wheel mounted onto a threaded shaft, and where this is utilised the operator rotates the thumbwheel until a gap greater than the depth to be drilled is provided. A pack of slip gauges or gauge blocks assembled to the depth of hole required is then inserted in the gap and the thumbwheel rotated back up the shaft until the pack of slips is trapped. The operator then locks off the thumbwheel (if a lock is fitted) and the hole drilled until the depth stop is reached.

Figure 4.23 shows the basic setting of a depth stop. However, when setting depths for twist drills, an allowance for the drill point needs to be made and the slip pack increased in height to accommodate this, and the operator also needs to ensure that the height of the table is not adjusted using the hand wheel on the machine knee before the hole is drilled.

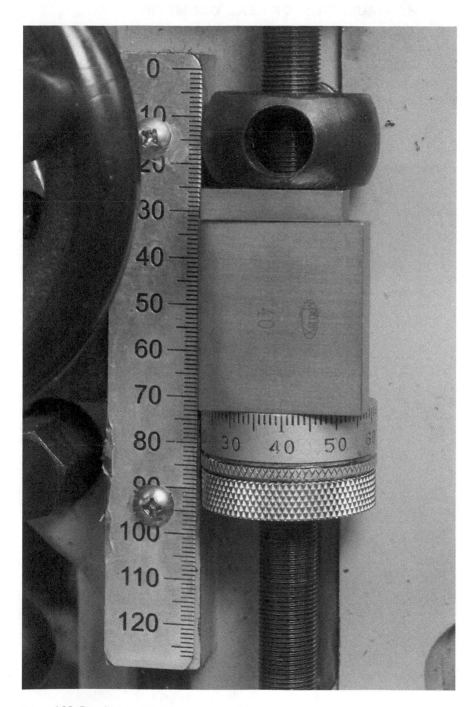

Figure 4.23 Depth stop adjustment

4.13 MANIPULATION OF HEAD GEOMETRY

The ability to rotate the cutting head of a universal milling machine in the x-axis, y-axis and z-axis, means that the cutter can be placed at an angle that will produce an elliptical form, or trough, when the workpiece is fed through a cutter. Figure 4.24 identifies the scales on the machining and shows the degree of rotation that can be achieved on each axis.

The shape of the ellipse, as in the width of the cavity machined with respect to its depth will change depending on the angle of rotation of the head in the

(A)

Figure 4.24 Turret mill head angular rotation scales

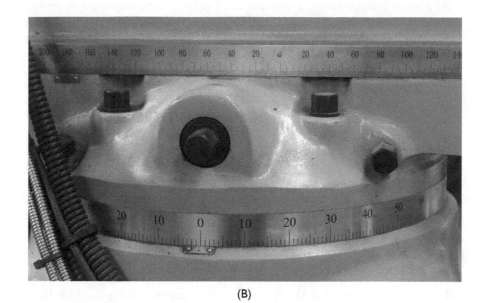

(B)

Figure 4.24 (Continued)

y-axis with the depth of ellipse reducing as the head is rotated towards 90°. However, there is a limitation to the angle that can be achieved as the workpiece has to have a clear path to pass under the turret head and full depth of cut. By extending the quill to its maximum extent some additional vertical clearance can be gained to enable relatively shallow ellipses to be machined, however, depending on the configuration of the machine head, there will be a finite limit to the ellipse that can be machined into the surface of a workpiece. The machining of this type of ellipse is normally undertaken by rotating the head on the *y*-axis, however, a similar approach can be taken by 'nodding' the head by utilising the *x*-axis rotation capability. Given the reduced amount of machine table travel on the *y*-axis, and axis reduced amount of rotation, it is less usual to machine an ellipse using the *y*-axis, however, when producing some complex shapes such as an ellipse machined into an angle face, it can be useful especially if the operator does not wish to reorientate the workpiece within a machine vice or work holding device.

Manipulation of head geometry to machine developed angle cuts results in material being removed using the bottom corner of an end milling cutter. In almost all cases this type of machining requires the use of a long series end mill to allow for enough space between the cutting edge and any part of the cutting head which may collide with the workpiece. As the angle of the head increases the trans-axial load increases in proportion to the axial load. Where heavy cuts are taken the cutter will deflect in proportion to the angle. Therefore, the operator needs to be circumspect when setting a cut depth and a number lighter cuts may be required at more extreme angles to ensure that chatter isn't experienced. The direction of feed, and feed rate is also an issue. Feeding the workpiece onto the end face of the cutter places more of a vertical load on the cutter and up through its central axis, however, the cutter in this orientation is almost being used as a slot drill, and while is downcut milling, the way that the teeth of the end mill bite into the workpiece may cause it to try and raise out of the workpiece due to the ability of the cutter to defect ant extreme angles. The alternative is to feed the end mill the opposite way with the workpiece fed onto the side of the cutter. In many respects this would be a more normal approach for an end milling cutter, but will provide an upcut action from the tool and will have a tendency to bend the tool upwards where excessive depths of cut are used. The solution to this is not easily identified before attempting such work, as the angle of head rotation, the tool diameter and length, as well as the type of material and depth of cut are all factors that control a successful operation.

4.14 TRAMMELLING IN TURRET HEAD

The ability to manipulate the orientation of a turret milling machine head is one of its key advantages over many other types of milling machine. The ability to rotate the head through a large angle on the y-axis and to 'nod' the head through a significant arc on the x-axis creates a problem of realigning the head when returning it to its normal datum position. Figure 4.24 shows a typical example of the scales used to align the head to differing angles.

It can be seen that the scales are marked in whole degrees and while alignment can be relatively well controlled through use of longer spanners on the head drive mechanism, backlash in the gearing mechanism, and the resolution of the scale/pointer, create a problem for achieving the return of the head to a perfectly aligned datum position. Where there is any slight residual misalignment the workpiece will be marked where one side of the tool digs slightly deeper into the workpiece than another leaving scoring on the machined surface of the workpiece.

Often this marking is not a problem, and the machine scales are normally accurate enough to keep any error within a few microns, however if surface finish aesthetics are important or machining is being undertaken to tight tolerances, trammelling the head back in corrects the error.

Trammelling the head back in is a relatively simple and the process is undertaken to ensure the head is perfectly square to both the x-axis and the y-axis. A DTI is mounted on a tommy bar which is inserted into a collet mounted in the machine spindle. Most operators manufacture these bars themselves and are often made from simply bent round bar that fits the DTO clamp, however, the stiffer they are any spring is removed, which provides greater confidence and speed of use. Figure 4.25 shows a DTI mounted on a simple tommy bar and inserted in the chuck of a turret mill.

Figure 4.25 Dial test indicator mounted for trammelling head

The longer the tommy bar is the easier any error is to identify and remove, however as the bar needs to be swung through an arc, any guarding, and the ability to use to trammel the y-axis often restricts its length to 300mm.

To trammel the head to a datum position, the machine table is cleaned and moved to a roughly central position of the x-axis. The machine vice or other work holding device is removed and the raised using the knee control handle until the DTI plunger has been displaced by a couple of millimetres and the dial marker set to zero. The table should also have been aligned on the y-axis such that the DTI plunger is roughly central on the ground upper surface lands of the machine table. The tommy bar with DTI on the end should also be well aligned with the x-axis.

The process of trammelling is very similar to clocking in a vice, in that once the DTI is set to zero and in reasonable alignment with the machine table x-axis, the tommy bar is swung through 180° until the plunger is resting on the ground top of the table opposite its starting position. Any deviation from a zero reading on the DTI indicates that the head is out of alignment on the z-axis. A positive or negative reading indicates which direction the head needs to be rotated. The head clamping bolts are slightly slackened to a point where rotation of the head is possible. Loosening too far may allow the head to drop slightly, therefore the minimum amount of slack possible is advised. The head is rotated until the reading on the DTI is half the displacement observed, when the head retaining bolts have been re-tightened. The DTI is set to zero and the process repeated. When there is no difference between the two positions the head is perfectly square to the x-axis.

The process for ensuring the head is square to the y-axis is exactly the same. The distance that the DTI can mounted away from the spindle is reduce as the width of the table is a constraint, and it is harder to be as accurate than when the length of the table is used, however, the process remains the same.

When adjusting the head and trying to make a fine adjustment backlash in the gearing mechanism may be an issue. If during any previous adjustment the head had been driven the opposite way to that required, it is likely that making a small adjustment will be difficult as the gearing can be quite coarse and the removing the back lash and making a fine adjustment can be difficult. Where this is experienced, the best method is for the operator, to rotate the head past the position required and then drive it back up until the DTI reading is acquired.

Once the head has a zero deviation from both the x-axis and the y-axis the head will have been perfectly aligned on the z-axis, and tool marking by heel drag eliminated. It is worth the operator closely examining the pointer and scale setting scales to identify if there is any misalignment once the head has been trammelled in, and noting any error to assist with subsequent realignment. Notes written on the side of the machine close to the scales provide a reminder of this as do electronic photographs, and can save a considerable amount of time.

4.15 FEEDS AND SPEEDS

Setting of feeds and speeds can be a complex area, and there is a vast amount of information to assist the operator. However, the machine operator only needs a few pieces of information to successfully configure his machine. The key information the operator needs is as follows:

- surface machining rate for the workpiece material;
- cutter material (i.e. high-speed steel or carbide);
- diameter of the cutter; and
- number of teeth on the cutter.

The surface machining rate varies for material types and is either quoted in surface metres/min (SMM), or surface feet/min (SFM). Material suppliers or stockholders rarely have this information, and unless the operator can access manufacturers data the internet is often a good source of values for differing materials. As a guide, typical values for common engineering materials are shown in Table 4.1.

These values are a guide and the achievable rate varies with cutter design, material and the behaviour of the material. As a general rule, softer materials can be machined more quickly than harder materials, but some higher feed and speed rates on softer materials can cause tearing of the material and lead to poor surface finishes. Some experience and experimentation to balance the materials, cutters being used and machine dynamics assists.

As can be seen in the table, the cutter material (i.e. high-speed steel or carbide) has a fundamental effect of the machining rate. While the table suggests that it is always going to be more efficient to use carbide tooling, that is not always the case as some high-speed steel (HSS) tools can lead to better surface finishes on some materials. The values mentioned stated in Table 4.1 are those for roughing cuts. There is a difference between rates for roughing and finishing, which while not significant, are generally higher and can be a factor in tool selection. If machining a workpiece that requires a high surface finish, or optimum cutting conditions, if the machine is unable to achieve the optimum spindle speed, then utilisation of a cutter material that requires a lower speed can be of use. Many modern carbide cutters are optimised for CNC machines which may have spindle speeds in excess of 10,000rpm, which few workshop machines can approach.

Of crucial importance is the diameter of the cutter. Machining rates are calculated for material removal at 'X' surface meters/min. This provides a standard factor accommodating cutters of differing diameters. A simple formula for calculating can be seen in Table 4.2.

Once the spindle speed has been calculated the operator changes the machine gear settings for vertical mills to the closest speed setting or, on a

Table 4.1 Typical values for surface machining rates using high speed steel and carbide cutters

	High-speed steel (SMM)	Carbide (SMM)
Mild steel	27	60
Carbon steel	15	30
Stainless steel	18	60
Cast iron (soft)	24	75
Cast iron (hard)	15	60
Aluminium	76	150
Copper	46	70
Magnesium	91	170
Bronze	46	70
Brass	46	70
Acetal co-polymer	400	800

Table 4.2 Calculation of feeds and speeds

	Calculation of spindle speed	Calculation of feed rate
Data	S = spindle speed D = tool diameter (m) Vc = surface machining rate (SMM)	F = feed rate N = no. of teeth on cutter S = spindle speed Ft = feed/tooth
Formula	$S = Vc \, \pi \, D$	$F = NSFt$

turret mill with the machine running, rotates the motor adjustment wheel until the correct speed is set.

Feed rates can be slightly more complex. Vertical milling machines often have variable feed rates which are changed using a gearbox setting. Where this is experienced the closest setting to that calculated is utilised. On turret milling machines feed rate is often controlled by an electrical control unit and a control knob is rotated to adjust the feed rate. These can often lack any values for feed rate. In these circumstances the operator should start with a slow rate of feed and increase until efficient material removal is observed.

4.16 SURFACE MILLING

One of the first and most common operations undertaken on a milling machine is surface milling. Surface milling is undertaken to either reduce a piece of stock bar to a known size, or to remove scale. Alternatively it is a precursor to more complex manufacture of shapes in the workpiece.

There are two approaches to milling a surface. One is by using a face/shell mill type cutter, and the second is by using an end mill. Modern face mills consist of a steel holder into which a number of carbide inserts are fitted as shown in Figure 4.26.

Figure 4.26 Modern face mill

These are relatively inexpensive and damage to any of the cutting inserts is simple and cheap to replace. There are, however, still a number of solid HSS shell mills found in workshops. There is little practical difference in their application, however, machining rates tend to be slower and if the cutter sustains damage, it either requires replacement or, to be completely re-ground. Shell mills will machine a wide area of the workpiece surface in one pass. The second approach is to use an end mill to machine a surface. Figure 4.27 shows a selection of end mills illustrating the choice an operator has.

While there are a variety of end mills covering differing numbers of flutes, helix angle, chip breakers, and diameters the operator can select, it is almost inevitable that to machine a surface a tool path will be followed to generate

(A)

Figure 4.27 Common end milling cutters

(B)

(C)

Figure 4.27 (Continued)

a single flat surface. The result of following a tool path is marking of the surface. Following each pass the table is moved across on the y-axis a distance just less than the cutter diameter using the hand wheel, and the workpiece fed back through the cutter in the opposite direction. Any slight error in the orientation of the z-axis will result in a series of tracks, and irrespective of this the differing direction of cut will leave a pattern on the material surface. It is not unattractive, and the surface will appear flat, but it will be visible. One feature of using end milling cutters to surface mill is that a high degree of wear can be experienced on the corner of the cutter leading generation of a ragged burr at edges and poor surface finish. Operators who can select an end milling cutter with a slight corner radius can avoid this problem, however, depending on the size of the radius will have a reduced width of tool path for each pass.

The selection of cutter type, the number of teeth, depth of cut, and feed rate are all linked to cutter availability, the material to be machined, and the capacity of the machine being utilised. Many face mills, especially those of the carbide insert type are designed for high productivity situations often involving the use of high feed rates, and depth of cuts that may not be achievable on common workshop machines. The selection of cutter type, speeds and feeds will be discussed in detail in a specific section, however, the operator of a turret mill fitted with an R8 collect system will often only have a seven insert general purpose cutter of approximately $\varnothing$70mm. When using an end mill on a machine with an R8 collet system, generally collets cannot accept a tool shank bigger than $\varnothing$20mm although tools with a greater diameter are often fitted with sub-diameter shanks giving some flexibility for larger tools, however, it is unusual to see solid end mills with a diameter greater than $\varnothing$25mm.

Insertion and retention of milling cutters for end mills and turret mills is almost identical. Modern turret mills either have the R8 type, ore ISO morse type collets fitted to them. Older end milling machines often have a traditional morse taper into which a locking tool holder is inserted. Both types of machine require the cutter to be inserted into a collet, or tool holder, and this is then inserted into the machine spindle. Some modern machines have an air locking attachment, however, most still require the operator to reach up to the top of the machine head and rotate the draw bar until it is fully engaged in the thread on the top part of the collet or tool holder. Using a spindle brake, the draw bar is then tightened with a spanner drawing the collet or tool holder tightly within the spindle taper and locking the tool in place such that it will not slip when the loads imparted by cutting are applied. If the machine has a shaft cover it is important that the operator replaces this prior to operation of the machine to ensure safe working conditions.

Face milling example

Following secure mounting of the workpiece and selection of cutter the workpiece is moved, either by hand or by using rapid feed until it is located under the cutter. The operator has a choice of either positioning the cutter in the centre of the workpiece or at one end to find the 'touch down' or initial datum point. The selection of start point is often driven by the type of cutter being used (i.e. end mill or shell mill), or whether the operator considers the workpiece to be generally flat. In ideal conditions the diameter of the cutter should be about 20% greater than the width of the workpiece, however, where this is not possible a tool path will have to be followed to machine the whole surface.

The operator starts the spindle rotation, and raises the knee to bring the workpiece up to the cutter to establish contact. In cases where there is a significant amount of material (i.e. in excess of 1mm) to be removed, there is no need for probing or use of cigarette papers, and a light cut can be applied. Where there is less material to be removed, or there is concern that the workpiece may not be flat, carefully establishing the touch down point and feeding the workpiece through the cutter, is a conservative method for checking that the workpiece is in contact with the tool, but there is no significant material removal. Irrespective of the requirement for precision and the approach taken for establishing an initial datum, the operator should now zero the z-axis datum. On machines that have a DRO, with a z-axis readout, this is set to zero. However, the majority of workshop milling machines tend to have a DRO with only x-axis and y-axis readouts, or none at all. In these cases the scale attached to the vertical knee control handle or wheel should be set to zero by releasing any knurled locking ring or screw, and rotating the dial to zero, before retightening any ring or screw.

The operator should ensure that the cutter has been moved off the workpiece and the initial cut depth applied using the knee control handle. The depth of cut is linked to the material being used, the cutter type and number of teeth and the feed rate. The operator will need to have calculated the correct spindle speed for the type of material being cut, however, the operator needs to have an understanding of the relationship between the depth of cut and the feed rate. The manufacturers of the tooling will identify a value for 'feed per tooth', however, on many common workshop machines the control mechanism for the feed often does not have any calibrated values, or has a selection which often does not correlate with the tooling values. While a rough guide is that the depth of cut should not exceed the diameter of the tooling for roughing purposes and 10% of the diameter for finishing cuts. This would be impossible when face milling with a shell mill. Moreover, undertaking a significant depth of cut at any appreciable feed rate may challenge the rigidity of the machine. Therefore, there is a balance between the depth of cut and the feed rate that is achievable without damaging the workpiece, stressing the machine or breaking the cutter. For face milling, where significant amounts of material are to be removed it is not unusual for an operator machining mild steels to apply an initial cut of 2.5mm engage the feed at a slow rate and then increase the rate until an optimum balance of material removal, feed rate and swarf clearance

Is achieved. Successive cuts can be set deeper or shallower, depending on the performance of the machine.

It is worth noting that the depth of cut directly affects the size of chip produced and the deeper the cut the larger the chip. While the type of cutter also has a significant effect on this, with 'ripper cutters' designed to produce small chips the depth of cut will always be a fundamental aspect of chips dimensions. The operator needs to be aware of this as surfaces that have high surface finish requirements can suffer from marking especially if a tool path is being followed when downcut milling, and there is a degree of recutting of chips. The flow of coolant and its direction can have a significant effect on ensuring the tool has a clear path and it can be used to 'wash away' swarf from the workpiece.

When face milling the operator either makes successive cuts, or makes an initial cut, the removes the workpiece form the vice, or work holding device, de-burrs and having turned the workpiece over reinserts it in the vice or work holding device to machine the opposing surface and repeats the operation. Successive cuts are taken at known pecks, by reading the depth of cut on the knee control handle scale. While there is no requirement to measure the remaining thickness of workpiece after every pass, it is worth periodically confirming that the depth of cut being applied on the scale correlates with the material depth actually removed. Measurement of a workpiece can be awkward on some machines especially if the operator is using vernier callipers as the beam of the calliper can often be prevented from being held upright by the spindle or machine head. Where this is experienced the operator needs to traverse the table clear of the interference. This does take a little time; however, it is essential when approaching the final pass to ensure the workpiece is within the required dimensional tolerance and requires accurate measurement. Following the machining of the workpiece face, or faces, the workpiece is either left *in situ* for further operations or removed and de-burred.

4.17 SLOT MILLING

Slot milling is an operation that produces a groove in the workpiece. These can be narrow (i.e. just the width of the cutter) or wider and be formed from multiple passes. They can have closed ends, or open ends, however, if they are closed, the feature could be described as a pocket.

Slot drills are generally used for slot milling, rather than end milling cutters, and if the feature to be machined is a pocket then they have to be used, as workpiece penetration will represent a drilling action. The majority of end mills do not have an overlapping cutting edge on the bottom of one flute of the cutter and while an elliptical tool path for surface penetration is built into most CNC machines it is highly unlikely to find a workshop machine that can do this with any ease.

Slot drills tend to have just two or three flutes, and are widely available in HSS or solid carbide. The come in three lengths in a similar manner to twist drills, with stub, standard and long series cutter available. The lengths are in proportion to their diameter, and some very small cutters will have short lengths of cutting edge on the end of a taper which leads to a much greater diameter. This simplifies mounting small diameter cutters in the spindle collet, but restricts the depth of slot that can be machined. The key difference between an end mill and a slot drill is the overlap of the cutting edge of the lower surface of the cutter. This overlap allows the cutter to be used in exactly the same way as a drill and plunged directly into the workpiece. Figure 4.28 shows a selection of slot drills in which the end cutting edge overlap can be seen.

When considering how to machine a slot or pocket the operator, will need to know the length of the useable section of tool, to ensure they have enough to machine down to the bottom of the cavity. Having some consideration of how the swarf will be removed from any closed pocket is a consideration as re-cutting of chips has the potential to damage the cutter and will affect surface finish. Peck depth is another consideration, as it is not uncommon to see a series of parallel lines like strata in a rock, which reflect each successive cut, however, this can be addressed by undertaking a light finishing cut at full depth, but the tool needs to be stiff enough to prevent deflection, despite what is normally a very light cut.

When plunging into a workpiece for the first cut, one side of the slot is going to be upcut milled and the other side downcut. However if the slot or

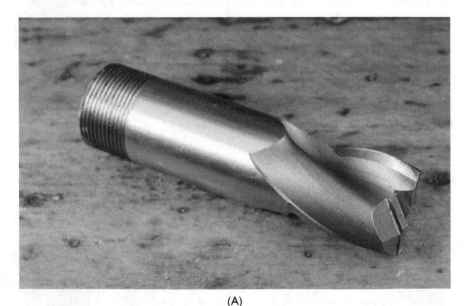

(A)

Figure 4.28 Typical slot drills

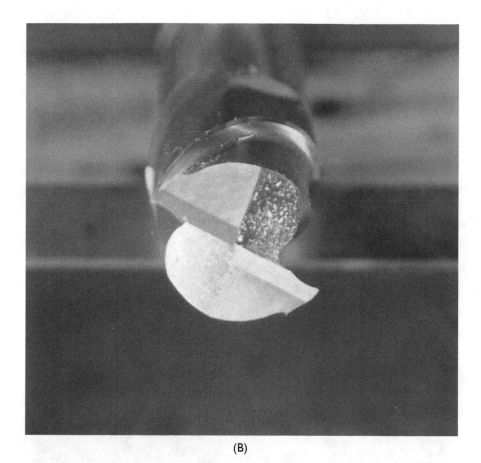

(B)

Figure 4.28 (Continued)

pocket to be produced is wider than the diameter of the cut then the operator will have a choice. The preference would be for downcut milling, not just because of improved surface finishes, but the swarf is ejected being the cutter on chip formation instead of being deposited ahead of the cutter with the risk of it being picked up and brought through the cutter again.

The first cut into a workpiece into either a closed or open-ended slot is going to have a cutting angle of 180°. If the slot drill only has two teeth then forward feed is controlled by the feed per tooth the manufacturer states. Irrespective of this when advancing the cutter into the end of a workpiece to form an open-ended slot, feed rates should be reduced until the 180° cutting angle of the tool is achieved to allow the loading on the tool to stabilise. Once into the workpiece this can be advanced to the most efficient feed rate. When

(C)

Figure 4.28 (Continued)

machining a pocket, this is less of an issue as the slot drill is fully supported having been drilled down into the workpiece prior to any feed being applied.

One of the constraints when machining a narrow pocket is swarf removal. When producing an open-ended slot this is generally less troublesome as a good supply of coolant tends to wash it away, and where any residue is left it is not difficult for the operator to quickly use a swarf brush to remove any residual debris. If machining a pocket, especially one that is deep in proportion to its width, it is common to get the machined pocket chocked with swarf chips. While much will be ejected on successive cutting operations, it is not unusual to hear crunching of swarf getting recut. If the pocket is intended to penetrate the entire thickness of the workpiece it can be a good idea for the operator to use the slot drill to penetrate through the workpiece in the centre of the feature to provide a swarf egress route as subsequent machining takes place. Where this is not possible it is normal practice for the operator to stop the machine when the pocket gets choked and brush out much of the swarf before continuing subsequent cuts.

Close attention needs to be paid to spindle speed and feed rate when cutting slots, as it is easy to generate chatter. Tool length, diameter, material hardness, spindle speed and feed rate all factor in this, as does the cutting angle. While cutting a slot that is the width of the cutter (i.e. having a cutting angle of 180°) means that the centre line of the tool is in the centre of the cut, the result of this is a fairly even chip thickness and thick chips can cause vibration as they leave the tool. Therefore, where it is possible to offset the cutter such that thick chips are formed as the cutter enters the workpiece becoming thinner as the chip exits, will tend to reduce vibration and also reduce temperature at the cutting edge which increases tool life. Generally shorter stiffer cutters, set to the correct spindle speed for the tool diameter/material surface machining rate, suffer less than longer tools being pushed harder.

Where a slot to be machined is wider than the tool width, the operator can make an assessment of the optimum cutting angle on the tool.

Slot milling example 1 (open-ended slot)

Having inserted the cutter into the spindle collet and firmly mounted the workpiece in the work holding device, an initial datum is to be set. The approach is very similar to face milling in that the cutter is brought down to the workpiece and the DRO scale or vertical knee control handle set to zero. It is normal for the upper surface to have been face milled or to be of a bright finish rather than a slightly undulating 'as cast' or hot rolled finish. Slots are most often dimensioned from an upper surface to a depth, although this is by no means certain. Where it is not the operator must calculate the depth at the maximum material condition, and its' allowable variation to give the upper and lower limits for the slot. Therefore, precision in determining the surface datum requires care, and the use of probes or cigarette papers is normal.

Having established a datum and moved the cutter off the surface of the workpiece the initial cut depth is to be set. While the material and machine power and stability are controlling factors, the depth of cut should be less than the cutter diameter. Of concern is the size of chip, and one side of the chip formed will be the length of the depth of cut. If swarf clearance is an issue the multiple shallower cuts will produce smaller chips which can be flushed away more easily than larger ones. There is a balance between depth of cut, feed rate and machine dynamics, as a deep cut and feed rate set to the correct feed per tooth, may be theoretically possible the noise, performance of the machine, and surface finish achieved may suggest otherwise.

Once the depth of cut has been set by raising the machine knee, the cutter is advanced to engage with the workpiece. It is worth using a slow feed until the full cutting angle is achieved to allow the tool to stabilise. Once the cutter is fully engaged the feed rate can be increased to its optimum setting. Coolant should be delivered to the cutter in as larger amount as possible, to ensure maximum cooling of the cutting edge is achieved and swarf is washed away. The rate that coolant is delivered must be tempered by the amount being sprayed, as it is easy for the operator to get both contaminated and wet, and also for both spray and spillage of the machine table to leak onto the floor producing a significant safety hazard. Adjustment by the operator of the flow rate and the direction will provide the optimum setting.

When producing an open-ended slot the operator has little to do until the cutter exits the workpiece. If the slot is being produced to the width of the cutter the next peck is set by raising the machine knee and a repeat of the first cut is undertaken while being fed in the opposite direction. This process is continued until the desired depth is achieved. The use of a depth micrometer, depth gauge or extending end of a vernier calliper is the common method for establishing depth, however, it is important to remove any burr from the upper edges of the slot that may lift any instrument being used to check depth. This burr removal is undertaken with the workpiece remaining in the vice or work holding device, and should not be undertaken off machine as workpiece the datums and squareness will be lost.

Where a slot is to be produced that is wider than the cutter, at the end of the first pass, the workpiece should be offset in the y-axis using the cross feed handle mounted on the machine knee. Assuming that the first cut was from the right hand of the operator to the left, the elegant method for producing wider slots is for the first plunge cut, to be down the centre of the slot, and the second cut taken by moving the table away from the operator in a +ve y-axis movement. This means that the return cut will be down cut milling, with its benefits to surface finish and chip formation. The distance that the table is moved will be that the takes the cutter to the desired edge distance or to the next tool path cut, assuming a 20% cutter overlap. The machine is restarted and feed engaged. As the cut exits the end of the workpiece the table is moved towards the operator in a −ve y-axis direction for the desired distance. If the cutter diameter is such that the entire slot is produced in two passes, then it is usual for the DRO to be zeroed, setting one edge as a datum. Any movement of the table across the y-axis will be that of the width of the slot MINUS the cutter diameter, and any material allowance for a finishing cut.

Use of a larger cutter usually provides for a stiffer cutter, rotating proportionally slower and one that is generally longer giving the operator a clearer view of the cutting process. Larger diameter cutters are going to reduce the number of tool paths, required making the operation more efficient and the ideal solution is generation of the width of a slot in two passes. As each pass is generated the increased width allows for more effective chip generation that is more easily removed especially where there is a 20% cutter overlap. Each peck will then only have a small section of closed slot and swarf will be able to climb up the cutter flutes and be cleanly ejected away from the cutting edge. This process then continues until the desired depth is reached.

The series of pecks often leaves parallel marks on the walls of the slot, and it is not unusual to take a finishing cut. A typical finishing cut will only be about 0.1 mm but it will be the full depth of the slot. Some tooling manufacturers suggest that the surface machining rate can be substantially increase for this type of operation, however, it is worth the operator checking the tooling manufacturers data before doing this. Generally knowledge of the machine and experience will show if this is worthwhile.

Slot milling example 2 (closed pocket)

The approach to milling a closed pocket has much similarity with milling an open slot, but there are a number of aspects that require the operator to pay close attention to. Set up of the machine and establishing a z-axis datum is identical to slot milling. However, two additional datums need to be set on the x-axis and y-axis in order to establish where the pocket is to be milled. These datums are set in a similar manner to identifying the z-axis datum i.e. through use of papers, or probing. In addition to this touching the cutter onto a non-significant edge, or using a centre finder (wobbler) to establish an edge datum. Machines that are fitted with a DRO are ideal for this type of operation. Machines that are not able to have their scales set to zero, and due to multiple rotations of the scales it is generally essentially to have marked up the workpiece with the outer edges of the pocket to be generated.

Once the datums have been set for the x-axis, y-axis, and z-axis, the limits for the maximum and minimum material conditions need to be established. It is easy for the operator if they are dimensioned off edges being used as a datum, but where they are not the operator will need to calculate them.

Once datum edges are established the machine table is lowered to allow the tool to pass over the top of the workpiece to a point that is roughly central to the pocket, or at least a few mm off the edges. The table is then raised until the slot drill penetrates the surface of the workpiece. The table continues to be raised by elevating the knee until the desired depth of cut is reached. It is not desirable to commence machining in one corner of the pocket as it is not unusual to have some movement when the cutter is plunged in and can result in a slightly offset corner.

Invariably the cutter selected is of a diameter that reflects twice the corner radius value stated on the drawing. This is significant as the operator needs to make a translational movement in the x-axis and y-axis and reset their datums as the edges of two sides of the pocket to be machined. it makes sense to set the x-axis datum on the side it was established on and similarly for the y-axis, however, this is not essential. When setting these datums using a DRO and setting a datum on the edge of the pocket closest to the axis datum, the value to be identified will be the distance from the datum edge PLUS half the diameter of the cutter. When setting a datum on the opposite side of the pocket, the value to be achieved will be the distance from the datum to the side of the pocket MINUS half the diameter of the cutter. As each edge is identified the DRO axis is set to zero. The distance to be milled will be the length and width of the pocket MINUS the full diameter of the cutter. It is often useful for the operator to make a note of these translational movements.

The identification of the x-axis datum and y-axis datum for the pocket places the cutter in one corner of the pocket. The operator then feeds the workpiece through the cutter with the tool path progressing in an anti-clockwise direction to downcut mill. The operator has a choice of continuing to peck and leave an island in the centre of the pocket or whether to remove the centre island with each pass such that the pocket progresses at an even rate. There are benefits and issues with both approaches. Continuing in a stable path to leave an island, is perhaps simpler for the operator, often quicker and reduces the possibility of machine marks where a cutter has been moved slightly too deep into the pocket sides. It also creates greater problems with swarf removal as the space is more congested.

Machining out the pocket fully with each peck reduces the swarf removal issue and ensures a flat bottom. This approach is often more successful when leaving an allowance for a finish cut. Finish cuts are again often only generally 0.1mm deep, however, if the pocket is deep and the required radius of the pocket corners sub 4mm then there is a possibility that the tool may try and deflect leaving an unsatisfactory finish. When undertaking a finish cut in a pocket, the tool is also run into the corners. This results in a cutting angle of 270° and care needs to be taken not to over push the tool or lobed corners are generated.

Depending on the material, cutter type, size of pocket and coolant flow, cutting operations may need to be periodically stopped to clear swarf with a small brush, however, once datums have been established machining pockets are fairly straightforward when using a DRO.

Where the machine is not fitted with a DRO, it is important for the pocket boundaries to be scribed on the edges prior to machining. It is possible to set Dial Test Indicators to indicate a zero-edge datum, however, this is time consuming. An issue for scribing lines is that it is not unusual for a burr to be generated which obscures at least 2 of these lines. It is recommended that the pocket edges be marked at the correct extremities, but the majority of the pocket be machined inside within 0.5–2.5mm of these lines. While leaving the workpiece in the vice or work holding device, any small burr is carefully removed using a de-burring tool or small file, while retaining the edge marks. The dimension from the workpiece edge datum to the inside pocket edge is established and the x-axis, or y-axis control handles are adjusted using the scales on the handle to move the cutter to the desired edge distance and each successive cut taken. The reason for limiting this to between 0.5mm and 2.5mm is that the lower value could be considered a finish cut, and the upper figure would be one to be undertaken in pecks but could allow for a subsequent finish cut, however, either approach would require less than one rotation of a control wheel which generally provide a lateral displacement of 2.5mm. If the operator's machine is less or more than that then the allowance can be adjusted, however, by leaving a small additional cut to be made a clear scribed edge remains in view, and final adjustments require less than one rotation of a control wheel, which reduces the potential for error.

4.18 END MILLING

End milling shares many of the applications that slot milling does, however, one key difference is that that end mills are unable to plunge into a workpiece in the same manner that slot drills can. While elliptical entry paths used on CNC machines get around this issue, it is exceptionally rare for this to be an option for general workshop machinery.

The advantage of end milling over slot drilling is the productivity and improved surface finish that can be achieved. A traditional end milling cutter has four flutes, and while there can be some overlap, where some slot drills have three teeth, as do some end milling cutters, it is normal for an end mill to have four or more. Figure 4.29 shows typical examples of end milling cutters.

Figure 4.29 Typical end milling cutters

Those cutters with more than four teeth are almost exclusively of the larger sizes as the space between teeth and the space corner radius is an important factor in swarf chip ejection. The greater number of flutes and greater flute helix angle than slot drills allows for better surface finishes, and often less vibration and chatter. This is a direct result of there being more a more consistent shearing action along the length of the tool, and a rough aphorism is that the greater the helix, the greater the shearing angle, and the lower the chatter and vibration. That said difficult materials require lower angles for cutter strength, therefore, most standard end mills have a helix angle of 30° which provides a cutter capable of machining a wide variety of materials.

Feed rates values for cutters are normally stated by manufacturers as a 'feed per tooth' and the increased number of flutes allows for an increased overall feed rate compared to slot drills. While it is easy to suggest that a four fluted cutter will have twice the feed rate of a two fluted cutter, manufacturers data will provide specific values as feed rate is not just a factor of the amount of material each tooth can dig into, but a combination of that, the angle of shear and ability of the tool to eject each chip. However, in general feed rates for end milling cutters are higher than those for slot drills and the surface finishes that can be achieved are better. Ensuring a satisfactory performance is observed requires the operator to pay close attention to calculation of correct spindle speeds and feed rates.

The design of end milling cutters vary and it is very common for the cutter diameter to be smaller than the shank diameter. Where this is present the depth of cut the operator can achieve is limited to the cutter maximum length of cut. This is shorter than total length of flute and represents the maximum length of cutting edge along all the flutes, and does not reflect the apparent length which includes any flute run-out into the cutter shank. The operator must also ensure that the shank will not impact on any edge as the cutter progresses into the workpiece.

Where the shank diameter is equal to the cutter diameter it is possible for the plain shank to progress into the workpiece following sequential cuts. This normally results in rubbing between the cutter shank and the vertical feature formed by the sequential progression of the tool down through the workpiece, and leaves a mark. If the tool has been re-ground at any time in its life it is likely that the shank diameter will be slightly larger than the cutting diameter, and will either damage/bruise the workpiece as it rubs it, or more likely deflect the tool out from its cutting face leading to an imperfect cut. The operator should take care to ensure that the tool selected has the correct length to complete the cut. Shorter tools tend to be stronger and capable of a heavier cut, but long series tools have the reach. The operator needs to decide whether he uses one tool or two to complete an operation.

End milling cutters are rarely used in situations where the cutting angle is anything approaching 180° and this affects the depth of cut and feed rate the operator can use. When using the preferred down cut milling approach a heavily loaded cutter, especially of the long series type can try and climb out of its groove leaving a tapered edge with a poor surface finish, and tool with a reduced life. The reduced cutting angle of perhaps less than 120° does not provide any support to the cutter, therefore its resistance to deflection comes from the inherent stiffness of the cutter, which is related to the tool length, number of flutes and helix angle, coupled with the depth of cut and feed rate. The operator needs to place close attention to the feed rate and depth of cut to ensure deflection does not occur. Resolving these issues by correct cutter selection and selection of feed rates and cut depths that are efficient, generally address any chatter and vibration issues.

Ejection of swarf chips are less of an issue for end milling cutters, as they are rarely employed in closed slots. The increased number of flutes, and in the case of ripper cutters the profiled cutting edge allow for good chip ejection, however, once again a good flow of coolant assists this and reduces the temperature of the cutting edge such that it ensures a good tool life.

End milling example 1 (using the side of the cutter)

End milling cutters are often used to square off the ends of a workpiece, and machine them to a set length. It is very common for a workpiece to be sawn off a length of bar stock, and the end condition is often neither especially square or having a good surface finish. While some workshops have efficient chop saws, any donkey saw will undoubtably deliver an undesirable finish. This requires the ends of the workpiece to be machined to provide a clean square edge, and a workpiece that is within dimensional limits stated on the component drawing.

Following placing the workpiece into a vice or work-holding device and establishing its square ness and addressing support issues for any significant overhang, the operator will insert a tool into the machine spindle. Generally for this operation cutter diameter is not crucial, however, small diameter tools can be fragile and cannot be 'pushed' without sustaining damage. Cutter length is more of a concern as the cutter has to be able to reach the full depth of the cut, and the length of the cutter may drive the diameter that is used.

It is normal for each end of the workpiece to be machined to 'clean up' each end and adjust to the correct length. Bar stock that has been sawn down will have an allowance for cutting and subsequent machining, however, this amount can vary significantly. The operator will need to measure the workpiece to establish how much material is to be removed. Following measurement this value should be divided by two to ensure an equal amount is removed from each end. While it is assumed that the workpiece is secure the operator should not consider taking a very light cut off one end that may be unsupported, and a heavier cut on an end that is held rigidly, as this rarely works, and the surface finish produced is either poor or the workpiece becomes distorted. Providing an equal cut for each end normally ensures that the tool has the same angle of engagement at each end.

There are two approaches that can be taken for machining each end. The first is generally employed where there is very little material to be removed, and in this situation a single light cut, or series of light cuts are taken at full depth using the side of the cutter. This approach is perhaps not best practice as it does not use the corner of the cutter to bite into the material and grip the workpiece as shearing action progresses, but it is a common approach and has its applications especially on thin workpiece sections. Figure 4.30 shows this approach being taken on thin section material.

The operator starts the machine and then brings the cutter into contact with the workpiece. It does not especially matter which part of the face comes into contact, however, if the edge is visible angled it makes sense to contact at that point. Small cuts are then applied by rotating the x-axis control handle or wheel located on the end of the machine table. The amount applied will depend on the length of engagement with the cutting edge and the diameter of the tool, however, in general this depth of cut is almost always sub 1mm. The cutter is then traversed along the y-axis by rotating the cross-feed control handle or wheel. Successive passes are undertaken in

Figure 4.30 End milling cutter side being used to machine thin section workpiece
edge

the same manner until the edge is cleanly machined. At this point the DRO
x-axis scale is set to zero and the cutter moved clear of the workpiece either
by dropping the machine knee or, traversing the table on the y-axis to move
the cutter clear of the workpiece, vice and anything that may impact it as
the machine table is traversed on the x-axis. The machine table is traversed
a distance equal to the length directed by that identified on the component
drawing, PLUS the diameter of the cutter being used. If the depth of cut is
shallow enough to allow a single pass then the process used on the first end
is repeated and the opposite end machined. Where this requires a number of
passes, it is useful to re-zero any DRO, traverse the table to a distance that
provides the depth of cut that is suitable and a series of passes undertaken
until the table is moved back to the zero datum at which the workpiece
should have two square ends that are cleanly machined and the workpiece
is within its dimensional limits.

Where the amount of material to be removed is more significant and would require a high number of passes using the side of the cutter, it is more efficient to undertake this operation using the end milling cutter in a more traditional manner, and this avoids any overloading of the tool or deflection that would result in a machined edge that was not square. The selection of cutter diameter should also be such that the operation will not require a slot to be milled at one or both ends.

The operation commences in a manner similar to that used when using the side of the cutter with the exception that the cutter is placed over the upper surface of the workpiece. With the cutter rotating the machine knee is elevated using the knee control wheel or handle until contact with the cutter is made. The table is then traversed on the y-axis using the hand wheel or control handle located on the machine knee. Generally this would be a translational movement in the negative direct (i.e the table is moved towards the operator, to ensure that the following cut is down cut machined); however, control of cutting direction is not essential if a full depth finishing pass is to be undertaken. A series of passes is undertaken with the knee being raised each time.

The peck depth of each cut will depend on the material being machined and the cutter diameter. The machine operator should have an awareness of the number of cuts to be taken and having a thin piece of material left for a final cut, can result in it tearing and becoming entangled with the cutter, leading to damage to both the workpiece and cutter. The operator should adjust the peck depth in penultimate cuts to leave an edge thickness that is capable of being machined through in one pass, but of sufficient thickness to prevent distortion or tearing in the penultimate pass. A rough guide for this would be leaving about 1.5mm of thickness for most metals. Stronger alloys may require less and softer ones especially plastics may require more. Following machining, it is not uncommon to utilise a full depth finishing cut of approximately 0.1mm to remove any residual machining marks.

Following the machining of one end exactly the same process as for thin section materials is undertaken where the table is traversed in accordance with the workpiece dimensional requirement, and an additional allowance made to accommodate the diameter of the cutter. The DRO is the zeroed and the series of passes repeated, along with any finishing to leave a clean machined edge and a workpiece within dimensional limits.

Where the machine being used is not fitted with a DRO, the operator is well served to accurately scribe the workpiece with a line to an accurate length. It is useful to do this once the first edge has been machined as this provides a clean datum edge. The use of a good-quality engineer's rule, eyeglass, and scriber should allow the scribing of a line to an accuracy within 0.25mm. Where the dimensional limits of a workpiece demand greater precision than this, the operator should machine to within 1–2.5mm of this line ensuring that the remainder to be removed is within one rotation of the table traverse scale. The workpiece is then accurately measured, which often requires the use of long scale vernier callipers, which typically have an accuracy of 0.02mm. Once the length has been established the final cut, or cuts, are taken.

End milling example 2 (stepped edge feature)

The first issue for the operator to select a cutter that has the length of cutting edge to reach to the maximum depth of cut. In the majority of case bigger can be better, however, if the feature to be machined is not the full length of the workpiece then any corner radius dimension will drive the selection of cutter diameter. Once this is established the tool is inserted into the machine in same manner as any other tool.

As for slot drilling it is most common for edge milling to produce is a straight parallel groove, however, the difference between a slot and a groove, is that a groove will have one open side. The effect of this open side is to reduce the cutting angle from 180° to perhaps 120° or less. This allows for much enhanced ejection of chips from the cutting area and allows for faster feed rates.

Where a continuous edge is to be generated the operator only needs to establish two datum surfaces. It is useful for the operator to establish the upper surface z-axis datum, and the y-axis datum carefully, using probes, cigarette papers or centre finders and subsequently zero and DRO readout or scales. However, where a significant amount of material is to be removed and initial cut can be taken, the workpiece measured and a datum size established from this. If this option is to be utilised then it helps the operator if the workpiece is marked with a scribed line as a visual reference marker showing the workpiece feature limits.

If a substantial amount of material is to be removed it is worth the operator considering using a ripper cutter for roughing out the majority of the material. The nature of the cutter allows a greater degree of material removal and chip formation is such that it is easily ejected from the cutting area. Once material is removed to within a finishing size the cutter can be changed to a smooth fluted cutter for the remaining cut or cuts. When a cutter is changed it cannot be assumed that they will be identical diameters and the datums need to be checked or re-established. It is also not certain that differing tool diameters will cut exactly the same amount of material. Therefore, it is useful for the operator to leave more than a single finishing cut thickness on the workpiece to allow for changing between ripper cutters to those used to finish the workpiece feature.

The size of the feature both in the y-axis and z-axis will to a degree control the peck depth, and it is not unusual for a series of passes to be undertaken down to a depth and then returning to the zero datum making a translational movement on the y-axis by rotating the hand wheel or handle on the machine knee and repeating a series of cuts to depth. The use of a larger diameter cutter can reduce this number of passes, however, machine size and power can limit this.

Irrespective of machine size once the majority of material has been removed a light finishing cut of approximately 0.1mm should be undertaken to remove any machining marks. However, where multiple passes have been used to establish the depth of cut, it would be normal to undertake a finishing cut against the vertical feature and to depth first, and then undertake the subsequent finish cuts to depth over the remaining surface. An end milling cutter will always leave a pattern at the bottom of any cut and an elegant approach is for the operator to ensure that equal translational movements are made such that the surface marking pattern is equal.

End milling example 3 (recessed edge feature)

Where a recess is to be machined into the edge of a component, the approach has little difference than machining a plane step along an edge. However, the two key differences are that the selection of cutter will be controlled by the corner radius dimension identified on the component drawing, and the requirement to establish a datum on the x-axis as well as the y-axis and z-axis.

It is quite normal for roughing cuts to be undertaken by ripper cutters when assisting in the generation of a recessed edge, and the diameter of these do not necessarily need to be of a diameter that accommodates any specific corner radius as they can be larger, leaving the finishing operations to address accurate corner formation.

Having established x-axis, y-axis, and z-axis datums, the cutter is lifted clear of the workpiece and any work holding arrangements and traversed to the start point of the cut. The operator should select the peck depth and having started the machine advance the cutter into the workpiece. The operator needs to undertake this relatively slowly as the initial plunge will involve the cutter experiencing a cutting angle which increases to 180° if the intrusion into the workpiece is equal to, or greater than, 50% of the cutter diameter. Slot drills only have two flutes to allow effective swarf chip clearance when operation in this manner, so operators need to have a degree of circumspection when plunging into a workpiece with a multi-fluted end milling cutter to allow time for chip ejection. Assuming that the feature to be produced is greater than half the cutter diameter then the depth that the cutter is inserted into the workpiece horizontally, should be to depth minus an allowance for a finishing cut, or ideally be no more than about 80% of the cutter diameter. This provides the balance between a reduced cutting angle of approximately 120° and room for chip clearance, and ensuring the number of passes is reduced to that which balances tool performance with time on machine.

Once the operator has advanced the tool into the workpiece the machine feed can be applied, and the tool loading rapidly reduces its set cutting angle. The cut is applied until it approaches the length required, which is established by observing the DRO readout or observing the reducing distance between the cutter and any scribed line. This observation can be difficult depending on the coolant flow and it is not unusual for scribed lines to be obscured. Following initial cuts this becomes less of an issue as the operator will see the machined step left from preceding cuts, and can stop the feed accordingly.

Where a ripper cutter of larger diameter than corner radii has been employed, it is unlikely that a plain helix end milling cutter can be employed to merely undertake a single, or low series of finishing cuts. If the difference between diameters is minimal it is possible to position the cutter over the centre of the radius less any allowance for a finish cut and raise the knee to utilise the cutting edges on the end of the cutter to remove unwanted material, and then subsequent finish cuts undertaken in a normal manner. Where the difference in diameters is such that the width of material to be removed exceeds the width of cutting edge on the bottom of the end mill, a series of localised roughing operations needs to be undertaken in each corner, followed by a finish pass. A third alternative is to fit a slot drill and machine down to remove unwanted material, although this adds an additional setting operation and is seldom worthwhile.

4.19 TEE SLOT MILLING

Tee slot milling is a milling operation that is more congested than most and requires some care in set up by the operator. Tee slots are normally manufactured to an ANSI/ASME specification or an ISO specification. As such need a cutter that is matched in diameter to the headspace width, and less or equal to the head space depth. Figure 4.31 shows a typical tee slot milling cutter.

Figure 4.31 Tee slot cutter

Tee slots are manufactured in two key stages, the first being machining of a slot with a width equal to the throat thickness stated in the standard for the size of tee slot being manufactured. This operation is normally undertaken using a slot drill on a turret or vertical milling machine and a side and face cutter on a horizontal mill. The second stage is to cut the headspace using a tee slot milling cutter. Figure 4.32 shows the basic form of a standard tee slot.

A slotting operation is undertaken to form the initial slot in a similar manner to which any slotting operation would be undertaken. The width of slot to be achieved may be able to be generated by using a cutter of the exact width, however, where a slot needs to be widened it is worthwhile the operator ensuring that the initial cut is taken on the required centreline position for the slot. The slot should be cut in pecks down to full depth (i.e. full depth to the bottom of the headspace as determined by the drawing or standard dimensional requirement). If the slot needs widening and has been cut down the centreline, the operator has the opportunity not only to continue to down cut mill on either side of the slot, but also has the option to change cutters to an end milling cutter if this is preferred. Once the slot has been manufactured to the correct width and depth all swarf should be cleared from the slot and the machine readied for generating the tee slot.

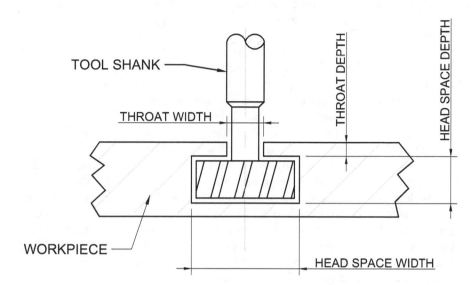

Figure 4.32 Tee slot nomenclature

Tee slot cutters tend to be supplied with 6 or 8 teeth, and are traditionally made from HSS. However, more modern examples allow for carbide inserts and staggered tooth configurations. The cutting edge configuration ranges from those with no cutting edge rake through to examples that have cutting edge rake and tooth stagger. Cutters with stagger and rake tend to be able to be pressed a little harder than those with vertical teeth, as the rake creates a shearing action and the load progresses across the tooth, and swarf chips are moved from the cutter more easily.

Once the initial slot has been cut to depth the tee slot cutter is mounted in the machine spindle. If the centre line of the slot was set as a datum then the cutter is merely aligned with this at one end of the slot. The datum height of the cutter is set using a probe, or papers in the same manner that and end milling cutter or slot drill *z*-axis datum would be set.

Tee slot milling requires the advancement of a cutter into a blind face relieved only by the central slot. It cuts across its diameter (i.e. on both sides of the cutter simultaneously); however, it does not have a 180° cutting angle as the central slot reduces this to approximately 90°. The depth of headspace required can be the depth of the cutter head, or may be deeper in which case two passes are required. One to cut the headspace and the second to deepen it to the required depth. Where depth of headspace height is greater than the depth of the cutter the operator needs to choose whether to cut at the full depth first and then raise the cutter to increase the height or work the other way around. Anecdotal evidence suggests that most operators cut at full depth first as the second raised cut is generally relatively light and space below the tee slot cutter allows coolant to and chips to flow away from the cutting surface.

Having set the correct spindle speed for the cutter diameter, it is slowly fed into the workpiece. This initial feed rate needs to be undertaken slowly as the tool is unsupported as it enters. Once fully engaged the feed rate can be that for the manufacturers stated rate for feed per tooth. However, the operator needs to closely observe the behaviour of the tool.

Cutting tee slots is a process undertaken in a confined space with limited space for chip removal. A significant flow of coolant is advised to ensure correct cooling of the tool and assist with the removal of swarf chips. If a cutter with straight teeth is being used there is no oblique shear action. the loads that the tool experiences during cutting will be significantly higher than that for staggered teeth and those with a rake that generates a shearing action. While a feed per tooth may have been correctly calculated, the performance of a straight fluted cutter than one that has a rake, may differ, and feed for a straight tooth cutter may be less. A second cut to machine the cavity to the correct throat depth is generally more easily achieved than the first given that the majority of material has been removed with the blind cut.

Tee slot cutters have a relieved neck to allow it to pass through the throat width, and while cutters are normally supplied to machine a standard tee slot, widening of the headspace is possible, but it is limited to the difference between the cutter neck diameter and the throat thickness. Allowing the shank of the neck to rub on the sides of the slot will damage the surface finish. An excess pressure will try and bend the shaft. This excess force will deflect the corner of the cutter into material and generally leads to breakage.

Where a square section undercut is to be formed, a tee slot cutter can be used rather than remove a workpiece and mount it at 90° to its original orientation, and machine using a slot drill. The operator only needs to make an assessment of the time it would take to reposition the workpiece and insert a slot drill compared to that of utilising a tee slot cutter. Utilising a tee slot cutter can be a good option if work holding issues are a significant factor.

4.20 DOVETAIL MILLING

Dovetail milling has much in common with tee slot milling and the initial approach is very similar with a slot being machined into the workpiece to a required depth before a shaped cutter is advanced into the workpiece. However, the difference is that, although entirely possible, it is less common for both sides of a dovetail to be formed in a single pass, and not uncommon for only one edge to have a dovetail machined. Dovetail cutters for engineering purposes generally come in two angles, 45° and 60° and examples can be seen in Figure 4.33. Other angles are readily available, but the operator needs to be aware that the angle stated is an 'included' angle. It should also

Figure 4.33 45° and 60° dovetail cutters

be noted that most dovetail cutters have straight flutes, and can have a narrow neck diameter, which reduces rigidity.

The approach to forming a dovetail slot is almost identical to Tee slotting, as a groove is first machined using a slot drill to form space for the dovetail cutter to work within. The selection of size of cutter to be used is controlled by the size of the initial slot, whether both sides are to be generated in a single pass, and the depth of the dovetail to be produced. Given that the cutters are standard angles then the largest possible cutter that can fit into a slot either by penetration from the end or by a vertical displacement into the slot followed by a horizontal movement into the workpiece. It is efficient to produce a dovetailed cut using one cutter, therefore the height of the cutter gives its cut depth and the operator selecting a cutter needs to ensure that the height of the cutter is equal to, or exceeds the height required on the workpiece.

Advancing the cutter into the workpiece is generally undertaken from one end, however it can be brought in from the side. Care needs to be taken until the cutter is fully into the workpiece at which the feed rate can be increased to the maximum for feed per tooth. The cutting angle is always going to be less than 90° to allow for the tool shank diameter, and this both reduces the

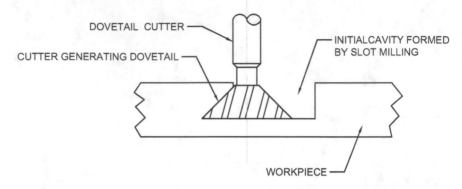

DOVETAIL CUTTER

CUTTER GENERATING DOVETAIL

INITIALCAVITY FORMED
BY SLOT MILLING

WORKPIECE

Figure 4.34 Typical dovetail configuration

load on the tool compared to a dovetail cutter and provides more room for swarf ejection. As with many cutting operations in restricted applications, high rates of coolant assist with chip dispersal.

Measurement of relatively square feature such as slots, Tee slots, and shoulders are simple to establish values for. Dovetails are an angled face that when produced in pairs, need to have a fairly well controlled size to ensure a mating component fits, or to prevent the generation of a razor-sharp edge as the angled face generated breaks through the top surface of the workpiece. It is normal for a dovetail to be cut such that the angled face breaks though into the wall of the slot previously cut leaving a small land, as shown in Figure 4.34.

It is possible to machine a dovetail in one pass, however, in harder materials, or those requiring a high degree of precision multiple cuts are used. To achieve this the cutter is set to the depth required and then the cutter is advanced into the workpiece in incremental amounts. The slot is measured as it progresses and the translational movements calculated until the cut is complete.

4.21 DIVIDING HEAD USE

A dividing head is an attachment that is mounted on a milling machine table and most commonly oriented along the x-axis. Dividing heads can be right- or left-handed, however, the most common type is left-handed with the attachment mounted to the left of the operator and the workpiece protruding to the right of the head. The mounting of a dividing head to the machine creates another degree of freedom allowing more complex shapes to be generated. A wide variety of types can be sourced, with some fixed in one plane and some that are described as universal, however, the universal type (i.e. those that can both rotate and move their axis of rotation through an arc) are very common in engineering workshops. Universal types are required for the milling of cams, tapered components and some gear cutting operations. An example of a common dividing head which can often be found in general engineering workshops is shown in Figure 4.35.

When coupled with the moveable head of a turret milling machine, five degrees of freedom are enjoyed on the machine, coupled with the ability for the workpiece to be moved through an arc, and the head of the machine being able to be tilted in two axis. With the freedom to position the cutter in a significant number of positions, shapes can be generated outside of the normal flat-bottomed slots and grooves that milling produces. The use of specialist cutters also allows for gear cutting, and hobbing to be undertaken without specialist machinery.

Dividing heads are often quite heavy and lifting them onto the machine table requires a degree of care to prevent injury to the operator. It is normal for the dividing head to have registers, or tenons located on its base which fit into the tee slot neck on the machine table. On good quality dividing heads these will have been mounted to an axial alignment of approximately 0.006mm, so any gross area will relate to the clearance between the registers

Figure 4.35 Traditional dividing head

and the width of the tee slot neck. These registers provide initial alignment, and if the head is supplied with a test certificate, only requires checking for work that has the closest of straightness or parallelism tolerances. When mounting the head operators should push on the head to ensure the registers bear up against one side of the slot if there is any excessive clearance in the tee slot neck. Retaining the dividing head only requires the fitment of standard tee bolts, however, in some cases the space to slide the front tee bolt in place is restricted and bolts of the shortest length that gives full thread engagement for the nut is advised. Figure 4.36 shows the front retaining bolt of a common type of dividing head.

Dividing heads are used for milling slots, grooves, splines, or teeth accurately around the circumference of a workpiece. They are also particularly useful in milling flats and polygons onto round workpieces. The workpiece is either clamped into a three-jaw chuck, or placed between centres in exactly the same manner as a workpiece is mounted in a centre lathe. Longer workpieces are supported by use of a tailstock, which have been manufactured to the same height as the dividing head centreline. It is also not unusual for longer workpieces to have an additional support, often a screw jack, placed centrally between the dividing head and tailstock to prevent deflection of less rigid workpieces while being machined. Figure 4.37

Figure 4.36 Dividing head clamping (front)

Figure 4.37 Long workpiece supported by tailstock and screw jack prior to milling

shows a long workpiece being supported by a tailstock and central screw jack prior to the commencement of machining.

Dividing heads have a number of features for controlling the orientation of the workpiece. While simpler dividing heads have a fixed axis, universal types allow the axis of rotation to swing through an angle. Typically this angle of swing is 90° and most universal heads will allow a negative angle of swing, often around −10°. The restriction on this angle is the point at which the depression of the chuck is impeded by the machine table. This is rarely a problem as machining a component in the opposite orientation will address this issue. However, the operator needs to consider the angle the workpiece needs to be rotated to, and which way round the workpiece will need to be, in order to generate the feature within the limitations of the dividing head.

While designs vary, almost all dividing heads have the same general features. The dividing head will have a hollow spindle that allows a workpiece to pass through, in a similar fashion to a lathe. The diameter of the hole will vary dependent on the manufacturer and model size. The end of this spindle will have an external threaded portion and an internal taper. The external thread is for mounting a chuck, or drive dog, and the internal taper is for mounting a dead centre used when mounting a workpiece between centres. If the head is able to rotate there will be a graduated scale and datum mark in degrees to indicate the amount of rotation. While rotation to elevate the dividing head from its horizontal orientation to vertical, or a proportion between the two normally is undertaken by hand, and axial rotation is always undertaken by mechanical means.

There are two methods for providing axial rotation; direct indexing or rotation, and indirect indexing. Direct indexing is the simplest method and the amount of workpiece rotation is either read of a scale marked in degrees, or by the engagement of a plunger into a machined slot on a circular plate.

Direct indexing plates for direct engagement by a plunger are often reversible with each side being machined with differing grooves to provide alternative increments. Typical examples of this are one side having 60 increments (i.e. minutes), and the reverse side having a slightly coarser arrangement with perhaps 48 increments. To change these plates the chuck or drive dog is unscrewed from the spindle the plate removed and reversed or an alternative mounted and the chuck re-fitted. Alternatively there may be a plain scale marked in degrees and a pointer rather than plunger utilised. This is used in a similar manner, and gives more flexibility, however, is less precise as operator error or parallax may become an issue when rotation a number of degrees. This type of dividing head indexing also requires a powerful retaining clamp as the lack of positive engagement by a plunger allows for slippage when under load if not tightly clamped, or will be using an internal worm drive to accept the load which is undesirable.

Fitted to the side of all dividing heads is a circular plate with multiple rings of holes drilled into it, two small moveable arms, and a handle which can be moved up and down a crank. This is the main control for rotating the workpiece and the key to Indirect indexing. The handle is connected to a worm drive which is used to rotate the spindle, and by default the workpiece. It is designed such that any number of sub-divisions can be produced depending on the workpiece diameter. Figure 4.38 shows a standard

Figure 4.38 Dividing head indexing assembly

indexing plate and handle assembly on a dividing head. It should be noted that other indexing plates with alternative hole configurations are available to ensure that any rotational increment can be achieved.

While differing manufacturers will have differing worm drive and gear combinations, it is standard for forty revolutions of the indexing handle to equal one full rotation of the spindle, and by default, the workpiece. While direct indexing is useful, the real strength of a dividing head is being able to rotate a workpiece through a closely controlled arc either as a discrete angular displacement or to achieve a number of equal increments on the periphery of a known diameter.

To rotate the workpiece through a known arc, the handle is pulled to disengage the peg inserted in an indexing plate hole, fully rotated a calculated number of turns, and normally a number of holes around on the indexing plate ring, after which the handle is released and the peg of the handle reinserted into the correct hole. To assist with ensuring the handle is inserted into the correct hole, two movable arms are used to identify the start and finish positions.

Some older, or more complex models of dividing head also have a drive shaft protruding from the rear. This is to allow a set of gears to be installed that links the milling machine table drive to the dividing head. The gear ratio that is configured will provide a proportional rotation while a translational movement takes place. This allows for helical milling along the axis of a workpiece.

Dividing head indexing example 1 (direct indexing)

A common operation for a dividing head is generating simple polygonal shapes from round stock bar. The classic example being the manufacture of a hexagon nut. Simple shapes with common angles between features can often be quickly and easily generated using the direct indexing capability of the dividing head.

The most efficient approach is to use a three-jaw chuck mounted on the dividing head rather than machining between centres, and to use direct indexing. First the operator needs to check that if the dividing head has a plunger that engages in a machined slot on the indexing plate mounted on the spindle of the dividing head as shown in Figure 4.39.

Where this is found the operator then needs to check that the divisions machined into the indexing plate are divisible by six (i.e. the number of sides on a hexagon). Where this is not found, the indexing plate can often be reversed with an alternative set of divisions found, or replaced with a suitable alternative. Where the dividing head is not constrained by fixed divisions and merely has a scale marked in degrees this issue is removed. Having attached the chuck to the

Figure 4.39 Direct indexing plunger engagement

dividing head and bolted the dividing head to the machine table the workpiece is clamped into the chuck in an identical manner to that used for a lathe. Any gear lever used to engage and disengage the spindle worm drive needs to be fully engaged to prevent free rotation of the dividing head chuck.

Prior to tightening of the workpiece in the chuck, the operator needs to assess how far out of the chuck the workpiece should be based on how well supported it is. Where an end milling machine or, turret milling machine is being used the operator needs to check that no part of the milling machine spindle or head assembly will impact with the dividing head during machining operations. The use of a longer series end milling cutter or extension of the spindle quill can often overcome this, but the operator needs to balance the length that the workpiece can protrude from the dividing head, with the likely bending moment, or vibration that may degrade the depth of cut that can be used, the surface finish and the dimensional accuracy. Where the workpiece is quite small it is not unusual for a tailstock to be required as the small size results in close proximity of machine castings, with the main body of the dividing head. Figure 4.40 shows the set-up of a workpiece clamped in a dividing head fitted with a direct indexing plate with 60 divisions, ready for machining to commence.

Selection of cutter diameter is a matter of availability and personal choice. If machining a hexagonal nut that is of a width less than an available cutter diameter then it makes sense to use it as it reduces the number of operations to be performed, however, this is not crucial.

Having released the dividing head index plate locking plunger, or any clamping device where a plunger is not used, rotate the head to a start point. This start point can be any selected by the operator, however, it is normally much simpler as machining progresses if the 0° position or 0 increment point on the direct indexing plate has been used as the start point. Many dividing head chucks have a line scribed onto the surface to show alignment with top dead centre, and give the operator a visual marker of the degree of rotation undertaken since commencement of machining. The operator should check that any engagement/disengagement leaver fitted to the dividing head is correctly engaged or the chuck will freely rotate It should only

Figure 4.40 Hexagon generated using direct indexing

rotate using the worm drive when the operating handle on the worm drive disc is rotated.

Having traversed the machine table such that the cutter is positioned behind the workpiece, the machine is started and the first operation is to set the datum positions. While many milling operations need careful identification of x-axis, y-axis and z-axis datums, producing a basic hexagon, or other polygon, is less critical of x-axis and y-axis datum identification. This is because if the width of the cutter is greater than the width of the feature to be generated then there will be no movement along the x-axis during machining. It is only where the length of flat to be produced is wider than the cutter, or must be manufactured to a known length would any datum be set. Where this is required, the standard techniques of touching the cutter on the end of the bar, use of cigarette papers or direct probing would be used and a datum set. When producing a flat surface for manufacture of a hexagon, the y-axis datum is not required to be set, however, the z-axis is always crucial when generating any feature to a required dimensional size. Given the type of operation to be undertaken when machining a round workpiece blank the first few cuts are rarely critical, in which case probing or use of cigarette papers is, unnecessary, and bringing the cutter into light contact with the workpiece is entirely appropriate. This is best undertaken with the cutter rotating to prevent cutter damage. Once this has been undertaken the scale on the knee elevation handle is set to zero or DRO scale zeroed to provide the z-axis datum.

The cutter is then moved off the workpiece, normally by traversing the y-axis hand wheel if the feature is narrower than the cutter diameter, and a depth of cut set. The operator needs to know what the depth of cut is to be. For even sided polygons this is a simple calculation as it is common for dimensions of these features to be given as a distance across flats (A/F). This only requires the operator to subtract the A/F dimension stated on the component drawing from the workpiece diameter, and divide this by 2. This gives the depth of material to be moved from each face, and if the cutter has only been lightly touched onto the workpiece such that it is only marked then the depth calculated will be the depth required to be removed.

When using a dividing head to produce a polygon the workpiece is rotated around a central axis. If an identical cut is not taken from each face the distance central between two opposing machined faces will not be on the central axis of the component. It is therefore crucial when machining features on a dividing head to provide opposing flats, that the same depth cut is taken to generate each face.

Once the depth of cut has been set, the y-axis hand wheel is used to advance the cutter into the workpiece. If a longitudinal cut is to be undertaken the x-axis feed is applied and the cutter advanced down the workpiece towards the dividing head chuck. Once the required length of cut has been achieved it is usual for a cross axis cut to be undertaken by using the y-axis hand wheel to pass the workpiece across the x-axis and provide a square end to the length of cut. The cutter is then returned to the start point and place off the workpiece. While a second depth of cut, and sequential cuts, can be applied and the face taken down to its full depth this is not advised. This approach requires the operator to have carefully set the z-axis datum and have calculated the exact depth of cut that needs to be applied to generate the feature. The dividing head then requires to be indexed and if the feature cannot be generated in a single pass then the knee will have to be dropped to reduce the depth of cut and sequential passes undertaken to generate the next, and sequential faces. This introduces a significant potential for error and the generation of all faces for each cut depth is recommended.

With the cutter off the workpiece the dividing head is rotated to allow machining of the next face. The direct indexing plate plunger, or indexing clamp is released, and the dividing head chuck rotated. It is generally unimportant which direction it is rotated, however the angle of rotation is. In this example the direct indexing plate has 60 graduations machined into it for the direct indexing locking plunger to engage with. This means that to produce a hexagon the dividing head chuck needs to be rotated and the plunger re-engaged with the tenth mark. Where a scale marked in degrees is utilised the chuck is rotated by 60°. This rotation is undertaken by rotating the handle and indirect indexing place assembly either clockwise or anti-clockwise. This shaft is connected to a worm drive with, typically, a 40:1 reduction, therefore a number of rotations of the handle to revolve the dividing head chuck by the angular displacement required, at which the plunger is reinserted into the correct increment slot, or the datum line aligns with the correct angular rotation required (in this case 60°) and any clamping screw tightened to prevent inadvertent rotation during machining.

Without changing the depth a cut, a second and, sequential cuts following rotation and are undertaken until all sides have a flat feature generated. The distance across flats is then measured accurately using a vernier calliper or micrometer. This process of measurement following generation of an initial flat is why it is not necessary to be overly precise when setting a z-axis datum, and just touching the tool onto the workpiece is entirely adequate. The dimensional distance across flats stated on the component drawing is then subtracted from the value derived from the measurement. The result is divided by 2 to give the total depth of cut required for each face, in order to provide an A/F distance that is symmetrical across a central axis.

Where this depth is greater than the depth of cut that is to be taken on each pass the operator needs to decide on the increments to be take. Differing cutters will produce surface finishes that vary. Some cutters like to have a more substantial cut than others and operators may wish to consider what the final depth of cut is to be, and while a heavy final cut is often undesirable, a depth of approximately 0.9mm can often produce a better surface finish than 0.1mm with some cutters. Experience with the types of tooling available, and their sharpness will provide the operator with the information required to make an informed decision. Calculation of the remaining depth to be removed from each face, and awareness of the final pass cut depth to be taken, provides the operator with the knowledge of how many cuts at a specific cut depth are to be taken to the penultimate cut. Once the operator has established the number of roughing cuts to be undertaken at a particular peck depth, there is no special requirement to re-measure. It is recommended that the operator checks the workpiece size, prior to undertaking the final finish cut to ensure that no adjustment is required, in order to create the feature to within the required dimensional tolerance.

Once all faces have been machined the workpiece can be removed, or subsequent operations undertaken. However, the process described will result in a machined polygon (hexagonal in this case) with flat faces that are equi-spaced about the central axis of the workpiece, as can be seen in Figure 4.40.

Dividing head example 2 (indirect indexing)

Where features of a less convenient increments or an odd number of slots or grooves are to be machined, the use of indirect indexing is required. Direct indexing, especially when utilising a slotted plate engaged with a plunger is limited to the resolution it has been machined in. The direct indexing example utilised an indexing plate divided into 60. This gives an angular rotation of 6° for each increment. Where a finer resolution is required, or the angle between features is not divisible by 6° or any other arc driven by the direct indexing plate then indirect indexing is required and typical accuracies come down to 6 seconds of arc. This also applies where a dividing head does not have a plunger type mechanical stop to limit rotation but merely a scale marked in degrees and datum mark precision may become an issue. This is especially relevant when machining gear teeth, or other components that are sensitive to angular accuracy.

Indirect indexing is where the circular plate drilled with holes in differing numbers and arranged in concentric rings of holes is used. The indexing plate has a series of concentric rings of holes, and each ring has a different number of holes. It is normal for two different plates to be supplied with each dividing head, which have differing complementary rings of holes to provide greater possible combinations of angular rotation. While rotating the handle a number of turns will provide a rotational movement of the spindle the series of holes allow further sub-division of the angular rotation achieved. The worm drive handle operates a shaft which passes through the centre of the circular plate, and itself is mounted within a machined slot on the crank of the shaft, allowing adjustment for the handle location pin to engage with each concentric ring of holes. To assist with accurate repetitive handle location setting a pair of sector arms are also mounted to the worm drive shaft and against the indirect indexing plate. Figure 4.41 identifies the key components of the indirect indexing controls that are used in precise radial movement of the workpiece.

Of key importance when determining the number of full revolutions and part revolutions of the control handle is knowledge of the dividing head worm drive reduction ratio. By far the most common reduction drive is 40:1, however, to ensure an alternative ratio is not fitted to the dividing head the operator uses the first check is to count the revolutions required to revolve the entire head through 360°. This is where a scribed mark on the dividing head chuck assists. With the mark set at top dead centre, or temporary mark placed on the chuck using a marker pen, the handle is wound through as many rotations as is required, to return the mark to top dead centre. The number of revolutions

Figure 4.41 Indirect indexing key components and controls

required to revolve the spindle through 360° is the reduction ratio, which is most likely to be 40:1.

The method for milling using indirect indexing is essentially identical to that used when indirect indexing, and it is the method for applying an angular rotation. While generating flat surfaces and polygons is often the preserve of direct indexing, the manufacture of splines, gears and slots requiring a precise angular relationship between features almost certainly requires indirect indexing. To establish the number of turns to be used to provide a given number of increments a simple formula is used:

$$A = \frac{\text{worm drive ratio}}{N}$$

where A = movement of the crank arm and N = number of divisions required. If it is required to machine a series of slots into a workpiece such as cutter a gear that has 23 teeth then the number of turns of the crank to give 1/23rd of the workpiece rotation needs to be calculated. The worm drive reduction ratio is then divided by the number of increments required.

Example calculation to determine rotation for known number of increments

In this case the worm drive reduction ratio is 40:1 so the following calculation is undertaken:

$$\frac{40}{23} = 1.739$$

However, to identify what 0.739 turns means in terms of numbers of holes to be engaged with on the indexing plate the operator needs to identify an angle of rotation or simply turn the figure into a fraction. This can be achieved by multiplying the numbers of holes in each concentric ring by 0.739 until a whole number is achieved. Alternatively the dividing head manufacturers handbook, or good quality machining handbooks often have tables which identify the numbers of handle full turns, hole circle, and fraction of the hole circle to be used to achieve the angular rotation to generate an exact number of equi-spaced divisions.

Therefore to produce 23 divisions on a workpiece, the diving head handle must be rotated through 1.739 turns or 1 full turn and 17/23 of a turn. The fraction 17/23 = 0.739 and this fractional rotation is achieved by inserting the handle pin into the 17th hole on the 23-hole ring.

Standard indexing plates have concentric rings of equi-spaced holes in groups numbers which can be seen in Table 4.3. The increments may be split across two plates and the operator may need to change the indexing plate to provide

Table 4.3 Standard indexing plate concentric ring divisions

15	16	17	18	19	20
21	23	27	29	31	33
37	39	41	43	47	49

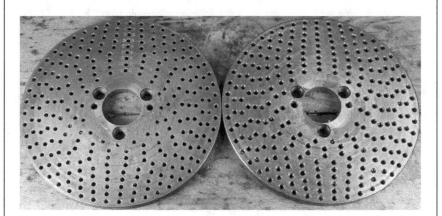

Figure 4.42 Common indirect indexing disc

the desired ring of holes. Figure 4.42 shows how each different ring of holes is identified such that the operator does not have to count them and can directly select the correct ring.

By adjusting the radius of the handle such that the handle pin fits cleanly into a hole on the 23 ring the handle is locked in position on that radius. Sector arms with opposing chamfers are mounted on the surface of the indexing plate and below the crank handle. These arms are then adjusted. This normally requires the loosening of a clamping screw or other locking mechanism. The trailing arm is then rotated until the chamfered edge presses up against the handle pin. The arm forward of the pin is rotated so that 17 holes are visible and both sector arms have their locking clamp tightened. To rotate the spindle through the required distance of arc the operator pulls the handle to disengage the pin in the indexing plate rotates it through one complete revolution, plus 17 holes as indicated by the forward arm, and releases the handle to re-engage the pin. The arm assembly is then rotated to bring the trailing arm back in contact with the pin on the indexing handle and the forward arm rotates with it ready for the next cut. The operator takes the required cut, and the process is repeated until all 23 cuts have been made.

Care must be taken to avoid over rotating the dividing head when applying increments. If the handle is wound too far, then it needs to be reversed at least one turn. This ensures that any backlash in the dividing head worm drive is eliminated and that the workpiece is rotated to its precise location for the subsequent cut.

Where an indexing distance is not controlled by the number of cuts to be made such as manufacturing a toothed gear, but the component detail drawing gives a dimensioned angle, an alternative approach needs to be taken. It is relatively simple to calculate the angular displacement of one rotation of the indexing handle, with the gear with the following formula used:

$$\text{Angle} = \frac{360°}{\text{worm drive ratio}}$$

So for a dividing head with a 40:1 worm drive reduction ratio the rotation angle is $(360 \div 40) = 9°$. This also means that 1/9th of a turn = 1°.

Example calculation to determine rotation for given rotational angle

If the operator was required to rotate the workpiece through an angle of 38° then the following calculation is undertaken:

$(38° \div 9°) = 4$ and 2/9 turns of the crank handle. Or 4 turns of the handle plus another 2° of rotation.

While there is no ring of holes with just 9 positions, there is a ring with 18, therefore, to rotate the workpiece through an arc of 38° the handle would be rotated through 4 full rotations and 4 holes of the 18-hole ring in the indexing plate. The calculations are not difficult, however, the operator needs to take care when establishing increments to be machined on a diameter, or angular displacement as the approach differs and the, use of the indexing plate

Where standard plates do not correlate with a required rotation a process called differential indexing can be applied. Differential gearing is used when a required number of divisions cannot be established by simple indexing. When this is encountered the use of a dividing head that allows direct drive of the dividing head spindle from the shaft of the dividing head disc. This is almost certainly limited to more advanced 'universal' dividing heads, and simple or semi-universal dividing heads are unlikely to be able to undertake differential indexing.

The connection is established through the fitment of a series of gears and idlers between the spindle and shaft, with the gear tooth combinations and intermediate wheels being the method used. Figure 4.43 shows a dividing assembled with gears mounted on the spindle, and disc shaft, with intermediate gears mounted on a change gear bracket. This combination allows the production of numbers of increments that cannot be produced by standard indexing.

To achieve this sort of set-up, a more sophisticated dividing head is required, and the operator must have access to the correct accessories for it to work. However, once the calculations to determine the gear combinations have been undertaken, attaching the change gear bracket, change gear spindle, and mounting the spur gears is a simple assembly task. The flexibility of the majority of dividing heads and the requirement to manufacture a component requiring the milling of features in increments that do not match with any of the direct and indirect indexing plates is unusual. Where the operator has access to a dividing head that is complete with its attachments, it remains an option.

Figure 4.43 Dividing head set-up for differential indexing

Dividing head example 3 (long workpiece support)

Long workpieces generally need support, either because they will deflect when being machined, or because the mass of the workpiece will cause it to be torn from the jaws of the dividing head chuck, especially when an additional down-force is applied during material cutting. Any operation the does not employ a chuck as the work holding device will be undertaken between centres to provide horizontal stability as well as vertical. While it is possible to support a long workpiece with a chuck and screw jack, this provides little lateral support and there will be significant vibration experienced at the workpiece end that is opposite the chuck, and where possible a tailstock should be employed.

The tailstock is normally supplied with a dividing head along with indexing plates and a gear set, however, they can be sourced separately. The operator must check that if the tailstock being used is not supplied with the dividing head and has been procured separately, the centre height is matched to that of the dividing head. If the centre height of the dividing head and the centre height of the tailstock are misaligned it will not be possible to machine a feature that is parallel to the axis of the workpiece. Where there is a difference in height the operator will need to insert packing shims under either the dividing head or tailstock to bring them level.

The tailstock normally has registers or tenons fixed into its lower surface in a similar manner to the dividing head. This assists the operator with locating the tailstock onto the same axis as the head itself. The tailstock is mounted using exactly the same method as for the dividing head, with a fore and aft tee bolts used to fasten it down. The tailstock itself will have a screw fed quill that allows for the centre to be inserted withdrawn from a workpiece to allow removal and replacement. Figure 4.44 shows a typical tailstock mounted onto a machine table ready for use.

The requirement to use a tailstock is highly dependent on the diameter, length, length of workpiece engagement in chuck jaws, stiffness of the work-piece, and the operation being undertaken. The operator needs to make a judgement when workpiece support is required, but the loads imparted by machining can be high, and any unsupported work will be subject to deflection and induced vibration if any significant cut is being undertaken.

Once the tailstock has been mounted the workpiece needs to be inserted into the dividing head. The workpiece will need to have a morse centre hole drilled into the end of the workpiece. Ideally this operation is undertaken in a lathe for ease of ensuring it is on centre, however, it can be drilled. With the workpiece being retained in either the dividing head chuck, or being held on the spindle centre by the operator who should also ensure that the driving dog be

Figure 4.44 Dividing head tailstock

fully inserted within the fork of the driver, then rotates the tailstock hand wheel to insert the point of the centre into the morse drilled hole. As there is no rotational speed that would cause significant heat build-up in the centre cavity, there is no requirement to lubricate the centre point, however, a small smear of oil or grease will assist in wear prevention. Following mounting of the workpiece the operator needs to check that the dividing head and tailstock are aligned.

The dividing head manufacturer will have checked its precision, and the operator does not need to determine axial errors of machine, but does need to establish the y-axis and z-axis alignment. This is normally achieved using the workpiece, however, if the workpiece is of an irregular shape or does not have a consistent profile then a test piece will need to be used. The simplest method to use is by locating two dial test indicators on the workpiece. The dial test indicators located at the tailstock end should be positioned at 0° and 90° (i.e. on the y-axis and z-axis) and the table then traversed. The operation can be undertaken with a single DTI but the use of two simultaneously identifies both height and alignment errors which can be of use. Figure 4.45 shows a simple set-up for this.

The dial test indicator located near the chuck or driving dog, is there to establish any ovality of the workpiece. By rotating the workpiece through 360°

Figure 4.45 Checking axial alignment of dividing head and tailstock

the operator will be able to observe any significant deviation especially when at the 180° point. Readings from the dial test indicator at the opposite end will establish the workpiece ovality and this can be factored into any reading identified at the tailstock end. Where a workpiece is found to have significant ovality or other deviation in profile, it is sensible for the operator to replace the workpiece with a test specimen such as a steel ground bar that has a morse centre drilled with which the operator has confidence in its central location. Once alignment has been established at the tailstock end the dial test indicator adjacent to the dividing head is removed and the table traversed such that the *x*-axis alignment is checked.

Where the workpiece is of substantial length, or lacks stiffness due to its configuration or mechanical properties, a screw jack can be used to support the workpiece. For excessively long workpieces this may involve a number of screw jacks. The type of screw jack employed should be such that it does not interfere with any rotating part such as the cutter or spindle. Its only purpose is to provide vertical support and not lateral. The ideal location of the screw jack is mid span between the head and tailstock, however, the features being machined may make this difficult. In an ideal situation the screw jack is set up on a part of the workpiece that will not be subject to machining, or on a section where a machined cavity will not rotate onto the screw jack head. Where this is unavoidable the operator will need to check that the screw jack head remains in contact with the workpiece and provides support when machining its opposite side. Screw jacks should have a mounting hole to enable them to be bolted down to the machine table, and devices used that lack this feature are unlikely to provide sufficient support. Figure 4.46 shows a long workpiece being supported mid span.

When adjusting the height of the screw jack and bringing it into contact with the workpiece, the operator needs to take care that the height it is raised to does not deflect the workpiece upwards. Where this happens the cutter will

Figure 4.46 Workpiece supported by screw jack mid-span

effectively dig deeper into the workpiece providing an elliptical groove. The height should be adjusted and locked such that light contact is made, preventing deflection away from the cutter, but retaining the axial alignment.

Dividing head example 4 (rotation of head)

Universal dividing heads and semi-universal dividing heads are able to rotate about the y-axis, often up to 90° and down to −10°. It is not normal for this rotation to be achieved by any geared or driven rotation and is achieved by the loosening of a clamp, or retaining screws followed by movement by hand to the required orientation, and re-clamping. The majority of dividing heads with this function have an engraved scale to indicate the angle rotated to. The accuracy of this scale is depends on the diameter of the dividing head arc and the operator needs to undertake a check to establish if the correct angle has been achieved if accuracy of greater than 0.5–1° is required. Figure 4.47 shows a dividing head rotated to 45° for the milling of a bevelled workpiece.

Figure 4.47 Dividing head rotated to 45° and bevelled workpiece

To check the angle of dividing head vertical rotation two approaches can be taken. Where the head is rotated to a vertical position a DTI is mounted on a bar one end of which is placed in the in the dividing head chuck, and the DTI mounted on the reverse end such that the plunger is in contact with the machine table. the head is rotated through 180° and the reciprocal reading taken. If the reading is the same on both sides the chuck is at 90° to the machine table. Where a difference is observed the operator does not need to undertake complex trigonometry, but the head is elevated or depressed by half the difference between the two and the process repeated until a zero difference is achieved. The clamping screws are then full tightened.

Figure 4.48 Runout measurement

Where a bevelled workpiece is to be machined, such as a bevel gear blank, the head is rotated to the required angle as indicated on the dividing head scale. A DTI is then placed with the plunger in contact with the bevelled surface. The table is then moved backwards and forwards on the y-axis until the top dead centre is identified. The machine table is then traversed so that the DTI plunger runs from one edge to the opposite. If there is no runout, the angle set is correct. Where there is runout, the operator halves the error by applying incremental adjustments until no runout is observed. Figure 4.48 shows this process being undertaken.

Where the workpiece to be machined is a shape or configuration that does not allow the approaches described previously, then manufacture of a bevelled arbor for the sole purpose establishing set angles is worthwhile. However, if the operator is confident of the accuracy of the scale, and their ability to precisely adjust angles using the scale, then precise checking may not be necessary. However, gear manufacture and the precision that requires, or the tolerance applied to the workpiece feature, may drive the approach taken.

Dividing head example 5 (helical milling)

Helical milling, or helical interpolation are widely used in manufacturing undertaken on CNC machines to produce holes and slots especially in materials which are hard, or cause rapid tool wear that prevents the use of conventional drilling techniques. In conventional manual machining helical milling is less commonly used, however, by connecting a dividing head to the machine table drive a work piece can have a translational movement in the x-axis as well as a rotational movement also in the x-axis. This allows for the machining of helical gearing and spiral slots. This operation is only produced in one direction, therefore, It is not possible to produce a double helix in one operation, as backlash issues will introduce error. The actual process for milling operation differs little from slot milling, or spur gear cutting, however, establishing the correct ratio between translational movement, to rotational is key to success.

To link the dividing head to the machine table lead screw drive gear is attached to the end of the table drive. Access to this table drive may be require the removal of a cover, or traversing handle. Machines that do not have this feature will be unable to perform helical milling operations, or will require a complex separate drive, which has an encoder either linked to the spindle speed or to the table drive shaft. By changing the gearing ratio between the table and the dividing head indexing shaft the amount of rotational movement

Figure 4.49 Typical helical milling gearing set-up

Table 4.4 Standard dividing head spur gear set

24	24	28	32	40	
48	56	64	72	86	100

with relation to translational can be adjusted. Figure 4.49 shows a typical set-up of gear linkage of the machine table to the dividing head indexing shaft.

Dividing heads that are capable of being linked to the machine table drives are supplied, or have a dedicated accessory, with a standard set of gears, mounting frames and drive shafts to construct a gearing train the provides the correct lead. Table 4.4 shows the numbers of teeth on a standard set of spur gear supplied with a dividing head. However, differing manufacturers may have marginally differing gear sets and the operator needs to be cognisant of the changes to indexing.

Operations that require manufacture of a feature with a short lead (i.e. one with a helix angle of less than 45°) will need to use an end milling cutter or slot drill, as any diameter of any arbor mounted cutter will not be able to cope with the requirement to adjust for the helix angle.

The term 'lead' is used to identify the properties of a helix, and the definition of a helix lead is 'the translational distance moved along a workpiece axis for one full revolution of the workpiece'. The diagram in Figure 4.50 shows a development of a helix.

When interpreting design information or component drawings, the operator needs to be aware of the difference between a helix angle and the lead angle. The helix angle is the angle between the angular groove or feature being machined and the central axis of the workpiece and tends to be used in gearing. The lead angle relates to threads and is the perpendicular to the axis of the workpiece. The diagram shown in Figure 4.50 identifies the lead and helix angles.

Each milling machine will have a standard lead for is machine table screw. Depending on the age of the machine or its origin, this lead may be a metric value or imperial. It is important to know this, as it may not be possible to machine some features if a metric dividing head and gear set is used on an imperial machine, and vice versa. Many standard imperial machines of British or American origin were fitted with a ¼" lead screw with metric machines often having a 5mm lead. To determine what the lead is for a machine the dividing head is linked to the machine table by building a set of gears that have an equal ratio (i.e. there is no reduction or acceleration in the gear train). With the line on the dividing head set to top dead centre, or similar, and precise, visual reference mark applied, the operator then winds the table handle, or engages the x-axis feed and counts the number of table shaft rotations to one full rotation

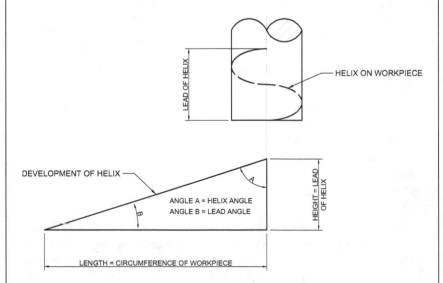

Figure 4.50 Development of helix to identify component features

of the dividing head. If the worm drive ratio is a 40:1 reduction then the table drive shaft should have undertaken 40 revolutions for 1 full revolution of the diving head chuck. If the ratio is not 40:1 there will be a substantially different number of turns, however, it will be a whole number. If the full revolution of the chuck has a number of turns and a fraction it is likely that metric and imperial drives are being mixed. Where this is experienced the operator needs to proceed with care to ensure that the correct values can be achieved, as there may be some combinations that cannot.

The distance moved by the table should be 10" for an imperial machine with a ¼" pitch lead screw and using a dividing head with a 40:1 worm drive ratio. For metric machines with a 5mm pitch lead screw this should be a translational movement of 200mm. By changing the gearing ratios between the machine table and the dividing head drive helix lead angles can be altered to that required. Once a helical groove is machined into the workpiece the table is dropped down so the cutter comes clear of the workpiece and returned to its starting point. This ensures that the backlash is eliminated and the cutter is clear of the workpiece while the table is returned so that workpiece damage is avoided. The workpiece can then be rotated using standard direct, or indirect indexing to make a series of sequential grooves. It is not possible to use differential indexing when helical milling, as the gearing is between the table and the dividing head and cannot also be attached to drive the dividing head spindle. This restricts the number of grooves that can be machined to those available from the direct indexing plates, or from the indirect indexing plates. Given the flexibility of these two methods it is rare for this to be an issue.

To find the helix angle required for a gear, milling cutter or similar large helix the following formula can be used, and will provide a value for angle A shown in Figure 4.50 above.

$$\text{Tan of helix angle} = \frac{\pi \times \text{pitch diameter of gear}}{\text{lead of gear tooth}}$$

$$\text{Tan of lead angle} = \frac{\text{lead of screw thread (pitch)}}{\pi \text{ pitch diameter of thread}}$$

If the lead of the helix is known as it is stated on the component drawing, the operator will not need to calculate the angles, however, if the angles are stated then the above formula will merely require transposition to determine the lead. Once the lead is known the gearing ratios can be identified. Manufacturers' handbooks and quality textbooks such as *Machinery's Handbook* (Industrial Press, 30th edition 2016) have detailed tables to identify gearing ratios, and the internet can be a good source of tabular information, which remove the need for calculation. In addition to this, nomograms can be used to determine unknown values, and an example of this is found in A.

The way that most gearing systems are arranged there are a set of driving wheels and driven wheels, the combination of this gives the ratio, however, idler wheels are required to ensure right- or left-handed rotation, and in some gearing combinations to make up any centre distance where a gap may occur. Two idler gear wheels or none at all will give a right-handed helix, and one idler will give a left-handed helix.

As an example using an imperial machine and dividing head with a dividing head reduction ratio of 40:1 and a ¼" pitch on the table lead screw, to achieve a required helix lead of 5¼" the following calculation is undertaken:

$$\frac{\text{Lead required}}{\text{Standard lead}} = \frac{5\frac{1}{4}''}{10''} = \frac{21}{40}$$

Whole numbers are required to calculate gear ratios so, 5¼" becomes 21 as the nearest whole number. Because gearing is a ratio the 10" denominator also becomes four times larger giving 21/40. However, the gearing system requires four gears. One that is attached to the table spindle, one that is driven, a second driving wheel and a final one mounted on the dividing head shaft. No idlers are required as the gears will mesh and a right-handed helix is being produced. To produce the four gear ratios required this is broken down as follows:

$$\frac{21}{40} = \frac{7 \times 3}{2 \times 2 \times 2 \times 5} = \frac{7}{8} \times \frac{3}{5}$$

Therefore, multiples of this are:

$$\frac{56}{64} \times \frac{24}{40}$$

This would give the correct gearing ratios, however, there are some problems in mounting the gears in this manner as there would be a clash between spacing collars and a nut. To solve this the order is transposed to give the following:

$$\frac{24}{64} \times \frac{56}{40} = \frac{\text{driven gears}}{\text{driving gears}}$$

The same ratio and lead would be achieved by using a gear set of 40 tooth (table spindle), 28 tooth (driven), 32 tooth (driving), 24 tooth (dividing head shaft) as derived from a manufacturers table of leads. If the feature was required to have a left-handed helix an idler gear mounted on a stub shaft could be inserted between the two sets of gears.

While the example used above was calculated for an imperial example, metric values could be used, with 10″ being substituted for 254mm and 5 1/4″ for 105mm. The principle for calculation of the gear ratios is exactly the same, with the operator trying to find combinations of gears within the set available to achieve the correct lead per revolution.

When milling features that have a very short lead as may be found when thread milling it would not be possible machine with standard gearing set-up, and feeds. Successful short lead milling requires the slowest feed settings, and generally an additional short lead attachment which allows the introduction of further gear reduction. It has the added benefit of reducing the torque load on the dividing head, as short lead milling the lead, tends to require a high angular rotation of the dividing head and gear reduction assists with this. One point for the operator to consider is that if short lead attachment systems are geared directly to the dividing head spindle, then the worm drive must be disengaged to avoid damage.

Where non-standard leads are required a close approximation can be achieved by utilising continued fractions. The calculations for this are quite mathematical and use terms that are no longer common in engineering, such as 'vulgar fractions' and 'quotients'. While the actual maths is straightforward it is fairly cumbersome, and while can lead to some machining error it would only be in the region of 0.04mm in 700mm, and as such could probably be ignored. Where non-standard leads need to be precise the internet is a good source of models to input data into. Manufacturers data and textbooks will also have pre-calculated ratios for a substantial number of leads commonly with 350–450 combinations identified. While it is entirely possible to calculate the gear sets for non-standard leads for manual machining, if the level of precision is required then use of a CNC machine with continuous path capability may be a better choice.

Dividing head example 6 (cam milling)

Machining of cams is becoming less prevalent than it once was with the replacement of machinery such as cam driven Swiss automatic lathes by modern CNC machinery, however, simple cam manufacture is another application for a dividing head geared to a table drive.

This operation needs to be undertaken on a vertical milling machine or turret mill that has a table gearing facility. The dividing head is rotated to give a rise that is proportional to the angle through which it is turned, and also through an angle in which the lead (or total rise) does not correlate with any table lead using standard gear wheels. The operator should select a standard table lead slightly larger than the lead of the cam as shown on the component drawing, and then set the angle of the dividing head and machine turret head to the same angle. The idea is that the result of this reduces the table lead to that required by the cutter (i.e. one complete revolution of the dividing head corresponds to exactly the cam lead). The principle is illustrated by the diagram in Figure 4.51.

The method for calculating the angle of rotation is as follows:

$$\cos\Theta = \frac{\text{Lead of cam}}{\text{Lead of table}}$$

The cutter selected for this type of operation should be of a similar diameter as that of the cam follower to be used when the component is installed. It should be of a length that accommodates the vertical rise and fall that will be experienced without adjustment (i.e. the lower end of the cutter will interface with the lowest part of the cam), and as it rotates the effective motion will show the workpiece moving up the cutter until a position at 180° is reached after which it will go down. It is also suggested that the cutter be used for down cut milling to ensure the cutting loads are applied towards the spindle nose rather than away from it to improve stability and reduce vibration.

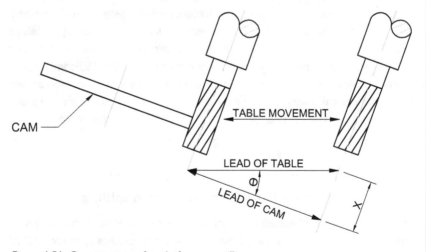

Figure 4.51 Components of angle for cam milling

Dividing head example 7 (spline milling)

Splines are commonly machined onto shafts and tend to take two forms. The first is a plain sided spline and the second is one that has an involute form. While plain splines can be cut with a standard slot drill, involute will require a bespoke profiled slot drill or end mill. It is also not unusual for a multi tooth cutter of either the 'side and face' type, or involute multi toothed cutter to be used either directly mounted on the arbor of a horizontal milling machine or to a stub arbor mounted in a vertical machine. Figure 4.52 shows a simple set up for machining a plain sided spline.

While it is common for plain splines to be machined using slot drills or end mills given their convenience, it is also not unusual for circular cutters mounted on an arbor to be used. This gives a radius 'ramp out' at the end of the cut and removes the vertical wall associated with the termination of slot machining. Removal of vertical walls be addressed to a degree by subsequent machining with a ball nose cutter. However, this is an additional operation, and can only

Figure 4.52 Plain sided spline milling

ever give a radius of half the cutter diameter, which is itself limited to the spline width. Therefore the operator needs to consider the approach to be taken where specific radiused features are not identified on the component drawing.

To machine a basic spline the dividing head is set up with the workpiece horizontal. It is unusual for splines to be particularly long and often the workpiece does not require a tailstock to be fitted, however if the spline is to be machined on a long shaft or is of a relatively small diameter this can be a sensible precaution.

Machining a spline is very similar to machining a spur gear in that there are a certain number of divisions for a given diameter. The rotation of the workpiece will controlled through direct or indirect indexing using the indexing plates in a standard configuration. It is unlikely that machining a standard spline will require any of the more bespoke differential indexing techniques.

Once the workpiece has been set up and the cutter mounted in the machine spindle the operator will need to set the datums. If using and end mill or slot drill this is simply a case of establishing the y-axis and z-axis datums using a probe or cigarette papers. While there are techniques for generating a semi-involute spline profile by milling off centre, it is usual to present the tooling on centre. If the spline is being produced to a standard class of fit, the tolerances involved will make carefully establishing datums crucial.

Where a horizontal rotary cutter is being used from the side of the workpiece establishing the y-axis datum only requires knowledge of the cutter diameter and probing, or use of papers. However establishing the z-axis datum requires a less direct approach. The tooth profile of the involute cutter does not lead itself to identification of a central point, therefore, the process is to align the top surface of the workpiece with a plain upper surface of the cutter. this is achieved by use of a DTI acting as a comparator.

The operator visually aligns the upper surface of the workpiece with the upper surface of the cutter. A reading is then taken between the two with the DTI, and an adjustment made by raising or lowering the machine knee. When there is no difference in value between the workpiece and the plain upper surface of the cutter they are aligned. The operator then raises the knee of the machine by half of the diameter of the workpiece, minus half the overall thickness of the cutter. This establishes the z-axis datum by aligning the centre line of the cutter on the centre line of the workpiece.

Machining of the spline is then approached in the same way that any other slot or feature is manufactured, with incremental or full depth cuts being taken depending on the size. There is generally no reason to drop the workpiece down when returning the cutter to the start for the next cut as backlash is not an issue, although some marking may be observed. If this is crucial the operator should drop the knee by a small amount and re-establish for the next cut. The workpiece is indexed through rotation of the dividing head indexing handle using the correct ring of holes.

Where larger splines are to be cut, the amount of material to be removed may be substantial and overload the involute cutter. Where this is likely it is usual to provide a 'gash' by machining with a gashing cutter, slot drill or end mill. This operation removes the bulk of material allowing the involute cutter to provide the spline profile. If the operator intends to use a slot drill or end mill to provide the gash, it would be produced on the top surface of the workpiece, and on an end milling machine the spline would be produced at the side. The operator must therefore check that the number of splines is such that there is a spline groove required at 0° and at 90°. Where there is a difference, the same technique must be employed (i.e. horizontally with a gashing cutter and then with the spline involute cutter). The size of the gash must be limited to ensure that it does not intrude into the workpiece beyond the point where material is removed by the involute cutter. Some care needs to be taken as it is efficient to remove as much bulk material as possible leaving less for the involute cutter to do, however, it is easy to overestimate.

4.22 ROTARY TABLE USE

Typically milling machines in general workshops are used to produce flat surfaces, longitudinal grooves, recesses, pockets, and edges. The common aspect is that these are all cut in straight lines. However, as with the use of dividing heads, more complex milling operations ca be undertaken using a rotary table.

A rotary table is another attachment which is normally bolted onto the machine table. It has a flat upper surface which is able to rotate and is itself fitted with tee slot grooves to allow the attachment of a vice or work holding clamps. A number of derivatives can be obtained with some that can tilt, those fitted with indexing plates and micro-adjustment features, however, the most common type is used in a flat orientation, are rotated using a manual hand wheel, and have a scale marked in degrees on the external periphery of the rotating plate. A typical example of a rotary table can be seen in Figure 4.53.

Almost all rotary tables have a central hole for the insertion of a central location register, used when aligning a workpiece. This generally requires a round plug to be machined on a lathe for use as a register for workpiece location. The hole can also be used to fit a variety of types of plug which can be used for accurate identification of the central datum point. These plugs

Figure 4.53 Traditional example of rotary table

are often a tapered pointed cone used with an optical eye glass inserted into the machine spindle, or a round plug base that fits into the rotary table with a square upper section used with a centre finder. Another common approach is to use a close-fitting cylinder and cigarette papers.

The key aspect of rotary table use is to ensure an accurate x-axis and y-axis datum is established exactly on the centreline of the rotary table prior to work commencing. Failure to establish this would result in an arc being machined rather than a true radius, and where an arc is required to be machined, the normal approach is to offset the workpiece on the table. By establishing a datum with the machine spindle directly on the centreline of the rotary table, the operator is fully aware of the relationship, between the machine spindle centreline and the workpiece. When this is established a workpiece can be offset by a known value to provide an arc centre that is off the rotary table axis.

The rotary table is bolted down to the machine table generally using standard tee bolts. Some tables are fitted with registers which mandate the orientation, however, the operator should ensure that they locate the rotary table on the machine table such that they can operate the machine controls and the rotary table from the front of the machine. The design of many tables is such that the

0° position is not at the front of the machine (i.e. central on the y-axis and at 90° to the x-axis); however, unless it is a more complex design which tilts, it will always be square with the z-axis. Not having the 0° position directly at the front of the machine provides some complication when not milling around a complete periphery and are machining a section of it and the operator needs to establish the start and finish angles, and operate between them.

It is unusual to fasten a vice to a rotary table because of the difficulty in establishing any workpiece datum when inserted by hand into the vice, while retaining the rotary table datum. This normally means that successive workpieces are either mounted onto a peg protruding from the central spindle, or by location on accurately set workpiece registers. Machine clamps utilising slots in the rotary table plate or direct bolting down are the most common approaches used. Figure 4.54 shows a component mounted and ready for machining.

Cutter selection for rotary milling is no different than that for other kinds of milling, the operator generally chooses between and end milling cutter and a slot drill depending on the feature to be generated and ease of swarf chip removal. Once the cutter has been selected and the correct spindle speed set, the z-axis datum is established in exactly the same manner as for end or slot milling.

Figure 4.54 Rotary table component clamping

The next task for the operator is to establish the radius of arc to be machined. Having previously set the machine datums on the centreline of the rotary table and set the z-axis datum using the upper surface of the workpiece, a translational movement in either the x-axis, and/or y-axis has to be undertaken. This is to position the cutter at the radius of arc required, with either the centre of the arc located in the centre of the rotary table, or with an offset applied which positions the centre of the arc off the centre of rotation. Where more than one axis is used to position the cutter the effect will be to produce an arc rather than a section of a circle when machining commences, while this may be the desired result, a circle or circular portion of arc will only be produced when the centre of arc, is located over the centre of the rotary table, and the effect of moving this arc centre will be to generate a feature that has a form similar to a section of an ellipse.

If a complete circle is to be generated by either slot milling or end milling, machining is relatively straightforward in that the centreline of the cutter is on a known radius, and in the case of a slot drill is moved into the workpiece to either the required depth or an appropriate peck depth, coolant applied and the table rotated until it has completed 360°. Similarly if end milling from an edge, the milling cutter is introduced into the workpiece by reducing (or in the case work out from an internal hole), by increasing the radius, and then the rotary cut achieved by rotating the table control wheel. The operator needs to understand the direction that rotating the rotary table hand wheel translates into the direction the table rotates. This is to ensure that the relationship between the direction of cutter rotation and table rotation is understood so a decision can be made as to upcut or downcut milling.

Where a section of arc is to be generated and additional datum is introduced which is locating the start point of the arc and its finish point. The finish point is relatively easy to establish in that it will be a number of degrees of arc from the start point, and this will either be directly determined from the drawing as a known dimension, or can be calculated if the drawing identifies arc start and finish centre points. Interrogation of component drawings should identify the arc start point, from the component drawing feature datum. The operator will need to established this datum in alignment with either the x-axis or z-axis prior to clamping down the workpiece. Where this datum is a straight edge, it is a relatively simple task to rotate the rotary table until it is at is 0° position, place the workpiece on the rotary table and align with either the x-axis or y-axis, and lightly clamp. Using a 'finger clock' or DTI, establish the workpiece in a parallel orientation to the machine axis and clamp down tightly. Once this has been undertaken the rotary is rotated through the required number of degrees until the start point is reached. This places the centre of the cutter on the centre point of the arc start. The wise operator records this value, or marks the start point on the edge of the rotary table platten. The number of degrees that represent the length of arc, are added to this figure and the value similarly marked or recorded. The

Figure 4.55 Arc milling with arc centre start and finish points visible on scale

cut is applied and the table rotated between the start point and end point. Figure 4.55 shows a radial slot being milled into a plate, and marks for the start point and end point put in place by the operator are visible.

Where the arc start point requires a clockwise rotation of the rotary table the number of degrees will need to be subtracted from 360° to give the value for the end point. Also when machining successive features that have a gap between them, the operator will need to establish successive start and finish points for each length of arc to ensure the correct size of land between each machined arc is correct. For this reason it is often simpler to operate with the rotary table moving in an anti-clockwise direction such that the operator down cut mills, and calculating arc centre end point value is a simple case of addition rather than subtraction.

Coolant flowing over marks placed on scales to allow the operator to identify start and finish points can be an issue when rotary milling. The tactical emplacement of tool steels, plasticine or similar dams, and diversions are the few approaches that can be taken, as a good flow of coolant assists which swarf removal.

When milling slots swarf build up can be a problem and re-cutting can lead to workpiece damage. The operator also needs to consider whether the rotary table is capable of absorbing the loads transmitted while cutting. Rotary tables are only retained by a central shaft and the standard machine shop rotary table is not large. They do not respond well to heavy cutting, as the large cutting angle can generate significant vibration, often causing machine offsets, and creating a damaged workpiece. When machining larger slots, the operator should consider cutting an initial slot using a significantly smaller diameter slot drill, often 50% of the desired diameter and the then once the cut is complete replacing with the full diameter. This approach does not require any adjustment to the machine datums as the arc start and finish points are all set on the tool centre line. However, it does significantly reduce the tool loading transmitted to the workpiece, and the volume of chips that need to be dispersed with each pass.

Where slots are to be milled through the component the operator will need to either place the workpiece on machine parallels, or place a sacrificial base material that the cutter can bite into. It is important that the integrity of the rotary table platten be maintained in the same way the machine table is protected. Where a sacrificial material is to be used, it merely needs to be flat and on a continuous thickness. However, rotary milling is an operation where the cutting point of the tool is obscured by coolant and swarf chips. While close attention to setting the z-axis datum and interrogating any DRO or z-axis control handle scale can protect the rotary table platten; one method to establish when the cutter has passed through the component and into a sacrificial material is to ensure it is made out of a different coloured material to the workpiece. If machining aluminium or steel use of a black or dark coloured engineering plastic to support the workpiece becomes very visible when machining and the operator knows they have passed through the component. The opposite applies when machining dark engineering plastics, in that sheet aluminium, or white coloured plastics can be used to provide a similar indicator.

Rotary milling can be combined with translational movements to produce much more complex shapes, such as offset rectangular pockets with radiused ends. These require some thought but the key aspect of manufacturing these correctly is establishing and accurate initial datum on the centre line of the rotary table.

Rotary milling example 1 (slot milling of arc)

Having mounted the rotary table to the milling machine table it is rotated using the rotary table hand wheel until the datum mark is adjacent to the 0° value on the edge of the platten. The centre datum position for the rotary table is the established. The simplest approach is either to use an electronic probe, or the use of cigarette papers. When using either method a short section of bar is inserted into the central hole of the rotary table allowing it to protrude at least 20mm from the platten. The section of bar needs to be a close fit with the central hole and be of a known diameter. A common approach is to use a section of steel ground stock bar as can be seen in Figure 4.56.

This section of bar is then probed, the offset applied in the x-axis and y-axis and the spindle is then located over the rotary table centre of axis. Where a probe is not available or the machine not configured for one, the pragmatic approach is to affix a paper around the stock bar and set the axis one at a time. When this approach is being undertaken the table is moved on the x-axis under rapid feed, or by winding the table until it is adjacent but not touching the ground bar and paper. The operator then incrementally moves the table using the hand

Figure 4.56 Stock bar used to establish rotary table datum

wheel until it is almost apparently touching the paper. The machine table is then moved forwards and backwards on the y-axis using the hand wheel. If the cutter removes the paper without damaging the ground stock bar, the table is dropped and a translational movement equal to half the cutter diameter plus half the stock bar diameter is applied and the DRO scale zeroed. This sets the x-axis datum. If the stock bar is damaged the table is backed off and with the cutter stopped the stock bar is rotated slightly and the operation repeated. Where the paper remains a small increment is applied and the process repeated until the paper is removed and the stock bar remains undamaged. The process for establishing the y-axis datum, is similar with incremental movements made on the y-axis and left and right passes made on the × until the paper is removed. Once this has been done and the correction offsets for stock bar diameter and cutter diameter made the spindle is located directly over the centre of rotation of the rotary table and both x-axis and y-axis datums identified.

The workpiece is then emplaced onto machine parallels or sacrificial pads and a datum point established. Often this is by directly mounting the component onto a hole which is the centre point of an arc to be machined. Where this is not possible positioning needs to be undertaken from a known point either through direct measurement from a known face of the machine to the central datum point on the rotary table, and offset applied or the use of registers employed. The x-axis and y-axis datums are applied such that the workpiece datum edge or face is oriented square to the machine axis, and the workpiece is clamped down.

Given that rotary milling often involves lighter cuts that traditional milling, and given the relative strength of the rotary table, clamping does not have to be as heavy. If a central hole is located on a peg set into the rotary table centre, often only one clamp is required freeing up space to allow the cutter to traverse through an arc without being impeded. The operator then firmly clamps down the workpiece. Where it is impossible to fully deconflict lengths of cutter arc and parts of the clamp it is not unusual to have previously machined sacrificial clamps and machine into them.

The workpiece is then moved until the start point for the arc centre is directly below the cutter. Where this is to be on an arc where the workpiece arc centre is located on the rotary table datum it only involves rotating the rotary table hand wheel until the desired angle is read out on the rotary table scale. The z-axis datum is then established by raising the knee until the cutter touches the workpiece, or paper, and the DRO and/or scale set to zero.

The operator should note the start angle or mark the rotary table scale with a marker pen and similarly mark the arc end angle. The operator should also check what direction of hand wheel rotation correlates to a clockwise or counter-clockwise rotation of the rotary table. Following this a cut is applied by raising the knee of the table until the correct peck depth is applied and the operator rotates the rotary table hand wheel until the first arc length is cut. Continuing cuts are undertaken until the required depth is reached or the operator witnesses sacrificial material in the swarf. It is normal to apply sequential cuts at both ends of the arc, and where surface finish is important having a cutter reground to slightly below the desired with and finishing with a full diameter cutter to apply a full depth finishing cut is one approach.

When rotary milling external features that have a defined start and finish point, the operator will need to consider the end radius size. It is not uncommon for rotary milling to be used to provide a rounded feature between adjacent linear features, and the diameter of the cutter may be such that it the side of the cutter away from the cutting edge may impact another part of the workpiece. The solution to machining into tight spaces like this are to use a slot drill to form the required end radius at the correct size, and use a smaller diameter end mill to form the section of arc required while providing a clearance elsewhere.

4.23 TURRET MILL QUILL USE

Turret mills and many end mills have a facility for feeding a quill down through the rotating spindle to allow for drilling and tapping operations that do not require the table to be raised for a cutter or tool to penetrate the workpiece. Many machines offer two options for achieving this with it being undertaken by hand or by using a power feed system.

All machines will have a clamping system to prevent the quill dropping when not in use, or to prevent it being pushed back up into the spindle if it has not been returned to its start position. The mechanism for this varies from machine to machine with some having a simple twist clamp, and others having more complex systems involving cams and locking dogs. Irrespective of the design of clamp, the ability of the quill to be locked in place gives the operator another degree of flexibility in selection of machining sequence.

The quill movement gives a degree of flexibility, however, the precision that can be achieved when using the quill feed is often less than that which can be achieved when raising the knee of the machine. For plain hole drilling using a Jacobs chuck mounted into an R8 collet system, or indeed mounting a plain shanked twist drill matched to a collet, is rarely a problem. Where a hole, or feature is being produced to an exact depth, the tolerance required as identified on the component drawing may preclude the use of the quill, which needs to remain locked in position and the depth of cut achieved by raising the table.

To drill a plain hole in a workpiece, the operator only needs to have set x-axis and y-axis datums, to ensure that the operator is able to locate the centre of the spindle over the exact position required. Given the rigidity of the workpiece when held in a machine vice, or clamped to the machine bed, it makes sense for the operator to centre drill the workpiece prior to drilling the hole, instead of using a centre punch. Using a centre drill provides a good clean start point for a twist drill, and also avoids any deflection of the workpiece in the vice if an overly hard hammer stroke has been used, and also avoids any inaccuracy from either striking a centre punch when it is slightly out of position, or from incorrectly marked out workpiece positions. By accurately establishing x-axis and y-axis datums and using a machine fitted with a DRO, the drill can be positioned directly over the centre point required normally to an accuracy of 0.0005mm. The operator should fit a centre drill into the chuck, and correctly tighten. The selection of the centre drill to be used is not crucial, however, it should not be larger than the drill to be used. Operations that employ a larger sized drill are better matched to a larger centre drill as the cutting edge configuration of larger drills may be such that it fails to engage in the centre hole that is provided by very small centre drills such as size 0 or size 00.

Where a plain hole is to be drilled right through the workpiece, the operator needs to ensure that the workpiece is lifted of the machine table using parallels, or has a sacrificial material placed under it that allows the operator to clearly see when the drill has passed through the workpiece and into the sacrificial material. It is helpful if this material is of a differing type and colour to the workpiece top provide a clear visual marker. The thickness of this material needs to be sufficient to allow for the drill point, and to allow the operator to become aware that the full width of the drill has passed through the workpiece. Where the workpiece is clamped in machine vice, the operator needs to ensure that any parallels that are being used to support the workpiece and ensure its squareness are not going to be impacted by the drill. This is not uncommon and the operator often has to tap a parallel clear of the exit position of the drill to ensure it is not damaged. Where a series of holes are to be drilled it is not unusual for parallels to be moved multiple times, or a third parallel introduced such that holes are drilled in a gap, and the workpiece is supported either side of the hole being drilled.

There will be a limitation to the size of twist drill used as this will be controlled by the machine collet size. Turret mills normally have a spindle fitted for an R8 collet, or ISO collet system. The machines are sized for cutters up to $\varnothing$20mm and for cutters with a parallel shank. It is unusual to be able to source twist drills with diameters bigger than 16mm, and Jacobs chucks larger than this. Therefore, there is a limitation to the size of hole that can be drilled. It can be tempting to take larger twist drills, especially those with a tapered shank and turn them down on a lathe to a parallel shank which can be directly inserted in a machine collet. This is poor practice as high loads can be generated by twist drills and these will be transmitted back up into the machine spindle. If larger holes are required then rotary milling, or having drilled a Ø12–16mm pilot hole on the milling machine, transferring the workpiece onto a heavy pillar, or radial arm drill to use a large diameter drill is a more prudent approach.

The speeds that should be used are directly related to the size of drill. While solid carbide drills are available, by far the most common tool material for standard workshop drills are HSS. Depending on the material being drilled, and its surface machining rate, spindle speeds are only likely to be in the range of 100–1000rpm for HSS twist drills.

To drill a hole to a known depth the operator needs to first establish a z-axis datum. This is normally achieved by unlocking the quill clamp and without the spindle turning use the quill manual feed handle to bring the drill point into contact with the workpiece. If a relatively small centre hole has previously been drilled, it is normal for the depth of drill point intrusion into this to be ignored, as plain drilled holes to a depth are rarely tightly toleranced. If the centre hole is of a significant size and the drill point intrudes well into it then the operator will have less depth adjustment to make for the drill point when setting the depth of hole to be drilled. It is normal for holes to be dimensioned for the length of full diameter required rather than to the point of the drilled hole. It is usual to see a dimension for the depth of a hole and there to be a separate arrow with 'permissible drill point' to recognise the further protrusion. To set the depth of hole, the operator merely needs to identify the relative position between the start point (i.e. the drill point touching the upper surface of the workpiece) and the intended depth the hole is to be drilled to, plus an allowance for the distance taken up by the drill point.

Almost all turret milling machines that have a quill feed operation will have a linear scale and quill depth stop ring, along with a rotating not, which may also have a micrometer scale etched into it. However, end milling machines may have a quill feed facility but this may not have any linear scale to clearly identify depth progression.

Figure 4.57 shows a typical linear scale and a nut with micrometer scale screwed down to the bottom of the threaded column.

This allows for full extension of the quill, and once the spindle has been started, the operator rotates the manual feed handle to press the drill into the workpiece, and merely has to observe the bottom edge of the quill depth stop ring until it is aligned with the depth required. The feed handle is then returned to the start position extracting the drill from the workpiece and the machine stopped. The problem with this approach is ensuring an accurate depth of hole where the component tolerance is less than 1mm. This is especially prevalent when the quill has been partly moved down to contact with the workpiece and is no longer set at 0mm. In this case the operator

Figure 4.57 Linear quill feed scale and quill depth stop assembly

has to add this value on to required depth of hole, and it is easy for errors to be experienced.

The use of the rotating nut allows for much greater accuracy. By rotating the nut up the threaded shaft and locking it in place with a lock nut the quill cannot be depressed beyond the set point, as the quill depth stop ring presses up against it preventing further advancement of the drill. Setting this visually by observing the alignment between the top surface of the rotating nut and the linear scale, then locking it off is one approach. The accuracy of this can be enhanced by utilising any micrometer scale which may be marked on the rotating nut as can be seen on Figure 4.57, however, operator parallax, the distance between the linear scale and rotating nut, and backlash in the threads, can still introduce error. A more precise method for setting a depth is to use gauge blocks (slips) as shown in Figure 4.58.

The operator builds a pack of gauge blocks (slips) set to the depth required, including an allowance for the drill point protrusion between the quill feed stop block and rotating nut. With the gauge block pack in place the rotating nut is locked tight, and the gauge blocks then removed. This provides a depth stop for the quill feed that is accurate and provides a physical stop to a set depth.

The approach described above can be employed where slot drills are being used, which has an advantage that no drill point has to be accommodated, but the disadvantage that slot drills are unable to reach anything like the depths that long series twist drills are capable of.

Many machines also have the ability to power feed the machine quill. This can take the effort out of repetitive drilling operations, and can have advantages when drilling difficult materials that may quickly work harden if a continuous feed is not employed. many machines, especially turret mills may have limitations to the size of drill during power feed operations, and this can be significant. The machine shown in Fig 4.3 is capable of using a ∅16mm drill with manual quill feed, but is limited to ∅9.5mm when drilling mild steel in power feed mode.

Power feed systems are operated in a number of ways depending on the machine type. On some end milling machines power feed uses the same handle as for manual feed, but are moved in or out to engage the feed. Turret mills generally have a separate set of controls. The machine user manual is normally the best method to identify the correct operation settings, however, there are generally some compromises that need to be made. I particular feed rate generally has a gearbox setting and the feed rate may not be precisely what is required. Irrespective of this, turret mills will generally have three quill feed settings, and a method of power feeding under operation control, where this may be a problem.

Following setting of the feed rate using the gear setting control, auto quill feed, and feed direction, all of which may be a quadrant lever or handle, and setting the depth nut to the correct setting, the machine is started and the feed engagement lever moved into position. The quill will automatically progress at the set rate into the workpiece, until the required depth is reached. Figures 4.59–4.61 show controls for engaging the quill feed on a typical turret mill.

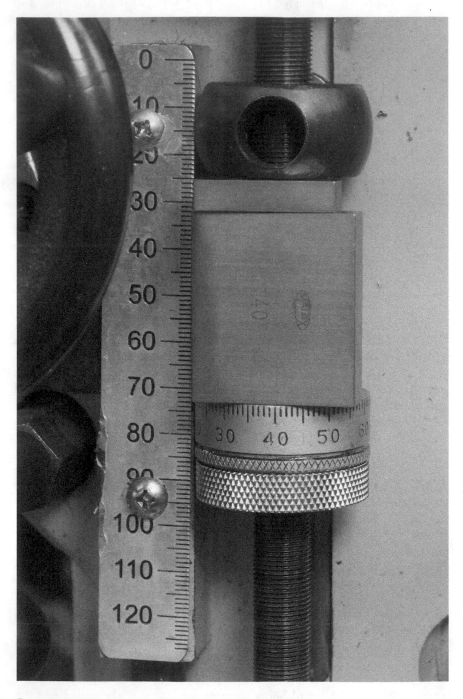

Figure 4.58 Gauge blocks being used to set quill feed depth

Figure 4.59 Quill feed rate controls on typical turret mill

Figure 4.60 Feed worm engagement lever on typical turret mill

Figure 4.61 Automatic quill feed trip lever, and fine feed hand wheel

Most feed systems have a 'trip' to stop the machine trying to progress deeper and damage the mechanism, however, machines do not always have a return mechanism, and this needs to be manually undertaken by the operator. Many machines also have a restriction on the maximum spindle speed feed can be used in which if using HSS tooling is unlikely to be reached, however, operators should be aware of the maximum spindle speed if damage is to be avoided. If the feed direction lever is placed in a neutral position the operator may have the option to use a small hand wheel to feed the tool into the workpiece. This hand wheel is often also used as a fine feed as quill progression with this can allow far finer feed than some auto feed settings or use of the quill direct feed handle. All other control lever settings remain the same with the exception of the feed direction, and the operator can continue to finely feed the tool into the workpiece until the feed trip lever engages with the depth nut.

4.24 DRILLING AND THREADING

Drilling is a relatively straightforward process and is discussed in sufficient depth in the section on quill feed use. However, one of the harder operations to successfully undertake is threading. Given the rigidity of the machine, the spindle and the fixed feed rates direct threading is almost impossible to achieve on an end milling machine or turret mill. While it is possible to set the quill to allow it to freely feed into a workpiece, and rely on the drag from the tap screwing into the workpiece to allow z-axis progression this is poor practice. it is also unlikely that the feed rates set for automatic quill feed will match the pitch of the thread required and effectively strip the thread as tapping is undertaken. Larger threads that may allow the use of thread mills are also constrained by feed rates and are generally the preserve of CNC machines with a capability of defining specific feed rates match to thread pitches.

Internal threads can be tapped using an auto reversing tapping head and the manual quill feed function of a vertical milling machine or turret mill. These attachments allow for the correct advancement of a tap into a hole and at the bottom of the hole automatically reverse to screw the tap back out of the thread. Their design allows a wide selection of thread sizes to be used, but again large threads may exceed the capacity of the machine. While it is possible to readily purchase these devices that can support an M30 thread size, the operator needs to consider if the machine has the capacity to produce this. Coolant and tap type are also considerations.

Where operators elect to use this type of device a reaction plate or post will need to be attached to the machine bed to provide a reaction post normally provided by a drill column when being used on a pillar drill. A spiral tap with a helix is also used. Hand tapping normally requires a periodic back rotation of the tap to snap off any swarf chips. Machine tapping using a reversing tapping head is one continuous process, and as such, requires a spiral fluted tap to assist with ejection of a continuous chip away from the cutting edge and out of the workpiece.

4.25 GEAR CUTTING

Gear cutting is closely linked to dividing head use, or (very uncommon) electronically controlled electronic rotation.

There are four principal types of gear that are commonly manufactured:

- spur gears;
- helical gears;
- bevel gears; and
- worm gears.

Spur gears are cylindrical and generally have a relatively short length of tooth, however, this is not a restriction or requirement. They have straight teeth which are machined parallel to the cylinder axis, and in use produce no axial thrust, however, can be noisy in use at higher rotational speeds. Helical gears have their teeth cut at an angle to the cylindrical axis and they have a higher load carrying capacity, are quieter than spur gears at higher rotational velocities, but produce some end thrust. They are more complex to cut, as unlike spur gearing which only requires simple indexing of the gear blank on a dividing head, helical gearing requires a rotational movement of the dividing head to form the gear helix during each tooth cut. Straight Bevel gears have teeth that are radially cut at an angle to the central axis. They are designed to mesh with a mating gear with its axis at a differing angle to the first. Typically they are cut at a 45° angle so that when the two gears mesh they provide drive shafts at 90° to each other. Worm gearing is similar to spur gearing and is designed to transmit power, and consists of a worm drive, and worm gear, with the shaft axis being at 90° to each other. All of these gear types are able to be produced on a manual milling machine, although helical gearing requires a dividing head to be geared to the machine table x-axis drive.

There are two methods for forming the gear tooth. This is by hobbing, or straight cut milling of one tooth at a time. Hobbing is normally undertaken on a specialist machine, however, it can be undertaken in the general machining workshop, if the milling machine has the ability to swing its table through an angle equivalent to the hob pitch, and connect the dividing head to the table drive. Given these requirements hobbing is often restricted to a horizontal or universal mill, but can be undertaken on a vertical mill, where the appropriate attachments can be made. The advantage of hobbing is productivity, as multiple gears can be manufactured simultaneously and the entire periphery of the gear is machined in one pass, however, where limited numbers of gears are to be manufactured, the use of milling is perfectly satisfactory.

The size and profile of gear teeth will vary, depending on the load, the gear ratio, and its function. In simple terms tooth profile is either cycloidal/hypocycloidal or involute. Cycloidal gearing tends to be restricted to clocks and certain types of pump, and the vast majority of gears has an involute profile. The profile of each gear tooth whether milling or hobbing is generated by a cutter which has been ground to the profile required and as such milling each tooth is generated by use of a form tool. The larger the tooth size the larger the cutter, but the involute shape remains the same, just increasing or decreasing in size depending on the gear size.

While the design of gear cogs involves many factors, and has some complexity to it, the information that is of most interest to the operator are the controlling sizes. In particular, is the number of teeth, and the depth of cut. It is not unusual for the gear component drawing only to have a stated pitch diameter, and number of teeth required, rather than all of the information that an operator may need. Therefore, some knowledge of gear terminology, and some simple calculations need to be undertaken. The diagram in Figure 4.62 identifies the principal features of a gear.

As previously mentioned it is normal for the number of teeth to be machined to be stated on the drawing, as its the pitch diameter. This means that the operator may need to calculate the outside diameter of the gear blank if it is not supplied, the whole depth of the tooth so that the operator knows how deep to plunge the tool into the gear blank, and the diametral pitch, or module so that the correct cutter is selected. Diametral pitch is commonly used in imperial sizes, and module for metric gears. Essentially they are the same thing in that they are a ratio between the number of teeth and the pitch diameter. The methods for calculating these sizes are as follows.

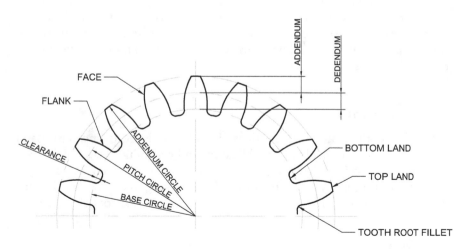

Figure 4.62 Gear tooth nomenclature

Diametral pitch/module

$$\text{Diametral pitch} = \frac{\text{number of teeth}}{\text{pitch diameter}}$$

$$\text{Module} = \frac{\text{pitch diameter}}{\text{number of teeth}} \text{ or module} = \frac{25.4}{\text{diametral pitch}}$$

The significance of this for the operator is that the cutter to be used will have DP or module engraved into it, therefore one or the other will need to be known to select the correct cutter.

Pitch diameter

$$\text{Pitch diameter} = \frac{\text{number of teeth}}{\text{diametral pitch}} \text{ or pitch diameter} = \text{number of teeth} \times \text{module}$$

Gear blank outside diameter

$$\text{Outside} \varnothing = \frac{\text{no. of teeth} + 2}{\text{pitch diameter}}$$

The gear blank is commonly produced on a lathe and supplied as a complete blank only requiring milling, however, there is no reason why this cannot be undertaken using a dividing head and rotary milling to the correct diameter. The outside edge of the gear blank forms the top land of the gear tooth, therefore close attention to accuracy and machining within tolerance is important as this face will be inserted into the mating groove such that the root clearance is maintained.

Depth of tooth

While the depth of cut for the tooth can be calculated directly it can be useful for the operator to calculate the addendum and dedendum sizes for the gear teeth, especially where they are given on the component drawing as a check. It also provides the operator with a method of checking to ensure that the type of tooth profile being machined does not have any specific root clearance.

$$\text{Addendum} = \frac{1.000}{\text{pitch diameter}}$$

$$\text{Dedendum} = \frac{1.200}{\text{pitch diameter}} + 0.002'' \text{ or } 0.05\text{mm}\,(\text{Min})$$

$$\text{Whole depth} = \frac{2.200}{\text{pitch diameter}} + 0.002'' \text{ or } 0.05\text{mm}\,(\text{Min})$$

Once the depth of cut, number of teeth and gear blank diameter have been established the operator can select the correct cutter. The shape of the tooth profile is dependent on the number of teeth in the gear. A small diameter pinion tooth flank, would have a much greater curvature than a larger gear. A gear considered to have an infinitely large diameter (i.e a straight rack) would have straight flanks. For this reason cutters are sold in sets of eight, to accommodate a range of gear tooth curvature, and Table 4.5 identifies typical applications for a range of cutters. Each set will be sized for differing gear profiles and a range of sets are required for large gears through to small gears, but each will be in a set to cover a range.

Where there is a concern that the tooth profile will be correct, it is worth the operator considering that the cutters will produce a perfect tooth profile at the lower value in the range and that the curvature becomes less accurate when approaching the upper figure. Therefore if the operator is to cut gears on the upper boundary between two ranges, it may be an option to use the next cutter in the set and have a more geometrically correct profile.

It is also worth noting that the cutters are sold for particular pressure angles. The gear pressure angle is the angle of obliquity between the gear tooth face and the tangent of the gear wheel. Its significance relates to wear and power transmission, however, typical values for this pressure angle are 14.5° and 20° although occasionally 25° pressure angles may be required. Differing cutter sets are required for gears with differing pressure angles. Gears in a set will always have the same pressure angle or they would not correctly mesh, and the operator needs to ensure that the right set is selected prior to machining. Cutters all have the required information engraved into their sides giving the operator the information he requires and Figure 4.63 shows an imperial series of cutters where the pressure angle, diametral pitch and tooth range are clearly marked. Metric

Table 4.5 Example of gear cutter set tooth ranges

No. 1	No. 2	No. 3	No. 4
To cut 135 teeth–rack	To cut 55–134 teeth	To cut 35–54 teeth	To cut 26–34 teeth
No. 5	**No. 6**	**No. 7**	**No. 8**
To cut 21–25 teeth	To cut 17–20 teeth	To cut 14–16 teeth	To cut 12–13 teeth

Figure 4.63 Imperial gear cutter set showing markings

cutters are almost identically marked other than they will have a module No. rather than diametral pitch.

Gears that have modified addenda, or non-standard diameters cannot be manufactured using a standard cutter set, and special purpose cutters need to be manufactured.

To cut a spur gear the operator selects the correct cutter according to the number of teeth to be cut and the pressure angle. This cutter is mounted on an arbor and inserted into the machine spindle. It can be advantageous for the orientation of the cutter to be such that it 'downcut' mills or climb mills, as this pushes the workpiece onto the arbor rather than lifts it off making the set-up more stable. The gear blank is mounted onto the dividing head. the bast method for this is between centres as it removes any inbuilt error in the three-jaw chuck. It is also normal for the gear blank to be itself mounted onto an arbor and needs to have a mechanism such as using a pre-cut keyway to prevent unwanted rotary motion around the axis of the arbor.

It is important that the gear cutter is aligned with the central axis of the workpiece. One method for ensuring this is to bring the side of the cutter into contact with the outer edge of the workpiece.

The workpiece is then moved clear of the cutter and the cutter and the machine knee (when using a turret or end mill) raised by half of the diameter of the gear blank, plus half of the thickness of the cutter. Precision is important when undertaking this task, and many turret or end milling machines do not have a DRO readout for the z-axis. This means that the z-axis scale needs to be zeroed prior to movement and close attention needs to be paid to the number of revolutions of the knee operating handle undertaken. Checking for the correct height setting is worthwhile, if time consuming, and the use of a comparator (i.e. finger clock) and stack of slips will provide a very accurate distance between the top of the gear blank and the side of the cutter. Failure to get the gear cutter on centre will result in the formation of a gear tooth profile that is asymmetric and will not correctly mesh with its corresponding gear.

Once the cutter is on centre the operator needs to set up the dividing head for indirect indexing as previously described to ensure that the workpiece is rotated the exact amount to ensure that a complete set of fully formed teeth are machined. Errors in establishing the correct diameter for the gear blank, or setting the cutter datum axis are clearly evident following indexing of the correct number of teeth. For gears cut in steel that have a diametral pitch no coarser than 8, or in cast iron 6, are often cut in one pass, however, it is often better to leave some material on and have a second light finishing cut. Some gears are hardened and ground, and the operator needs to establish if an allowance needs to be made for subsequent grinding, such that the full plunge depth will not be made, with the cut stopped short of full depth. Larger teeth will need to be machined in incremental cuts so that the cutter is not overloaded. The cutter profile is intended to produce a full form tool, therefore it is easy to overload it given the length of cutting edge. Once the depth for cutting has been set, either by raising the knee for horizontal machines, or making a translational movement of the y-axis for end milling machines or turret mills, the machine is started, and the cutter is fed through the material until it passes out the other side. It is important that when selecting or manufacturing an arbor the operator checks that there will be enough room for the cutter to pass through without damaging the dividing head. The head is then indexed.

While originally developed for universal milling machines 'Gashing' cutters were developed to rough out the tooth profile prior to using a gear cutter to take lighter finishing cuts to provide the full form. These cutters were either 'Vee' shaped, or a slit saw was used to remove a significant amount of material in either on single pass down the centre of each tooth, or with two angle cuts to remove material efficiently where large teeth were to be machined. The process is commonly used in specialist hobbing, however, is almost unused in traditional milling. There are significant benefits, in removing material prior to final tooth profiling in extended tool life, and surface finish improvement. For larger gear teeth removal of material using a slot drill or end mill can be a pragmatic solution, especially where harder workpiece materials are being machined.

For milling of helical gearing, the approach to be taken is the same for standard helical milling in that the dividing head needs to be connected to the table drive, using a set of gears that provides the correct x-axis rotation with respect to the machine tables x-axis linear travel. However, an adjustment needs to be made because the curvature of the pitch line of the helical gear is less than that given by the pitch radius. What this means in real terms is that the cutter that is selected to cut the gear is not the one selected for the number of gear teeth, but one that has been selected based on a calculation to give a notional number of teeth, that takes into account the effect on the gear tooth profile, of lead angle, diametral pitch and cutter diameter.

To ensure that the correct tooth profile is maintained the actual numbers of teeth to be cut must be must be multiplied by secant^3 Θ. A secant of a curve is the line that intersects a curve at two distinct points, if it only intercepted at one point it would be a tangent. The symbol Θ represents the helix angle, so to calculate the adjustment for the correct cutter the following formula is used:

Req'd no. of teeth × secant^3 Θ = no. of teeth for cutter selection

For example if the operator was required to cut a 15-tooth gear with a 20° helix angle the equivalent number of teeth would be as follows:

$$15 \times \text{secant}^3\, 20° = 18.077$$

Normally the requirement for a 15 tooth gear would require the use of a no. 7 cutter which has a range of 14–16 teeth, however, with the correction a no.6 cutter should be used which has a range of 17–20 teeth, and will give the correct tooth profile after making an allowance for the secant.

Example 1 (imperial)

To manufacture 15 tooth helical gear with a 20° helix angle and a diametral pitch of 5.

$$\text{Pitch diameter of gear} = \frac{15}{5} \times \text{Secant}\, 20° \quad (\text{note secant}\, 20° = 1.06418)$$

Pitch diameter of gear = 3.1926″

$$\text{Lead} = \frac{\text{pitch circumference}}{\text{tangent of helix angle}}$$

$$= \frac{3.1926″ \times \pi}{\tan 20°} = \frac{3.1926″ \times 3.142}{0.36397}$$

Lead = 27.557″

This now allows the operator to calculate the gear combinations for connection of the dividing head to the machine table lead screw. These can be calculated as previously described and the gear ratios identified, however, reference to dividing head manufacturers tables, or those found in engineers reference handbooks [1] can be much quicker. In this case it can be shown that the closest gear set would be as follows:

- Gear on worm (driven) **56**.
- First gear on stud (driver) **32**.
- Second gear on stud (driven) **44**.
- Gear on screw (driver) **28**.

This is based on a lead of 27.5″ rather than the calculated 27.557″. The difference is equivalent to 0.002″ per inch of gear face width. While for short gears or those mating at 90° this can probably be ignored, however, for gears of significant width this is a problem, unless the mating gear was matched, albeit with an opposite hand. One method for providing a solution to this problem is to calculate the exact number of teeth required on a dividing head spur gear to give the correct lead and manufacture it. Alternatively a method for addressing the problem of non-standard leads can be through use of continuing fractions to get a close approximation although this may still require the manufacture of spur gears.

One thing to remember is that helical gear milling is restricted to end milling machines or turret mills which can have a power feed to dividing head or universal horizontal milling machines where the table can be skewed.

It is entirely feasible to manufacture worm gears on a milling machine. Worm gears are designed to transmit power, and the gear is driven by a proportionately small diameter worm. The edge profile of a worm gear is concave and each tooth of the concave surface has a full profile. This means that a worm drive gear cannot be produced in the same way that a spur gear can, and while it is theoretically possible to plunge a gear cutter into the edge of the gear blank the cutter would have to have a diameter twice that of the radius of the worm gear profile. This is rather an unlikely possibility, so the manufacture of worm gears requires hobbing.

The gear hob must be manufactured with a form identical to the worm itself, except that the outside diameter is increased slightly to give the required clearance at the root of the worm gear groove. The allowances for this are stated in the standard being used for the design, however, it is likely that the component detail drawing will identify this. Where the operator cannot see this, or the worm and gear details are not fully described the operator will need to establish the clearance values.

It is normal for the gear blank to be 'gashed' prior to hobbing. This allows for the hob to drive the gear blank and it is not uncommon for a standard gear cutter to be used for this purpose. it is important, however, to ensure that this penetration does not go beyond any depth that hob is intended to reach as these gashes will be fully cleaned out by the hob to give the correct worm gear profile. Once the gear hob has been manufactured the wheel blank is set up on the dividing head. Where an end milling machine or turret mill is being used the dividing head will need to be rotated up to angle that is equal to the lead angle of the hob.

The gashed gear blank is mounted on an arbor that allows free rotation. The fit between the arbor and the gear blank, and any mounting mandrel needs to be close but still allow free rotation. The centreline of the hob needs to be carefully positioned such that it is exactly on the centreline of the worm blank and the table x-axis locked. The hob is then plunged into the workpiece by traversing the y-axis until the full depth is achieved. The cutter will then rotate the workpiece until the cut is complete.

The machining of straight bevel is in many respects relatively straightforward. It is probably more easily machine on a horizontal milling machine than vertical or turret mill as not only does the work piece need to be inclined such that the workpiece surface to be cut is set at an angle such that it is parallel to the machine table, but also oriented such that the tapered profile of the gear tooth will be achieved. Straight bevel gears are probably the most common type of bevel gear, with curved tooth zerol, spiral, and hypoid gears being more specialist, albeit produced in large numbers for applications where power, transmission, ratio, noise or smooth action are more important features.

The key aspect of bevel gears is that they are conical gears (i.e. they are shaped like a cone) and the tooth profile is not parallel. This means that there will be a difference on toe and heel tooth profile with it being narrower on the inside edge that the outside. The shape of each tooth means that the operator will need to undertake two cuts to form each tooth. Specialist bevel gear cutters are required, which again come in sets of eight, however they are narrower than sets of cutters for spur gears, as they need to pass through a relatively narrow gap on the inside of the gear than the outside. Selection of cutter for this operation is not dissimilar to that for spur gearing, although it brings in the added factor of the bevel gear angle. Tables for the selection of the correct milling cutter, and the tooling offsets can be found in good quality handbooks [1].

The important factor for the operator is ensuring that the relative angle between the cutter and the workpiece is 0° to ensure the bevel angle and dedendum angle is correct and the gear tooth does not climb or descend into the workpiece outside of its required limits. The operator also needs to ensure that the centre line of the cutter is set to the centre line of the notional gear cone, after which an offset is applied to allow each side of the gear tooth to be cut. For turret mills or end milling machines this will involve raising or lowering the machine knee. For horizontal milling machines it will be a horizontal movement on the y-axis. On vertical milling machines, or turret mills, the dividing head will need to have its location tenon's removed to allow it to be bolted down at the correct bevel angle. Determining this angle requires some ingenuity and the use of a sine bar against a plain side of the dividing head and DTI is one method that will have a suitable degree of precision. The dividing head will then need to be rotated vertically to give the correct dedendum angle for the gear tooth profile. On Universal (horizontal) milling machines this arrangement is perhaps simpler, with the bevel angle being achieved by rotating the dividing head vertically, and rotating the machine table through the dedendum angle, and therefore does not require any difficult clamping of the dividing head as the dividing head tenons can be left in place with clamping achieved in the normal manner.

Rack and pinion manufacture, is a relatively straightforward process. The pinion is essentially a spur gear, and the rack is a straight gear, which does not require the use of a dividing head. Cutter selection for the pinion will relate to the number of teeth to be cut, whereas a rack effectively has a straight sided tooth so would utilise a no. 1 cutter (i.e. one meant to cut 135 teeth to a rack). It is important that the workpiece blank be set at 90° to the cutter and be perfectly aligned with the spindle axis. When using a turret mill or vertical milling machine, the operator will need to check if the cutter diameter is large enough to reach the workpiece without the workpiece blank interfering with the machine head. Where this is the case the operator will need to investigate if the extending the quill to its maximum length will negate this problem, where this is not achievable the rack would either need to be manufactured in small sections, and accept the compromises to accuracy and assembly that this may cause or recognise that rack manufacture is not possible on these types of machine and may requires specialist hobbing or the use of a universal mill.

Measurement of gear teeth is in many ways as relatively straightforward as measuring a thread. Measurement of threads is normally undertaken, especially for smaller gears, using an 'over wire' method. measurement of gears uses a similar method. Spur gear can be simply measured using two wires of a known diameter, and an accurate measuring instrument such as a micrometer or callipers. For even toothed gears wires, or rollers are placed opposite each other and a measurement taken across the top of the wires. For gears with an odd number of teeth then the roller is placed in the gap that is closet to opposite. Wires or rollers often come in standard sets for the purpose of metrology, however, any roller or wire of the correct, and known, diameter can be used, with drill shanks often being the utilised as a last resort.

The diameter of wire used is relative to the module or diametral pitch of the gear and the pressure angle. What the operator is trying to establish is whether the gear pitch diameter is within its toleranced limits. This means that smaller toothed gears will require smaller diameter wires etc. Using too large a diameter wire will result in the wire bearing on the tooth involute above the pitch circle, and too small below it, and potentially below the top land of the gear teeth, giving a false reading. There are a number of approaches the operator can employ to determine the wire diameter to be used, and the dimensions to be achieved. Perhaps the simplest approach is to use tables in good quality machining handbooks [1] which have tables for wire sizes and over wire dimensions for a range of diametral pitches, pressure angles and numbers of teeth, they also have diametral pitch equivalents for module gears. This provides a simple reference guide for operators to check if the teeth cut depth is correct.

An alternative approach is to utilise a CAD system to draw the gear profile at its upper and lower limit and then identify the wire diameter to be used at the nominal pitch diameter. Manipulation of the CAD software will enable the operator to establish the upper and lower sizes across the wires where absolute precision is required. It should be noted that these sizes are for finished gear teeth, without any allowance for backlash. Where gears are to be hardened and ground in later operations the operator will need to make an allowance for the depth left for gear tooth grinding.

Another approach for gears with larger tooth profiles is to measure the chordal thickness of two or more teeth. Two teeth are required to ensure that the flats of the calliper or micrometer being used, bears upon the involute curve at the pitch diameter. Once again tables or CAD systems can be used to determine the values required, however grinding allowances also need to be compensated for. For smaller tooth profiles this method may be difficult given the potential for handling errors being incorporated into any valid measurement, and the larger the tooth profile is the easier this method is to undertake. For very large gears this may be the only accurate and practical method.

4.26 USE OF COOLANTS

Most workshop milling machines have a facility for, or actively use, coolants supplied through a pipe system to the cutting tool and work piece. The primary role of this flow of coolant is to reduce the temperature of the cutting tool and workpiece. The intention of this process is to extend tool life, increase productivity and help prevent any heat treatment of the workpiece as a result of heat inadvertently introduced as a bye-product of the cutting process. This amount of heat can be substantial as witnessed by the 'blueing' a swarf chips, and in extreme cases chips being expelled from the cutter bright orange from the heat. The energy used to cut metal is efficiently converted into heat. The heat degrades the cutting edge of the tooling rapidly, therefore, its removal by using a flow of coolant rapidly reduces the temperature at the tool point.

Most machining operations can be undertaken without using coolants, however, the use of them allows for increased feed rates, greater depth of cut and significantly increases the tool life. The application of coolant to the cutting area of tooling has a positive effect, although the degree of tool life extension is more marked for HSS tooling than carbide inserts.

The cutting action of a tool involves a shearing process where the material is deformed and converted to a chip, either continuous or as a series of small swarf chips. The majority of energy involved in this process is converted into heat, which is focused at the tool point and along the surface of the tool where swarf rubs over the surface of the tool. Some tools are more resistant to heat than others, and carbide tool inserts can withstand significant amounts of use without coolant, however, in time will develop thermal cracks which lead to cutting edge failure. High-speed steel tooling degrades quickly under significant load, and it is not unusual to see HSS tool cutting edges glow bright orange due to heat. Tooling under these conditions has a short life, and creates a poor surface finish. The addition of relatively small amounts of coolant rapidly quenches this heat build-up and substantially extends tool life.

The selection of coolant can be quite involved as they vary from specific cutting oils, solids, and dry powders to gas cooling. The selection of a cutting fluid or coolant can depend on the machinability of the material being cut, the cost and the effect that it may have on the health of the operator, especially long term. However, most machines found in a general machine shop use essentially one type for logistical ease and low cost. These tend to be from a group of water-miscible fluids. These fluids are an emulsion of oil, water, and an emulsifying agent. The proportion of mixture varies, but a mix of 1 part oil/emulsifying agent to 20–30 parts water is quite common for machining steels using carbide tooling. If using HSS tooling then a slightly richer oil mix of 1 part oil to 10–20 parts water is used. Machining stainless steels may again involve a richer mix of approximately 1:5, however, differing manufacturers products will have recommended mix values, and operators should consult product data sheets for the optimum mix. A broad 'rule of thumb' is that the harder or more difficult a material is to machine, the more concentrated the fluids become. This also applies to types of tooling, where HSS tooling would generally have a richer oil mix than carbide tooling machining the same material.

The advantage of this emulsion of oil and water is that water has excellent cooling properties, and the oil content provides good lubrication, with a secondary advantage of providing some corrosion protection to the machine. While there are a wide variety of oils which can be tailored to differing materials it is normal for a machine just to have one type which can be used across a wide variety of workpiece materials. The only exceptions to this would be where a machine was being used in a repetitive production environment, and machining one type of material. Alternatively some materials that are vulnerable to a chemical reaction such as copper, brasses, and bronze etc. may be subject to staining from chemicals within the coolant especially some sulphur or chlorine compounds. Where components that have aesthetic requirements, or high surface finish requirements, adjusting the cutting depth and feed rates to reduce cutting temperatures and not using coolant, or changing to a passive type are the options open to the machine operator.

Delivery of coolant to the workpiece and tool cutting edge is normally via a sectional/flexible pipe, or a rigid pipe consisting of an elbow joint and extending delivery section to accommodate differing orientations. Both designs have their advantages and disadvantages, in that rigid pipes are stronger, but can be more difficult to keep oriented in the right direction, while sectional pipes can be more easily oriented but are more vulnerable, when swarf wraps or knocks the pipe away. Irrespective of this the machine operator needs to ensure that coolant continues to be delivered to the tool point as the cut progresses. When considering what orientation to deliver coolant to the tool and workpiece it is worth assessing where the coolant delivery pipe, and coolant stream will be at the end of the cut to ensure that coolant does not impact the chuck in any appreciable volume. While this does not cause any harm to the machine it is normal for the coolant to be sprayed out onto the floor below the guard, and out of the back of the machine creating a safety hazard.

While coolant manufacturers and literature suggests that coolants should be delivered at 10–20lt/min, this represents a best case. In general it is not a requirement to deliver large quantities of coolant to the tool point which can often be a rather wet process, and depending on the type of coolant being used can be a long term hazard to the operator. However, the delivery of a stream to the tool such that there is little or no steam witnessed rising from the tool/cutting area is often a good indicator. Workpieces that are more than hand hot when checked at the end of each cut are an indicator of insufficient cooling. Poor surface finish and 'blueing' of swarf chips is also an indicator of insufficient cooling.

One advantage of using larger volumes of coolant when milling is the flushing action it provides. While excessive amounts of coolant can result in milling being a rather 'wet' process, and has safety implications for slippery floor surfaces, it can also be a very effective method of removing swarf chips away from the cutter, and assists in preventing secondary cutting of swarf chips. Similarly when machining a closed pocket, a good supply of coolant removes swarf chips, and while will never completely clean out a pocket, does provide considerable assistance.

When machining more exotic materials such as magnesium or metals that may react with water no coolant would be used. In general this should not be a problem for the operator, as the cutting speed and feed rate should not be so high that heat does not dissipate fast enough, and the operator needs to have an awareness of these marginal conditions.

4.27 TOOLING

As with turning tooling the selection of milling cutters depends on the operation to be undertaken the tooling available and the material to be cut. While carbide insert tooling is widely seen on CNC machines, even in comparatively small cutters, it is still common to see both solid carbide, and HSS cutters in a standard workshop environment. The one exception to this is the replacement of shell milling cutters and helical 'slab' milling cutters for horizontal mills with carbide insert tooling for face milling. For the majority of other types of machining where cycle times or time on machine is less critical the solid cutters tend to be the predominant type, generally as they are substantially cheaper than alternatives.

While almost all tools will cut material selecting differing tooling for different materials can have a substantial effect of surface finish and ease of machining. Tooling rake angles for the cutting edge of a tool will generally be lower for harder materials especially when HSS tooling is used, with softer materials allowing a higher rake angle with HSS. Carbide tools tend to have a lower rake angle to provide support to the cutting edge, and because the wear resistance is greater.

Many modern tools have an applied coating to extend tool life, to achieve sharper cutting edges, allow much finer cuts, or provide improved surface finishes. There are a significant number of types of coating, titanium nitride (TiN) having been in use since the 1960s. Differing manufacturers all make claims about the advantage of particular coatings, and there is clear evidence to suggest that coated tools have increased performance when machining difficult, soft but abrasive, or extremely hard materials. However, for general machining of a wide range of standard steels, aluminium, or engineering plastics the operator needs to consider if the performance advantage of using coated tools is worth the extra expense compared to the use of solid carbide or HSS tooling.

When selecting a tool the operator needs to consider how many teeth are required. Feed rates are normally driven on the basis of 'feed per tooth' so cutters with more teeth theoretically can have greater feed, however, the space between the teeth are crucial for swarf chip clearance and this is affected by the workpiece material properties. Very ductile materials are more likely to produce a continuous chip and fewer teeth with larger inter-tooth spaces provide an easier exit route for the chip thereby reducing the potential for 'chatter'. Thinner materials are more effectively machined using cutters with more teeth and a reduced inter-tooth space as this prevents any tendency for the tooth to dig into the workpiece. In general the operator should select a tool based on having no more than two teeth actively cutting at a time, and that the helix of the tool and rake angle be matched to the material properties of the workpiece.

The availability of differing types of tooling is significant across the world, and there are a substantial number of manufacturers. Tooling design is a discipline of engineering in its own right, however, manufacturers catalogues, data sheets, and technical sales teams are good destinations for operators to seek advice for machining specific material types and operations. However, a 'standard' set of end milling cutters, Slot drills and two or three sizes of ripper cutters in either HSS or solid carbide, to fit all R8 collet sizes (or similar chuck) are a good starting point. A ∅75mm carbide insert face mill to fit an R8 collet or similar chuck system is also an asset to a standard workshop milling cutter set.

End milling cutter

End mills differ from slot drills in that they have a greater number of flutes, and traditionally had a recess in the end and a centre hole to assist with cutter grinding. While many modern cutters now have on cutting edge that goes over centre on the end of the cutter allowing a plunge cut, the essential differences are the number of flutes and the general application. While the traditional end mill is a flat-bottomed cutter, as with slot drills there are a number of alternative profiles that have either a radius on the bottom corner or a completely round profile, often referred to as a 'ball nosed' cutter.

End mills also break down into two subsets. Cutters can have a plain or smooth cutting edge throughout the flute, or can have a secondary tooth profile along the cutting edge of the helix, giving it a 'wavy' edged appearance. These cutters are known as roughing cutters. Roughing cutters are used for rapid removal of material but provide a rougher surface finish. The secondary tooth profile acts as a mini cutting edge, and the length of tool in engagement with the workpiece is effectively longer. The profile generates smaller chips than a straight flute and tends to experience less vibration or chatter. In general ripper cutters are employed to remove substantial amounts of material quickly and then supplemented by finishing cuts undertaken with a plain flute cutter to provide a good surface finish.

The number of flutes on an end mill, and the helix angle is key to the performance of the cutter, as is the clearance angle. There are a multiplicity of designs of end milling cutter, and each is tailored to a different material or application, and close attention to manufacturers data sheets provides guidance on correct cutter selection.

The helix angle on a cutter can be between 15° to 60° with standard milling cutters having a helix angle of around 30°. The helix angle is significant as it provides a shearing effect at the cutting edge reducing the loading on the tool, and also moves the swarf away from the cutting zone. A straight fluted cutter would not be effective in moving swarf away and would have the impact of cutting transmitted along its length leading to vibration and chatter. This suggests that cutters with a high helix angle would be better, and in fact some premium cutter have a variable helix to further reduce vibration and chatter. Cutters with high helix angles, and consequent high shearing angles, have more consistent loading along its cutting edge as there is a progressive cutting action leading to improved surface finishes and good chip discard, however, they are less good on materials that are difficult to machine where the strength of the cutting edge and stiffness is important. The operator when selecting a cutter needs some awareness of this as cutting a softer aluminium, or thin edged material may benefit from a cutter with a high helix angle than a complex stainless steel that would benefit from the increased stability of a cutter with a lower helix angle. Standard workshop cutters will always cut material, however, if the operator has a selection of cutters to choose from, then there are benefits to the standard of cut produced.

The number of flutes is also a feature of end mills, and the key differentiator between slot drills and end mills. While there is some overlap, in that three fluted slot drills and three fluted end mills can easily be found, the traditional end mill has four flutes or more. End mills with more flutes tend to be of much larger diameters. The key consideration for this is the space between each tooth which needs to be large enough to allow free ejection or flow of the chip away from the cutting edge, and the space between the corners where the back of one tooth meets the front of the next does not have a corner, or significant turning angle that chips get clogged up preventing efficient ejection.

Face mills

Unlike end mills or slot drills face mills or shell mills are only for facing (i.e. producing a plane surface rather than a pocket). The cutting face is located at the edge, or periphery of the cutter so it is only used in a horizontal, or vertical plane (on horizontal milling machines) and cannot be used to plunge into a workpiece. It is normal for the entire surface of a workpiece to be machined to form a flat surface with a face mill, and creating a step should be avoided as many face mills to not have sufficient clearance between the body of the tool and the cutting edge. The depth of cut is limited by machine power, workpiece material and the insert size.

There are two principal types of face mill. The first is the carbide insert type. These have become cheap and widely available and enjoy a higher material removal rate than solid HSS shell mills. Figure 4.64 shows an example of a carbide insert face mill and a traditional HSS shell mill.

Carbide inserts which are indexable are inserted into the periphery of the cutter, and when damaged or blunt are turned to provide a new cutting edge. Given the ability of the tool to absorb head and load, they can be run without coolant, however, cognisance of workpiece material properties

Figure 4.64 Carbide insert face mill and HSS shell milling cutters

needs to be made, given that heat transfer can be so significant as to harden some alloy steels. Use of coolant will remove this problem. The tools can sustain significant depts of cut and the feed per tooth value stated by the manufacturer provides for a high rate of material removal. The swarf chips produced by these cutters, tend to be small and fast moving and as such are ejected from the cutting area. These can be hot and where hand feeding is being undertaken it is not unusual for the operator to be impacted by these hot swarf chips.

The cutter will mill a surface to the width of the cutter. Given the load on the machine that this type of machining imparts, it is not unusual for any spindle backlash to be taken up. While this is not especially significant, if the operator is trying to get a consistent surface finish, where backlash is present it is not unusual for rotational grooves to appear in the work-piece once the outer diameter of the cutter passes off the workpiece. This occurs because vertical load is removed from the spindle, as the cutter is only removing material up to the maximum diameter of the cutter. Once this passes from the surface of the workpiece the load on the spindle is removed and drops slightly moving the trailing radius of the cutter arc to dig into the workpiece scoring it. The use of backlash adjusters, and ensuring the spindle clamps are engaged is the primary method for preventing this. Poor surface finishes can also be because the head of the machine is not exactly perpendicular to the machine table, and having an accurately trammelled head is an important factor in achieving good surface finishes when using a face milling cutter.

Shell milling cutters are the traditional cutter for face milling. They are traditionally of the solid HSS type, although carbide insert versions are freely available. The key difference between a face mill and a shell mill, is that a shell mill can cut into a corner as they have a peripheral cutting edge as well as one on the lower face. Almost all shell milling cutters will need to be mounted onto an arbor that then fits into a machine spindle, however, this is not always the same for face mills, which quite commonly have a tool body profile manufactured to an ISO, or R8 style collet system. The ability to machine on the face and side of the cutter is its advantage, but any requirement to mount on an arbor increases its expense. Often machine shops are equipped with face mills and utilise slot drills or end mills for subsequent machining operations.

Use of HSS shell mills requires some care. The cutter will often have more teeth than the equivalent carbide insert type, with a right-handed helix and a square corner. The operator will need to consider the depth of cut and feed rate, in order to avoid swarf clogging up the inter-tooth spaces and to prevent significant wear to the corner of the cutting edge. The way these cutters are ground makes this corner vulnerable to damage if pressed too hard.

Slot drills

Slot drills are a type of end milling cutter that typically only has two flutes and a high helix angle to allow easy egress of swarf chips. Its key difference from an end milling cutter is that all slot drills can centre cut. By this there is a full cutting edge to the centre of the tool on the bottom face. This allows for vertical plunge cutting to a depth required and then move in a horizontal translational movement. Figure 4.65 shows a typical slot drill tooth profile.

The wide gap between the teeth and helix angle assist the ejection of swarf chips which is particularly useful when plunge cutting and machining out closed end pockets. The method of mounting and the selection of tool length is exactly that for end milling cutters, and often slot drills are used in applications that an end milling cutter could be employed in. Given that feed is a function of feed per tooth, feed rates can be slower and surface finishes can be more coarse than that achieved by an end milling cutter.

When plunging into a workpiece the hole depth is limited to the length of flute up the shank of the tool, as enough space must be left for swarf chips to be ejected. This generally means that the operator will need a long series set of slot drills to pierce through any thickness of material. In broad terms, a standard-length slot drill will have a flute length of approximately 2–3 times its diameter, with long series slot drills, being 3–4 times the length of its diameter. Differing manufacturers will have their own approaches, however, it is not unusual for long series slot drills to have three flutes rather than two. When drilling into a workpiece with a slot drill the surface finish

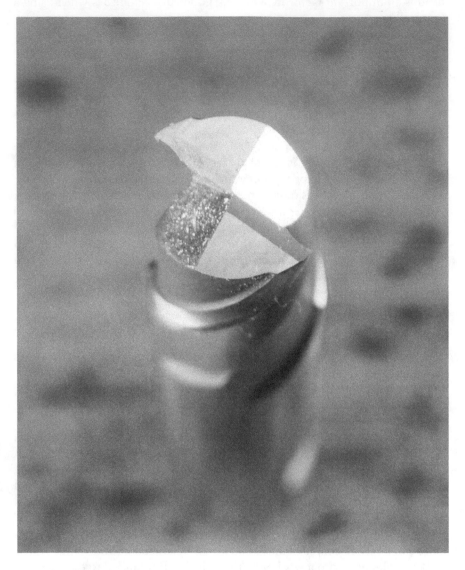

Figure 4.65 Typical slot drill

achieved can be quite impressive and operators may wish to drill holes with this type of cutter, rather than using a centre drill, twist drill and reamer to achieve the same surface finish.

For many years there has been a clear delineation between slot drills and end milling cutters. End milling cutters having three flutes or more, with four flutes cutters being the most common, and having a centre hole on the end of the cutter which prevents plunge cutting, and may have a profiled cutting edge of the ripper type. While slot drills traditionally had two flutes an over-lapping cutting edge on the bottom to allow plunge cutting and it was rare to find one with a profiled edge. This clear delineation has been eroded and it is not uncommon to find end milling cutters that one tooth on the lower surface that extends to the centre of the tool allowing plunge cutting.

Modern CNC machines also now have built in ramp on or spiral sub-routine that allows for end milling cutters, especially centre cutting tools, to pierce a workpiece and allow and end mill to be used further eroding the differences between the two types. However, for the general workshop this ramp on approach is not available, and having a set of slot drills to comple-ment a set of end milling cutters is essential.

Tee slot milling cutter

Tee slot mills are a bespoke cutter for machining of tee slots, typically those in a machine table. The cutter consists of a plain or morse tapered shank with six or eight teeth. These teeth are also staggered left and right. It is normal for the cutter to have a reduced diameter for a section between the cutting teeth and shank. The diameter of this section denotes the throat width that the cut-ter can pass through (i.e. the slot that will have been previously machined to depth using a slot drill or end milling cutter). The length of this reduced section defines the throat depth and controls the depth of material remaining above the tee cavity produced by milling. Figure 4.66 shows a typical for tee slot cutter.

Figure 4.66 Tee slot cutter

Tee slot cutters come in standard sizes and these are driven by the type of slot to be produced. There are a number of national standards that remain extant, for Tee slot dimensions, however, there is also an ISO standard. Tee slot cutters will be matched to the requirements of these standards particularly with respect to the cutter diameter, face width, which gives the breadth of the tee, the throat thickness and throat depth. Operators intending to machine tee slots will need to ensure that the cutter is matched to the size, as there is little latitude for operators to increase the internal width of the tee as the only lateral movement by the cutter will be that provided by the clearance between the initial slot machined and the throat diameter of the cutter. Similarly, the depth of the key is constrained by the length of the throat of the cutter.

Dovetail milling cutters

Dovetail milling cutters share a high number of features with tee slot milling cutters in that they have a shank and portion of the cutter with a reduced throat and throat length. This means that the same restrictions to depth of cut apply. The cutters have an angled cutting face with a 60° angle and 45° angle being the most common. Typically the widest part of the cutter is at the lower end, however, inverted dovetail cutters are freely available where the narrowest part of the tool cone is at the bottom of the cutter and widest part above this. Figure 4.67 shows an example of typical 45° and 60° cutters.

However, dovetail milling cutters have a number of applications. While all operations involve producing an angled feature in a workpiece, this can be as an equal slot similar to a tee slot, a parallel slot with a wide throat thickness and angled undercut, or a simple shoulder with a single angled undercut.

Dovetail milling cutters have a series of teeth which come to a sharp corner. This corner (especially on 45° cutters) can be vulnerable to damage, the depth of cut, feed rate and the way the tool is advanced into the workpiece needs care to preserve the corner. Unlike tee slot milling cutters, dovetail milling cutters do not have staggered teeth. It is relatively unusual to machine a slot like a form of tee slot and it is more normal for the cutting to be undertaken on one side. The operator therefore has the choice on whether to employ upcut or downcut milling.

Prior to dovetailing a groove has been machined with a slot drill or end milling cutter and the dovetail cutter is used to create the undercut. It is normal for the undercut to be created at full depth, meaning that the initial penetration of the workpiece is undertaken by the tool corner. The depth of cut that can be taken with this type of cutter is substantially less than that for and end milling cutter or slot drill and operators should have a degree of care when using these tools. Initial cuts can be light as the corner is advanced into the workpiece, but as the full length of the tool cutting edge

Figure 4.67 Typical 45° and 60° dovetail milling cutters

is engaged loads substantially increase and care must be taken to ensure that the cutter does not flex away from the feature being cut.

Indexing boring mills

Indexing boring mills are less common in general machining workshops, although if the workshop is one that regularly undertakes precision boring operations, then it is more likely they will be encountered. In broad terms an indexing boring mill consists of a cutting tool, which is similar to a lathe boring bar, but mounted vertically. This is inserted into a body that has the ability to move at right angles to the spindle axis, controlled by a precision adjustment. This allows the operator to increase the radius of cut very precisely.

The tools are only used to increase the diameter of previously drilled or formed holes, and are a direct alternative to the use of a rotary table. The ability to index the tool allows for the generation of hole diameters outside of standard drill or reamer sizes. It also overcomes any issue with drills may produce a hole slightly oversize, or with poor surface finish. The advantage of using an indexing boring bars as an alternative to rotary milling, is that a hole can be drilled, the table dropped on the z-axis and the indexing cutter inserted without losing a datum, or having to remount the workpiece.

The recommended cutting speed used for these tools is around 50% of the calculated speed and there are some recommendations that identify spindle speeds for these tools should not exceed 600 rpm. This is due to the unbalanced nature of the tool producing vibration and the cantilever effect of the boring bar protruding downwards. When boring deep holes such as cylinders this protrusion is likely to be significant, and as such lower speeds, quill feeds of the machine minimum setting, and peck depths of a maximum of 0.5mm, reducing to as little as 0.08mm for harder materials. When using carbide inserts within the boring bar the operator will need to select an appropriate insert for the material being cut. The criteria for this are not different from the approach taken when turning, with larger corner radii providing a stronger tool, and being good for roughing operations and smaller radii providing for finer cuts, a more square corner in machining to a step and a better surface finish, however, this is only if very fine feeds are available of the quill feed settings. Where this is not possible then a slightly larger tool nose radius may be a better solution.

These type of tools often have an adjustment scale, but these are for indication of incremental adjustment. The most efficient method for operators to determine incremental offset is to place a dial test indicator directly on the toll body and measure the incremental change as the tool adjustment screw is rotated. The operator also needs to remember that this offset is on the radius and the increase in diameter of the generated hole is twice the offset.

Fly cutter

A fly cutter is a tool used on a milling machine that generally has one cutting point and is often used in place of a face milling cutter. They tend to be cheaper than face mills, especially those with multiple inserts, and it is not unusual to see home workshop manufactured fly cutters, where a body has been turned and a left-handed carbide insert lathe tool mounted into the body. Fly cutters allow for a wider diameter of cut with the respect to tool loading, and less machine power is required to machine a width compared to that with a multi tool point cutter. However, the trade off is a reduced feed rate, increased spindle speed and depending on the tool design and machine rigidity, the introduction of a degree of vibration.

Surface finishes achieved with a fly cutter are normally better than those achieved with a face mill as there is not drag from tool inserts not engaged in cutting material, and it is quite common to see engine cylinder heads being cut with a fly cutter, rather than milled and ground. However, where larger diameters are being cut with a fly cutter, the greater the precision required when trammelling in any adjustable turret mill head, if a high surface finish is to be achieved.

Gear hobs

Gear hobs are a form of gear cutter and while extremely common tend to be the preserve of specialist gear manufacturers or workshops that have a universal milling machine. Essentially a gear hob is a helically cut gear form, that has a series of axial cuts made. These axial cuts provide a cutting edge and turn what is essentially a worm drive pinion into a milling cutter. Figure 4.68 shows a typical gear hob milling cutter.

The cutter is used to cut spur gears, helical gears, and worm drive gears, however, there use requires a dividing head to be connected to the machine table drive shaft for spur gear and helical gear manufacture and the angle of the cutter when mounted on the shaft spindle needs to be angled with an equivalent to the helix angle of the cutter. The manufacture of worm drive gears on a standard turret mill or end milling machine is relatively straight-forward with a hob, however, spur gear or especially helical gear manufac-turing is requires a more complex set-up for a general workshop and the use of this type of cutter is less common in a standard machining workshop.

Gear cutters

Gear cutters as described in the section on gear cutting come in sets of eight. This is because the involute gear profile changes depending on the number of teeth to be cut in the gear periphery. As with almost all milling cutters

Figure 4.68 Typical gear hob

carbide insert tooling is available, however, it is still extremely common to find solid HSS gear cutters as the predominant type found in general cutting workshops. This is partly due to gear cutting in general machining workshops to be undertaken more slowly than might be undertaken in high volume gear manufacturing environments. When cutting helical gears the requirement for a dividing head to be geared to the table drive to provide rotational as well as translational movement restricts the feed rate which can be applied, and removes any time advantage of using carbide insert tooling.

An additional advantage of using HSS cutter sets is that the module or diametral pitch and range of teeth to be cut are clearly marked on each cutter, whereas carbide insert cutter systems tend to have a standard tool holder into which a set of inserts for each gear tooth profile are inserted. Figure 4.69 shows a typical set of HSS gear cutters in which the gear tooth profile information is clearly displayed.

The operator needs to ensure that the correct insert is fitted into these types of toolholder and that a complete set is inserted prior to machining. Solid cutters clearly marked remove this complexity, however, carbide insert systems have all of the advantages of being able to accept higher loads, heat resistance, and ease of addressing cutting edge wear.

Figure 4.69 Standard gear cutter set (imperial)

Drills

There are no special drills used when milling and operators either use standard HSS twist drills inserted directly into a machine collet or into a Jacobs chuck fitted with an appropriate R8 or other collet system shank. Figure 4.70 shows a Jacobs chuck mounted in an R8 system collet.

Figure 4.70 Jacobs chuck mounted in R8 collet system

This approach lends itself to the use of morse centre drills, and drills of odd sizes such as those used for tapping, where direct insert into a collet can be restricted by the standard 'whole millimetre' sizing. The alternative approach being to drill with slot drills, however, there is again a restriction on this approach with diameters being limited to those of the cutter diameter, and fractional tapping sizes are often not available.

Large diameter holes are generally not achievable on standard workshop machinery. Plain shank drills over ∅16mm are relatively uncommon, as are Jacobs chucks above this size. Operators may have to consider rotary milling or drilling the largest possible hole on the milling machine as a pilot hole, and subsequently drilling to a required diameter as a separate operation on a heavy-duty workshop drilling machine or radial arm drill.

4.28 UPCUT AND DOWNCUT (CLIMB) MILLING

There are two directions that most materials can be cut, when milling one being 'upcut' milling and the other being 'downcut' milling, otherwise known as climb milling. The analogy being drawn the cutting edge is always trying to climb up the slope of the workpiece being machined. Downcut milling is where the table feed is in the direction of rotation of the cutter. Upcut milling is the reverse of this, where the direction of table feed is in the opposite direction to the feed of the cutter.

The primary effects on tool life are cutter speed, feed rate and depth of cut. However, the direction of cut also impacts tool life. When milling the operator needs to consider how much heat is going to be built up at the cutting edge, where the swarf chips are going to go and how stable the cutting process is going to be to prevent vibration. Irrespective of cutting direction, whenever the cutting edge of a milling cutter enters a workpiece it is subject to a shock load and the operator needs to take into account the cutter type, geometry, size and condition of the machine being used and operation to be undertaken.

There is a widely held aphorism that milling should always be conducted such that chip generation goes from 'thick to thin'. This is to ensure that as the cutter rotates the cutting edge exits the workpiece with the chip at its thinnest. The advantage of this is that the load on the cutting edge reduces through the cut, but more importantly does not experience frictional rubbing as the cutter enters the workpiece and increasing load which is suddenly released as it exits. While the helix angle increases the shearing action to reduce the sudden impact loads that would be experienced by a straight tooth cutter the rubbing action remains. By downcut milling the cutter does not have this frictional start as its initial cut forms the thickest section of chip and the depth of cut reduces to zero as the cutting edge exits the workpiece. The chip is then ejected behind the cutter removing the potential for secondary or over-cutting of the removed chip. This method pulls the workpiece into the cutter, but also away from the table spindle drive. On machines that have excessive wear or no backlash adjuster, vibration, sometimes severe, can be experienced along with a poor surface finish. However, where backlash adjusters are fitted or the table lead screw is in good condition, surface finishes are likely to be better than for upcut milling. The effect on tool wear, and chip ejection means that downcut or climb milling is the preferred choice for cutting direction.

Upcut milling generally produces a degree of work hardening as part of the cutting process and as the chip thickness increases during the cut produces a hotter chip, which reduces tool life. However, depending on the condition of the machine, or the work holding device upcut milling can be a preferred choice in some situations. For machines that have excessive wear, or no backlash adjuster, it is not uncommon for a cutter to bite into the workpiece and drive the machine table through, damaging, the cutter, the workpiece and potentially the machine. Where this is suspected upcut milling is recommended. Where precision operations involving the workpieces being mounted lightweight work holding devices such as dividing head arbors when gear cutting, upcut milling can provide a more stable workpiece.

In general operators should consider downcut milling as a preferred direction, unless machine dynamics or the work holding considerations prevent this.

4.29 TOOL CHANGING

When using milling cutters tool life is affected by a number of factors such as cutting speed, feed rate, depth of cut and the material being machined. Irrespective of cutter material or any coating eventually tools wear out. When cutting feeds and speeds have been correctly calculated and the depth of cut is set at a level that the cutter, and machine, are not overloaded then a substantial life can be expected. The first indication that a cutter needs changing will be a degradation in the surface finish of the workpiece. Once this surface finish goes below any roughness value stated on the component drawing, or begins to become an aesthetic issue, the operator should change the tool.

While changing the tool is a straightforward process (i.e. undoing an ISO or autolock type collet system), or slackening off the spindle drawbar for R8 collet systems, is straightforward the operator should examine the tool to establish whether the tool has worn out due to abrasive wear, or is exhibiting other failure modes such as built up edge, crater wear or plastic deformation etc. Where tool damage other than abrasive wear are encountered operators should review the cutting conditions for the workpiece material be machined to ensure that machining operations can be completed without any degradation of surface finish for the feature being generated.

4.30 COLLETS

There are four principal types of collet that are encountered in general engineering workshops. The R8 collet system and ISO collet system are becoming universal, however, their predecessors based on a morse taper spindle interface, especially the Clarkson autolock type which are still widely seen.

All systems rely on a collect being inserted into a spindle, and drawn up the spindle by a threaded drawbar screwed into the end of the collet. The drawbar is tightened using either a spanner applied to the upper end of the drawbar located on the top of the machine, or by actuating a powered system, which is normally pneumatic. This draws the collet into a taper and locks it tight which allows for the power to be transmitted to the tool holder without slippage. In the case of the R8 collet system this process also locks the tool in place.

The R8 system has a locking taper at the bottom of the collet, and this results in the collet being locked into the spindle taper close to its end and also providing clamping force in the area the cutting tool is inserted into. Therefore tightening of the drawbar not only locks the collet in place, but clamps the cutting tool within the collet. An example of the R8 collet systems is shown in Figure 4.71.

Most other multi-tool types, have separate clamping mechanisms and there are two principal approaches taken to this.

Autolock systems utilise a screwed shank tool that is inserted into a small collet body, which has a tapered bottom edge, and this is itself inserted into a screwed insert, as shown in Figure 4.72.

The system can require a little adjustment to ensure that the centre hole in the end of the tool is pressed into an internal centre point of the collet body, and that the screwed insert presses on the collet insert taper and not the bottom edge of the collet body. However, when properly adjusted the screwed insert is tightened onto the collet insert taper locking the tool tightly in place and transmitting the rotary motion with sufficient friction. The tool is also supported by a point in the collet body preventing rotation of the tool in the threaded insert. Differing manufacturers to Clarkson, had similar approaches, and many remain widely used. Screwed shank tools also

Figure 4.71   R8 collet system

Figure 4.72 Clarkson autolock system

remain widely available for common applications, however, more bespoke or niche tooling, tends to only be available with plain shanks recognising the gradual move towards ISO or R8 collet systems, and the obsolescence of screwed shank systems.

ISO collet systems use a similar approach with a collet body inserted into the machine spindle and a separate collet insert. However, while a male threaded assembly is often used holding the collet insert and clamping the tool, a grub screw can be used to lock the tool within the collet body. The collet inserts do not have a thread and plain shanked tools, with or without a flat portion on the tool shank are used. Figure 4.73 shows a typical example.

The collet inserts also have significantly more slits machined into them allowing for better and more parallel compression giving good stability. The collet insert body is also tapered giving a larger frictional area, however, the use of a C spanner, or similar to tighten up the tool in position is similar across bot ISO and Clarkson types.

A third type seen is a collet that is a simple tool holder. The collet body has an ISO taper and is inserted into the spindle as for all types but has a hole sized for a range of tool diameters. The tools are retained within the holder by tightening two or more grub screws which press onto a flat machined onto the shank of the tool. This flat is common to most plain shank tools allowing them to be fitted to multiple types of tool holder. The advantage of this type of holder is reduced cost and simplicity of the tool holder, along with speed of tool changing, however, to allow insertion of the tool there

Figure 4.73 ISO collet system

has to be some clearance between the tool and the tool holder body. While small, a gap will exist with only a small part of the tool circumference in contact with the tool holder cavity, and prevention of rotation achieved by point loading from the grub screws. In extreme situations the tool can have slight movement leading to surface finish defects. While in most machining situations, operators will notice little difference between the types of collet system, but where complete support is required, ISO, or R8 systems provide the highest amount of stability.

4.31 BEHAVIOURS OF DIFFERING MATERIALS

The majority of materials machined on milling machines are relatively benign, and once the workpiece has been well clamped down and the correct feed and spindle speed selected rarely provide and problem for the operator. However there is a class of materials that can be considered 'difficult to machine materials'. These are a group of materials which lead to poor surface finishes, deterioration or hardening of the workpiece sub-surface area, high temperatures at the cutting point and poor heat dissipation which leads to excessive or rapid tool wear. While some very ductile materials such as copper can be vulnerable to tearing, generally the correct selection of tool and close observation of feed and speed resolves this. However, a lot of titanium alloy- and nickel alloy-based workpiece materials can cause significant problems. Materials containing significant amounts of nickel, cobalt, tungsten, and chromium also present problems. While some of these materials are less commonly come across, stainless steels have been widely used for many years, but alloys that were considered exotic a few years ago are becoming more mainstream, and require careful selection of tooling to ensure successful machining.

Titanium alloys continue to become more commonly used given their high strength to weight ratio, corrosion resistance and relative toughness. However, because the material generally has a relatively low modulus of elasticity it deflects under load of the cutting point causing chatter and vibration and at the same time strain hardening. Not only does this degrade the cutting edge itself but also can lead to a built-up edge on the tool itself. Nickel based alloys are very robust materials and have high strength. The inclusion of carbide particles in its structure, high strain hardening and its high strength cause wear on tooling and again the formation of a built-up edge. Stainless steels suffer from a high rate of work hardening, and do not dissipate heat well, high amounts of force are required to machine the materials which leads to heat and causes rapid tool wear and often suffers from poor surface finishes if care is not taken. Cobalt chrome alloys have similar issues, however, magnesium-based alloys machine well, but are vulnerable to setting on fire at temperatures above 450°C. On this material only mineral neutral cutting oils can be used so often it has to be machined dry, reducing the depth of cut and feed rate that can be used and leads to extended machining times.

The principal problem with machining difficult materials is the heat produced at the cutting edge of the tool leading to rapid wear or built-up edge. There are a number of solutions such as using diamond tools or cubic boron nitride tooling. This has excellent wear resistance and hot cutting capabilities however, are also expensive. Diamond can in some circumstances chemically react with titanium or steel workpiece materials. The solution is careful selection of tooling, and the use of coated tools. Coatings that are applied to the outside of carbide or HSS tooling provides a barrier between the tool material and the workpiece. These coatings can often be extremely hard and have high heat a wear resistance. Seeking direct advice from tooling manufacturers will identify the correct tool type and tool material for the application being considered. Most machining is adequately undertaken using HSS or carbide tooling, however, it is becoming more common to see difficult to machine materials. Investigation and discussion with tooling manufacturers to determine the correct tool that will provide the balance of tool life, workpiece surface finish and material properties post machining.

4.32 METROLOGY

Practically any machining process involves the application of a cut, or a series of cuts to remove material, followed by measurement to determine if the correct size has been achieved. More commonly, a series of roughing cuts removes the bulk of material and then a measurement is taken. This measurement identifies how much more material is required to be removed to achieve, a depth within the limits of the dimensioned size, or more carefully to achieve the maximum material condition.

While many machining operations do not require complex measurement and removing material to a simple scribed line is entirely appropriate. However, where the scribed line is obscured by a burr, or precision greater than 0.5–0.25mm is required, measurement needs to take place to accurately determine the size of a component feature, or to identify the depth of material that remains to be machined away. There are a number of approaches, and measuring instruments that can be used to determine sizes, however, there are constraints that need to be accommodated, which can drive the selection of measuring instrument, and the accuracy to which the size needs to be determined is also a fundamental factor. Another common problem is removal of the workpiece from the vice or work holding device. As soon as the workpiece has been removed it is unlikely that that it can be returned to exactly the position it was in before. This is because coolant, swarf chips can be displaced, or that movement of vice jaws, is not consistent. The amount of movement can be quite small, however, if work is being produced to tight tolerances, not moving the workpiece is key in order to ensure that the cutter cutting edge, and face being cut, remain perfectly aligned. Similarly the frame or scales of some measuring instruments can be quite long and have the potential to clash with parts of the machine, especially the cutting heads. This type of clash can be quite subtle, and lead to measurement errors. Where it is not possible to move the machine table away from the head such that the workpiece is clear of any obstruction, a different measurement methodology may need to be selected. A common example of this can be measuring depths using vernier or digital callipers where the length of the beam containing the main scale and outside large jaws commonly protrude upwards by at least 200mm and impacting the machine head if the table has not been traversed.

One consideration the operator needs to make when selecting a measurement instrument to identify the size or depth of a feature is its resolution. Accurate measurement is a combination of the resolution of the measuring instrument and the competency of the operator. It is generally accepted that operators should use instruments that are calibrated to one-tenth of the tolerance band being measured. Therefore, if a feature had a dimensioned size of 20mm ± 1mm the overall tolerance would be 2mm and a vernier calliper with a resolution of 0.2mm could be used. However, if the dimensioned size was 20mm ± 0.05mm then a micrometer with a vernier scale and a resolution of 0.005 would need to be used. That said, the accuracy of the machine, and the operating environment also needs to factor into how the operator will achieve a valid measurement, and for machines commonly found in workshop environments 0.05mm can often be the limit of the machine's accuracy.

The skills of the operator and the method used to measure affect the accuracy of the measurement taken. In optimum conditions a skilled operator will have little problem in repeatably taking measurements to an accuracy of 0.02mm when using a digital calliper calibrated with a resolution of 0.01mm. However, vernier callipers are often only calibrated to 0.02mm or even 0.05mm. Parallax errors and differences between the axis the component is measured at and the reading axis along the callipers beam can often lead to much greater errors, as does the reach required by the operator to get to the workpiece feature being measured and any congestion around the workpiece. Where this may be an issue, and multiple measurements need to be undertaken it is common for a gauge repeatability and reproducibility study to be performed to identify operator variation.

One of the most versatile tools in the workshop is the vernier, or digital, calliper as it allows external sizes, internal widths, and depths to be measured using one instrument. It is also very common for vernier type callipers to have both imperial and metric scales on one instrument and, for digital models to have a simple switch between imperial and metric systems. This flexibility and low price make them an important instrument when machining. However, they lack the accuracy of micrometers, and while more expensive digital callipers have a resolution of 0.01mm they suffer from errors introduced by having the measuring axis some distance from the reading axis, and not having the callipers positioned perpendicular to the feature being measured. Close attention needs to be paid by the operator to ensuring that errors relating to Abbe's principle and, those relating to the positioning of the jaws perpendicularly across an axis or feature. Care also has to be taken to ensure that the beam of the calliper does not inadvertently foul and part of the machine that may lead to error. Having selected an instrument with the correct resolution for the feature tolerance, the solution to many of the problems when using a calliper is to undertake a series of repeated measurements and check that there is a consistency of reading. Where there is not the operator should investigate, until consistency is achieved. One other problem with a vernier of digital calliper is that they are limited in the maximum external diameter they can measure. While the beam scale may be 200mm+, the length of the jaw provides a constraint on the maximum external diameter that can be measured. The distance between the end of the calliper outside jaws and the beam provides the pocket that the workpiece has to fit within, effectively the radius of the workpiece. While callipers are hugely flexible when it comes to measuring linear features, depths, and internal diameters, they are restricted when measuring external curved surfaces or diameters and operators need to select a different instrument to overcome this.

Micrometers are often stated as having 10 times the accuracy of callipers, and while this is not always correct, as there are callipers that have a very fine resolution, it recognises errors that creep in as a result of operator method and instrument resolution. The alignment between the parallel measuring faces and the measuring scale of a micrometer removes many of the errors that can be found in measurements taken with a calliper, especially a vernier calliper. Standard micrometers have a resolution of 0.01mm or 0.001″. Micrometers are also commonly found with an additional vernier scale allowing a resolution to 0.002mm. Some digital micrometers take this down to 0.001mm or 0.00005″, however, in the general machine shop accuracies of this magnitude start to require a temperature-controlled environment. This accuracy, and the speed of use makes micrometers a highly effective instrument for measuring sizes of workpiece features during and after machining. Micrometers, while highly accurate, easy and quick to use have a limited range and only take one type of reading. The overall distance a micrometer can measure over is limited to 25mm (1″). This means that to have the capacity to measure over a range of sizes and numbers of micrometers is required. Therefore, to measure the typical 200mm range of a cheap calliper, eight micrometers would be required (i.e. 0–25mm, 25–50mm, 50–75mm etc.). Micrometers are also task specific in that standard micrometer is only used for measuring across external faces, and while this can be a quick and accurate method for determining a size, depths, internal widths, wall thicknesses require task specific instruments, such as depth micrometers, inside calliper micrometer, ball and anvil, internal microm- eters, and bore micrometers. These instruments again only measure over a range, although it is common for depth and bore micrometers to be supplied with a range of spindles which cover a wide range of sizes. This requires a significant investment in measuring instruments to cover the range of appli- cations and sizes that a digital or vernier calliper could achieve, and the operator needs to balance resolution and accuracy of the work performed when sourcing measuring instruments.

When machining slots, pockets and stepped features the use of gauge blocks (slips) is also a highly accurate and quick method for establishing the size of a feature. Gauge blocks can either be inserted into slots and subsequently wringing multiple blocks together until they cannot be inserted to derive a size. This approach, while extremely accurate, takes some time and can almost be considered an inspection method rather than something undertaken to determine a remaining cut width or depth. That said physical difficulties in getting a calliper or depth micrometer into a position to take a valid mea- surement of a feature can often be overcome by inserting gauge block packs into a slot to determine its depth. While use of a finger clock (DTI) used as a comparator alongside gauge blocks determines absolute values, using a finger run between the slip face and outside surface of the workpiece can also be sur- prisingly accurate, and has utility especially when undertaking roughing cuts.

Metrology is an engineering discipline of its own, however it is a key part of successful machining. All machining is undertaken to remove mate- rial to a known size and measurement is undertaken to establish whether that size has been achieved or to identify how much more material is to be removed. This operation can be important when machining difficult materi- als where cut depth for each peck, is critical given work hardening or feed constraints. The allowable error, or tolerance that is permitted for a feature is also something that drives the selection of measuring instrument, and can be a controlling factor. What tends to happen in practice is that a number of methods are employed from a simple scribed line being used to mark up the workpiece, callipers being used during the roughing phase and microm- eters often used to determine the depth of finish cut and final sizes. Once machining is complete inspection is normally undertaken off machine using micrometers, gauge blocks, a DTI used as a comparator. The approach an operator wishes to take, can be constrained by the equipment available, however, it is important for them to consider how accurate measurement of features will be achieved while the workpiece is being sequentially cut, that does not require its removal from the machine. Where this is not possible and the workpiece has to be removed, it may be that the tightness of the tolerance requires a subsequent machining operation such as grinding to achieve final sizes, and the operator, while free to use a less accurate method, must ensure a grinding allowance is left, and the depth of material left will be a function of the time required to grind to size and the accuracy of the measurement method.

4.33 USE OF SINE BARS

Sine bars are common in machining as a method for setting up workpieces and work-holding devices to an extremely precise angle. This can remove the requirement for a workpiece to be subsequently ground, or make subsequent operations more simple to execute.

Sine bars are relatively simple devices. They consist of a plane body with rollers inserted into two steps. The distance these rollers are apart are critical to the accuracy of the sine bar, and the bars themselves come in a variety of lengths, such as 100mm, 200mm, 300mm or their imperial equivalents. The length of a sine bar is that between the centres of the two rollers, and not the overall length of the sine bar. All sine bars will have their nominal length engraved or etched into the side of the body. Figure 4.74 shows a typical sine bar.

There is no difference between metric or imperial units sine bars other than imperial unit sine bars require imperial gauge blocks (slip gauges), an metric unit sine bars require metric gauge blocks. It can be argued that metric slip gauge pack heights can be converted to imperial units and vice versa, but this introduces a potential opportunity for error.

Most sine bars are used in metrology labs or inspection departments, however, they have practical uses in the workshop environment where accurate set-up is key. The principle behind a sine bar is that of producing a right-angled triangle. A pack of slips is inserted between a plane surface and one roller of the sine bar. The height of this pack of slips is a function of the sine of the angle required. The sine of an angle in a right-angled triangle is calculated as follows:

$$\mathrm{Sin}\,\Theta = \frac{\text{opposite}}{\text{hypotenuse}}$$

The hypotenuse of the triangle is known as this is the nominal length of the sine bar i.e. the distance between its two rollers and the required angle is generally as it is commonly a dimension on the component drawing known, therefore the unknown factor is the height of the opposite side, therefore:

Opposite = hypotenuse × sin Θ

The value from this gives the size a pack of slip gauges needs to be built up to in either mm or decimal inches. Once this has been established a pack of slip gauges is inserted between a plane surface, such as the machine table, or surface plate and the end roller to lift the sine bar up to an angle. Figure 4.75 shows a sine bar elevated by insertion of a slip pack.

The workpiece is then mounted onto the upper surface of the sine bar and clamped in place. The slip pack and sine bar are then removed and the

Figure 4.74 Typical sine bar

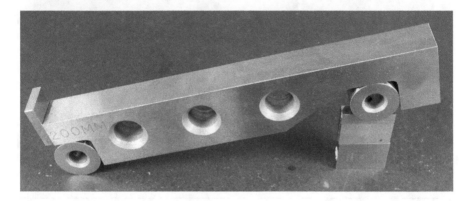

Figure 4.75 Sine bar set to inclined angle

workpiece remains at a precise angle. This method is only recommended for angles up to 60° given the pack height, and where angles in excess of this are required the operator should change the orientation of the machining operation i.e. considering milling on the side of a cutter rather than the end.

Example 1 (metric)

To identify the slip gauge pack height to produce an angle of 35° using a 200mm sine bar:

Sin 35° = 0.5736 (derived from calculator or tables)
Length of sine bar = 200mm
200mm × 0.5736 = 114.72mm (slip gauge pack height)

Figure 4.76 shows a 200mm sine bar with a slip pack of 114.72mm being used to raise it to an angle of 35°.

Figure 4.76 Sine bar set to 35° inclination

Example 2 (imperial)

To identify the slip gauge pack height to produce an angle of 35° using a 6-inch sine bar:

Sin 35° = 0.5736 (derived from calculator or tables)
Length of sine bar = 6″
6″ × 0.5736 = 3.4416″ (slip gauge pack height)

Sine bars work horizontally as well as vertically and all they require is a plane surface to work off. Figure 4.77 shows a workpiece being clamped to an angle plate prior to milling, with a sine bar being used to support the workpiece at a known angle.

Similarly sine bars can be used to locate work holding devices at a know angle to facilitate milling. Figure 4.78 shows an angle plate to be fastened

Figure 4.77 Sine bar used support workpiece at known angle while being clamped

Figure 4.78 Sine bar being used to mount angle plate at angle to machine table

to the machine table at an angle to the direction of feed in order to mount a workpiece such that an angled cut is achieved.

4.34 DIGITAL READ OUT SYSTEMS

Digital read out (DRO) systems are widely used in general machining workshops. It is unusual in recent times to purchase a machine that is not fitted with a system, and many machines have been retro fitted with DRO systems to assist both machining accuracy and productivity. A DRO system does not allow an operator to do anything that could not be accomplished by other means, however, it does significantly speed up operations, especially for accurate incremental moves are made which otherwise would involve packs of slip gauges and DTIs.

The most basic systems will give x-axis and y-axis displacements. They all have the ability to set a zero datum on the edge of a piece of work. More advanced systems will have built in calculators, sub-datum functions to allow multiple datum positions to be entered and stored, line hole positioning functions, inclined surface datum positioning, pitch circle diameter (PCD) position functions, and Arc functions. Some systems are also fitted with a z-axis readout, and almost all have the ability to switch between imperial and metric units.

The accuracy of the system is ultimately constrained by the type and resolution of the system encoders being used, however, the readouts typically will give values to 0.0005mm.

While the more advanced systems provide the flexibility for complex operations such as Arc function, the majority of operations require only the setting of an x-axis and y-axis datum on a selected datum edge of the workpiece and the simplest of systems allow this. To set a datum a centre finder or cutter is placed in the machine spindle. If a cutter is being used then it is normal for a thin paper such as damp cigarette paper to be attached to the datum surface. The table is moved such that centre finder touches the datum surface and displaces, or the cigarette paper is ejected from between the cutter and the workpiece. The table is the dropped down to avoid the cutter or centre finder impacting the workpiece, and the machine stopped. The table is then moved half the diameter of the cutter and thickness of any paper, or half the diameter of the centre finder ball, and the DRO x-axis or y-axis scale zero button pressed to zero the scale on the DRO. This zeroes the scale with the centreline of the machine spindle directly over the edge of the workpiece and is a set datum. Once both x-axis and y-axis datums are set the operator has a known position to make translational movements from.

An alternative to setting a datum on the edges of a workpiece, is to set the datum on the workpiece centre. The reasons for doing this are normally set by the component drawing and its dimensions, however, this can often be pragmatic where the workpiece outer edges are not clean or clearly defined. I this case the above process is essentially followed, however, each side is found either using papers or a centre finder. The first edge is touched and the x-axis or y-axis scale on the DRO is zeroed. The table is dropped, the cutter or centre finder moved to the opposite side of the workpiece, the table raised and the edge finding process repeated. The distance traversed displayed on the DRO readout is halved, either by using the calculator function on the DRO or operator's calculator, or by mental arithmetic. The table is dropped to clear the cutter or centre finder form the workpiece surface, and the table moved until the value calculated is displayed on the DRO. The spindle is then located on the centreline of the workpiece for that axis, and the axis display is zeroed by pressing the readout zero button. The process is then repeated for the other axis and when both scales show a zero value the machine spindle is over the centre of the workpiece.

Setting a z-axis for DRO scales which have this function cannot be achieved with a centre finder, however, the use of papers is a good alternative. The difficulty the operator will have with setting z-axis datums is that it will only apply to the tool length being set. If the operator changes the tool, it will almost certainly have a different length from the previous cutter and the axis datum will need to be re-set. This does not apply to the x-axis or y-axis settings as the datum line as established by the method above is based on the centre of the spindle being set on a datum edge, and will remain a datum irrespective of the diameter of cutter inserted.

The operation manual for the DRO system mounted on any machine will give details of how to set the more complex functions, such as sub-datums, PCD, and hole line offset features, and the approach varies depending on the model. Operators need to familiarise themselves with the functions and operations of each type.

When machining a pocket or feature within a workpiece the operator needs to identify the distance of translational movement that need to be made along the upper surface of the workpiece for both the x-axis and y-axis, and then the pocket length and width, having made an allowance for the diameter of the cutter. The translational movement that needs to be made from the datum edge will be the dimension stated on the component drawing, *plus* half the diameter of the cutter. The distance to be moved to form the pocket will be the dimension stated on the component drawing *minus* the full diameter of the cutter as half the cutter diameter is already in the workpiece and the datum is on the cutter centre line so half the cutter diameter has to be allowed for on the other side of the pocket. These parameters need to be established for both horizontal axis, whereas a z-axis datum only needs to be established, and the scale on the DRO zeroed or the calibrated dial on the machine knee control handle set to zero.

Example 1 – basic pocket (datum in pocket corner)

To machine a pocket 50mm long by 40mm wide set 25mm in from the x-axis datum edge, and 15mm in from the y-axis datum edge. Having established the x-axis on the right-hand side of the workpiece, the y-axis datum edge on the front of the workpiece, and zeroed the DRO readout scales. If using a ∅16mm slot drill the following translational movements need to be made to position the cutter ready to plunge into the workpiece.

1 For the x-axis, move the table a distance of:

   25mm (dist of pocket from the datum edge) + 8mm (∅16 cutter ÷ 2) = 33mm

2 For the y-axis, move the table a distance of:

   15mm (dist of pocket from the datum edge) + 8mm (∅16 cutter ÷ 2) = 23mm

The DRO readout can then be zeroed for the x-axis and y-axis scales, and the operator then calculates the pocket sizes.

3 For the x-axis, the table will need to be moved a distance of the following to form a pocket:

   50mm (Length of pocket on x-axis) – 16mm (∅16 cutter) = 34mm

4 For the y-axis, the table will need to be moved a distance of the following to form a pocket:

   40mm (dist of pocket from the datum edge) – 16mm (∅16 cutter) = 24mm

It is common for operators to write these translational distances on the machine using a wipeable marker, as most DRO systems do not have a readout for them.

This method is simple where only one pocket has to be machined, and re-establishing the original datums is a simple subtraction exercise, but also does not provide for any finishing cut, and has the cutter plunge into a pocket corner which can, depending on the material leave a slightly lobed, and potentially out of tolerance corner. An alternative to this approach is to find the pocket centre, and either set that as the datum or (ideally) set a sub datum where the DRO functionality allows.

Example 2 – Pocket (datum in pocket centre)

To machine a pocket 50mm long by 40mm wide set 25mm in from the x-axis datum edge, and 15mm in from the y-axis datum edge. Having established the x-axis on the right-hand side of the workpiece, the y-axis datum edge on the front of the workpiece, and zeroed the DRO readout scales. If using a ⌀16mm slot drill the following translational movements need to be made to position the cutter ready to plunge into the workpiece in the centre of the pocket.

1 For the x-axis, move the table a distance of:

25mm (dist of pocket from the datum) + 8mm (⌀16 cutter ÷ 2) + 25mm (50mm ÷ 2) = 58mm

2 For the y-axis, move the table a distance of:

15mm (dist of pocket from the datum) + 8mm (⌀16 cutter ÷ 2) + 20mm (40mm ÷ 2) = 43mm

The DRO readout can then be zeroed for the x-axis and y-axis scales, or a sub-datum set and the operator then calculates the pocket sizes.

3 For the x-axis, the table will need to be moved a distance left and right of the sub-datum of the following to form a pocket:

25mm (Length of pocket from centre of sub-datum on x-axis) – 8mm (⌀16 cutter ÷ 2) = 17mm

4 For the y-axis, the table will need to be moved a distance of the following to form a pocket:

20mm (Width of pocket from centre of sub-datum on y-axis) – 8mm (⌀16 cutter ÷ 2) = 12mm

---

It is common for operators to write these translational distances on the machine using a wipeable marker, as most DRO systems do not have a readout for them. The operator, then knows he has to traverse the table left and right of the pocket centre by 17mm on the x-axis, and forward and backwards by 12mm on the y-axis. The advantage of this approach is that where the DRO system has a sub-datum function it means the original datum's do not have to be recalculated, but more importantly the operator will find it simpler to leave material for a finish cut without having to calculate odd sizes, and avoid any issues relating to machining eccentricity when plunging into a pocket corner.

DRO systems provide the operator with an accurate real time readout of where the tool centre or cutting face is. They require skill and practice, especially where advanced features are available, however, they remove the need for mechanical forms of measurement, especially when checking the size of a workpiece between cuts.

Chapter 5

Surface grinding

The process of grinding can be defined as to 'sharpen or smooth by friction'. This is perhaps a little misleading as a grinding wheel is made up of a series of minute abrasive grains, each of which behaves like a cutting tool and removes material from the workpiece. These grains are retained in a revolving disc by a vitreous bonding agent which weakens during use until it cannot support the abrasive grain, which breaks away. The benefit of this is that the effective cutting edge is replaced, and to a degree grinding wheels can be considered self-sharpening, and leaves a fresh cutting grain to remove material from the workpiece surface.

The process of surface grinding provides a flat workpiece with a high surface finish, which is generated by passing the workpiece under a rapidly rotating grinding wheel. The depth of cut and effective material rate is not high, however, wheel material types, grain size and feed rate can alter the material removal rate, albeit with an effect of surface finish.

Surface grinding is the process by which a rotating grinding wheel has a workpiece passed under it, to produce a flat ground surface. The grinding wheels used are normally plain wheels, which are commonly ⌀200mm and have a width of approximately 20mm. While many types of wheel are available this type is the predominant one used in surface grinding.

5.1 THE SURFACE GRINDER

The traditional workshop surface grinder consists of a machine table that can rapidly oscillate in the x-axis, traverse in the y-axis, with a machine grinding head that moves in the z-axis rather than the machine table moving up and down a column on a knee. This is to ensure machine stability given the precision of the cutting operation.

The table x-axis traverse mechanism is controlled by use of a hand wheel with a relatively high gear ratio, or by an automatic feed system. The y-axis traverse mechanism is also by a hand wheel but is much lower geared than the x-axis. Where a power feed mechanism is provided for the machine the y-axis will also have a reversible traverse feed mechanism. While some

DOI: 10.1201/9780429298196-5

production machinery may have a feed on the grinding head this would be exceptional in a workshop environment. The head which mounts the machine spindle and grinding wheel is brought down incrementally using a low geared hand wheel, normally with a fine adjustment wheel attached. For machines that have a power feed system there is often a z-axis rapid feed mechanism, but this would never be used to feed the wheel into the workpiece. Figure 5.1 shows a typical workshop surface grinder for general grinding operations.

Workpieces are commonly mounted on a magnetic vice, however, non-ferrous workpieces need to be mounted in a vice or other work holding device. The difficulty here, is that any clamping mechanism raised above the surface of the workpiece being ground becomes a potential obstruction to the wheel or machine head assembly.

Grinding is a process that can be undertaken wet or dry. Machines, even the most basic often have a facility for liquid cooling, and while coolants bearing fat-based emulsifiers are generally not recommended for grinding, many operators use the same coolants used in their other machinery. It is only when, high volumes of work, special workpiece materials, grinding wheel types, or surface finish requirements demand the use of discrete coolants. Coolants are particularly useful when grinding thin sections as it prevents bending due to heat. It is also not uncommon for local extraction and ventilation (LEV) systems to be fitted where grinding is undertaken dry. This is to ensure that the operators do not ingest grinding residues, and over the long term prevent harm.

5.2 GRINDING WHEEL TYPES

A grinding wheel, is one group of a set of materials and products know as abrasives. In general workshops grinding wheels used for surface grinding have an abrasive substance held within a vitrified bond. The manufacture of the grinding wheels involves the abrasive and bonding materials being mixed and then fired in a furnace to give hard strong, but brittle structure. The wheels require dressing with a diamond to form a plane flat surface prior to use but hold their shape during use.

Modern grinding wheel materials are synthetic, with the vast majority being silicon carbide or aluminium oxide locked into a vitreous bond. These materials have almost supplanted carborundum or diamond for workshop grinding, although diamond is still retained for some more specialist grinding operations especially those involving grinding of very hard materials such as carbides. The advantage of silicon carbide and aluminium oxide grinding wheels is their hardness and brittleness which gives them a friability. It is this property that causes grains to break off presenting a new sharp edge to the workpiece, and they are the type generally used in surface grinding, with tougher abrasive materials and bonding agents being used for hand grinding or off hand grinding applications.

Figure 5.1 Typical workshop surface grinder

Aluminium oxide wheels are likely to be the most common type selected by an operator as they are best suited to grinding of a wide range of steels and can accommodate material conditions from fully annealed to hardened. Silicon carbide wheels are more likely to be used when grinding non-metallic materials, non-ferrous alloys, non-magnetic stainless steels, and cast irons. Silicon carbide is also known to chemically react with ferrous metals.

Grinding wheels are supplied in a number of grades, each of which have differing properties. There are five main properties that affect the performance of the wheel when in use, as follows:

- **Abrasive** – the type of abrasive material used (i.e. silicon carbide/aluminium oxide etc.).
- **Grain** – the particle size of the abrasive grain on a scale of 4 (very coarse) to 1200 (very fine). The grain size relates to the application it is to employed in. Coarse grains tend to be used for rapid material removal grinding operations such as plough grinding. The surface finish is proportionately rougher than that for surface grinding operations, where the grain size will normally be much smaller giving a slower material removal rate but a very good surface finish.
- **Grade** – graded soft to hard, representing the tenacity of the bonding material to retain the grain. The grade is an indication of the strength of the bond that is retaining the abrasive particles, with an A grade indicating the softest bond and a Z grade indicating the hardest bond. Where the abrasive material is less likely to fracture presenting a sharp cutting edge, a weaker bond is useful as it allows the grain which may have a dull cutting edge due to friction with the workpiece surface to break away presenting a new cutting edge. Where the abrasive easily fractures a harder bond is appropriate, to retain the abrasive and improve the wheel life.
- **Structure** – the porosity of the wheel, with higher numbers denoting higher porosity. The significance of porosity is that the grains that are held in place by the bond do not fill all of the space and there are gaps between each grain and its supporting bond. These gaps allow the heat to dissipate, allow for coolant to circulate and for the temporary storage of swarf chips. When selecting a coarse grade of wheel which will cut thicker chips, the porosity of the wheel will need to be greater.
- **Bond** – the bonding material used in the wheel construction. There are seven generic types, and while a resinoid bond cutting wheel is sometimes used on a surface grinder for cutting off very hard materials the normal bond for surface grinding is vitrified.

One limitation of all grinding wheels is their peripheral speed. All wheels are manufactured to a maximum speed which is expressed either as a peripheral surface speed in surface metres per min or surface feet per minute. Alternatively it will clearly state a value for its maximum RPM. Operators when selecting a wheel should ensure that the machine speed does not exceed the maximum RPM for the wheel.

There are a number of national standards and an international standard which relate to the marking of abrasive wheels, however, the information that is carried on them is universal. All wheels will have the following information marked on them:

- company or brand name;
- lot no. or traceability code;
- dimensions of the wheel;
- maximum operating speed, stated as RPM or peripheral speed in surface m/min; and
- wheel specification data.

The wheel specification marking gives the operator exact information about the wheel. All wheels have a series of letters and numbers to inform the operator of the abrasive type, grain size, grade (hardness), structure (porosity), and bond type (i.e. vitreous). Figure 5.2 shows a typical grinding wheel.

The label on the grinding wheel shows the manufacturer's coding and from this wheel it can be seen that 180 × 20 × 31.75 refers to the wheel sizes:

Figure 5.2 Typical grinding wheel for surface grinder

- 180 = wheel diameter in mm.
- 20 = wheel thickness in mm.
- 31.75 = spindle hole diameter in mm.

2 A 60 L5 VS 8 are the wheel property information:

- 2 is the manufacturers prefix for the exact type of abrasive.
- A is the abrasive type, in this case aluminium oxide.
- 60 is the grain size (on the border between being classed as medium and fine).
- L is the grade of hardness on a scale between A (soft) and Z (hard).
- V denotes the type of bond (in this case vitrified silicate).
- 8 is the manufacturer's marking to identify the wheel.

Grinding wheels are also supplied in a series of colours. Aluminium Oxide wheels are found in white, pink, ruby red, brown or grey colours. Each of the wheel colours indicate grinding characteristics with grey and brown wheels tending to be used in heavier production grinding or surface grinders on low to high carbon steels. Pink and white wheels are generally used on harder steels which need cooling and friable properties to avoid burns to the surface of the workpiece. Wheels with a ruby red grit are usually reserved for grinding of tool steels. Silicon carbide wheels are normally found in black or green colours. Black silicon carbide has a very sharp grit and is normally used to grind non-ferrous metals as well as plastic and some rubbers. Green silicon carbide abrasive is sharper than black and is commonly used for the grinding of carbides, titanium and materials that have be deposited via a plasma spraying process.

When an operator is considering selecting a wheel type for surface grinding, detailed information is available on manufacturers websites and in good quality workshop handbooks. However, in broad terms, aluminium oxide is used for grinding high tensile strength steels and materials, with silicon carbide being used for lower tensile strength steels, cast irons and non-ferrous metals. Coarser grain sizes are used for rapid removal of materials and finer grits are used for improved surface finishes. Softer grades of bond are used when the machine power is low or good surface finish is important and harder grades where maximum wheel life is important and finish requirements are less crucial.

5.3 WHEEL MOUNTING

Mounting of a wheel on a surface grinder is not especially complex but requires some care. Nearly all surface grinders require the dismantling of the wheel guard to expose the existing wheel and machine spindle. It is good safety practice to electrically isolate the machine to prevent inadvertent operation while holding the grinding wheel. Wheels are normally mounted on a spindle clamped between two spindle retaining plates or cheek pieces, as shown in Figure 5.3.

The outer spindle retaining plate is unscrewed using a peg spanner and being careful not to dislodge any balance weights. The old wheel is removed and any residual dust brushed out. Grinding wheels will have a centre hole matched to the machine spindle size and they are lightly pushed into place. It is important for the operator to check that the paper discs with manufacturers details and wheel grades remain attached on both sides as these are a pad between the hard metal surface of the spindle cheek pieces and the hard brittle wheel substrate. The operator should also check that the maximum RPM for the wheel being mounted is more than the maximum RPM of the machine.

Where the removed wheel is worn out or damaged it should be smashed to prevent it being reused. The operator should spin the wheel by hand to check for free rotation, and the guard reassembled. The wheel is the ready for dressing, and use.

Figure 5.3 Surface grinding wheel spindle assembly

5.4 WHEEL DRESSING

Dressing of a grinding wheel on a surface grinder refers to the process of moving an industrial diamond across the periphery of the grinding wheel to provide a plane surface that is concentric with the machine spindle axis. When a wheel has first been mounted on a machine it will requires dressing, effectively a precise 'truing' process to ensure that it is perfectly concentric with the machine spindle it has been mounted on. After this the wheel will require periodic dressing.

In a perfect environment a wheel would never require dressing, as the friability of the abrasive, or the loosening of the bond would constantly provide a sharp fresh cutting edge to the workpiece, until the wheel diameter was reduced to a small diameter, and the peripheral speed reduced below that required. However, the properties of the material being ground, the depth of cut, the use or failure to use coolant, the grinding wheel type being used and the type of grinding being undertaken often means that the wheel requires periodic dressing.

In particular using the wrong type of wheel for the material being ground has a significant effect on the regularity of dressing, as does operations where surface scale or heat treatment debris needs to be penetrated, and in this situation it is not uncommon for the operator to have to undertake frequent dressing of the wheel until clean material is exposed.

Dressing, and truing, is undertaken using a single point diamond, the size of the diamond. A diamond, normally purchased with it retained in a mild steel bar is itself mounted in a steel housing. This housing should be angled approximately 10–15° down in the direction of the rotation of the wheel to maximise its life, and ideally, also 10–15° to the axis of the machine spindle. It is clamped onto the magnetic vice, or other work holding device ready to be used. The table is positioned such that when the diamond touches the wheel it has a 10–15° angle. Severe angles in relation to the curved surface of the wheel and diamond run the risk of the wheel surface interfering with the diamond retaining bar rather than the diamond. Figure 5.4 shows a diamond dressing bar presented to the wheel for dressing.

Prior to starting the machine it is essential for the operator to check that any automatic table feed has been left engaged. The result of this being engaged is to immediately drive the diamond into the grinding wheel leading to either a wheel that requires significant dressing to return to condition, damages the wheel requiring its replacement or ejection of the dressing device. Either of the last two outcomes represent an unsafe condition, and may result in the operator experiencing a wheel burst at some future point, or being impacted by the ejected dressing device.

Having started the machine the diamond is advanced into the grinding wheel. For newly mounted wheels it is normal for the operator to hear that the wheel is on being dressed for a portion of the periphery, and as the

Figure 5.4 Surface grinder wheel dressing

dressing process advances it will become apparent that the diamond is in contact with the wheel all around its periphery.

The process of dressing a grinding wheel takes care, and time. Effective dressing will ensure that the wheel is in an optimum condition, for material removal and production of a high-quality surface finish. The depth of cut for each pass should be no more than 0.025mm (0.001″) and for fine grain wheels or those being used for high finish grinding operations limited to no more than 0.010mm (0.0004″). The cross feed rate is similarly controlled as it is important to avoid having a spiral pattern dressed into the wheel. I broad terms coarser grain grinding wheels can cope with faster cross feed, and a finer grain wheel will require a slower cross feed. What limits the speed of cross feed when dressing the wheel is the grain size. When wheel dressing the operator is trying to ensure that there is a plain flat debris free surface across the diameter of the wheel. The diamond needs to engage with all of the grains around the periphery of the wheel across the whole surface to ensure it is ready for grinding a workpiece. This means that the cross feed velocity when dressing is going to be a function of the wheel diameter, machine RPM and grain size. The general rule is that the cross feed rate should not exceed the diameter of a grain, or be less than half of the grain size. This can still be relatively quick on standard ∅200mm surface grinding wheels where spindle speeds of 2500+rpm are normal and data derived from workshop handbooks suggest that for a 19mm wide wheel the diamond should be fed across the width of the wheel in 1.6 seconds. This is probably hard for the operator to achieve by hand, however, the important aspect of wheel dressing is being gentle. The diamond should be carefully introduced to the wheel using the table hand wheel until it is just at the surface of the grinding wheel. The cut is applied by bringing the head down by rotating the z-axis fine adjustment wheel. When the diamond can be heard to touch the wheel the diamond is passed across the surface of the wheel using the y-axis hand wheel. A series of very light cuts are undertaken with each one conducted in one smooth pass across the face of the wheel.

Sequential cuts are taken until the wheel is clean, and any grinding debris is removed. This is quite easy for the operator to see and as the diamond passes over the surface of the wheel the colour lightens until a clean, new surface is presented. On initial use, or after any impact it is likely that the wheel will not be true. In this case dressing should continue until the diamond can be heard to be cutting continuously around the periphery, and the surface is clean.

5.5 THE MAGNETIC VICE

Secure work holding is an important aspect of successful surface grinding. The most common form of work holding on a surface grinder is a magnetic chuck or table. These are precisely ground pieces of tooling that have been produced to a high squareness tolerance, and a high flatness tolerance. They have a permanent magnet that can be switched on or off, normally through turning a simple lever through 180°. Figure 5.5 shows a common magnetic vice fastened to a surface grinder.

The advantage of using this sort of work holding device is that the work-piece has no clamping equipment that may temporarily distort the workpiece, restrict the area to be ground or have a potential collision issue with a part of the grinding machine while the workpiece is being passed under the grinding wheel. When the magnet is energised it provides a substantial clamping force, which for most grinding operations is more than enough to firmly retain the workpiece while grinding is taking place. The upper surface shows a series rectangular sections, and it is over these sections that provide the magnetic clamping force.

Most magnetic vices have a skirt that can be raised or lowered on at least one side or end. This allows for a workpiece to be simply aligned with the edge of the vice. When mounting a magnetic vice onto the bed of a surface

Figure 5.5 Magnetic vice/table

grinder, it is crucial that the alignment be as accurate as can be achieved. Surface grinding is a precision process, and normally undertaken when significant value has already been added to a workpiece. When grinding to a shoulder, and error in alignment of the work holding device will result in the

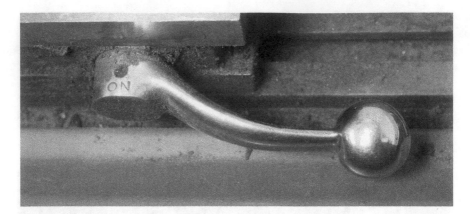

Figure 5.6 Magnetic vice operating handle shown in 'on' position.

grinding wheel digging into the shoulder. With typical workshop magnetic vices being 400–600mm long, it is crucial that the vice is 'clocked in' to the limit of accuracy of the DTI being used. Once this is achieved the operator has confidence that any component butted up against the vice skirt will be aligned with the machine table axis.

The upper surface of the magnetic vice needs to be carefully looked after especially to ensure that it does not suffer from raised burrs due to impact as these prevent a workpiece from being flat and lead to grinding errors. One normal method for the upper surface to suffer damage is by the magnetic not being energised by the operator not rotating the magnet control handle. It is a good idea for operators to check that the handle is shown in the 'on' position, to prevent the workpiece from being ejected from the machine, as shown in Figure 5.6.

Other than protecting the upper surface of the vice, these work holding devices require no maintenance other than light corrosion protection. They are, however, only useful for ferrous materials. Work holding of non-ferrous materials, often still utilises a magnetic vice but the magnetic vice is used to clamp down small steel vices, such as a grinder universal vice, or other work holding devices.

5.6 SINE BAR USE

Sine bar magnetic vices are widely available and commonly used in surface grinding. The precision that can be established for workpiece angles, and compound angles is essential when gauge making, or undertaking other types of precise work. The use of magnetic vices that can be set to a precise angle are an order of magnitude more accurate than other types of work holding device.

Magnetic vice sine bars consist of a standard body with a plane mag-netic surface which can be switched on an off. The lower part of the body is hinged in one or more commonly, two directions allowing the vice to be tilted to a known angle on both the x-axis and the y-axis. This angle is set by inserting a pack of slip gauges between each axis roller and its base. Where two axes are used a compound angle is formed, however, they can be used to provide a workpiece angle individually on the x-axis or the y-axis.

The distance these rollers are apart are critical to the accuracy of the sine bar, and the bars themselves come in a variety of lengths. Typical sizes for grinding tables are 250mm × 125mm or their imperial equivalents. The length of a sine bar is that between the centres of the two rollers, and not the overall length of the sine bar. All sine bars will have their nominal length engraved or etched into the side of the body. Figure 5.7 shows a typical (imperial) sine bar sine table.

There is no difference between metric or imperial sine bars other than imperial sine bars require imperial gauge blocks (slip gauges), and metric sine bars require metric gauge blocks. It can be argued that metric slip gauge pack heights can be converted to imperial units and vice versa, but this introduces a potential opportunity for error.

Most sine bars are used in metrology labs or inspection departments, however, they have practical uses in the workshop environment where accu-rate set-up is key. The principle behind a sine bar is that of producing a right-angled triangle. A pack of slips is inserted between a plane surface and

Figure 5.7 10″ × 5″ magnetic sine bar grinding table

one roller of the sine bar. The height of this pack of slips is a function of the sine of the angle required. The sine of an angle in a right-angled triangle is calculated as follows:

$$\text{Sin}\,\Theta = \frac{\text{opposite}}{\text{hypotenuse}}$$

The hypotenuse of the triangle is known as this is the nominal length of the sine bar i.e. the distance between its two rollers and the required angle is generally as it is commonly a dimension on the component drawing known, therefore the unknown factor is the height of the opposite side, therefore:

Opposite = hypotenuse × sin Θ

The value from this gives the size a pack of slip gauges needs to be built up to in either mm or decimal inches. Once this has been established a pack of slip gauges is inserted between the plane surface on the machine table body for the axis, and the end roller to lift the sine bar up to an angle. Figure 5.8 shows a sine bar machine table elevated by insertion of a slip pack on the x-axis and y-axis for grinding a compound angle.

The angle clamping mechanism is then locked and the pack of slip gauges removed. The workpiece is then mounted onto the upper surface of the sine bar and the magnet switched on. This method is only recommended for

Figure 5.8 Sine bar set to inclined angle on x-axis and y-axis

angle up to 60° given the pack height, and where angles in excess of this are required the operator should consider side grinding, or mounting the work-piece in a vee block which is then located onto the sine bar machine table.

Example 1 (metric)

To identify the slip gauge pack height to produce an angle of 35° using a 250mm machine table sine bar:

Sin 35° = 0.5736 (derived from calculator or tables)
Length of sine bar = 250mm
250mm × 0.5736 = 143.4mm (slip gauge pack height)

A slip gauge pack of 143.4mm would be inserted between the machine table roller and base to elevate the magnetic surface by 35° the table locked in place and the slip pack removed. The process for the y-axis is similar although the length of the sine bar will almost certainly be different unless the sine bar machine table is square.

Example 2 (imperial)

To identify the slip gauge pack height to produce an angle of 35° using a 10-inch machine table sine bar:

Sin 35° = 0.5736 (derived from calculator or tables)
Length of sine bar = 10″
10″ × 0.5736 = 5.736″ (slip gauge pack height)

5.7 MACHINE OPERATION

Common workshop surface grinding machines will have an on/off switch for the wheel, and where an automatic feed is available, there is often a separate switch to power this. This is because the wheel head is often driven by a belt connected directly to an electric motor, whereas it is common for the feed mechanism to be hydraulically powered with a separate motor powering a pump. As an additional safety precaution, it is considered good practice to start the wheel, but not start the feed power until the wheel has been dressed to avoid any inadvertent initiation of feed.

Once the wheel has been dressed, the workpiece is placed on the magnetic vice, or retained with a work holding device. For tall workpieces this may require the head to be raised, and if complex support blocks are to be used the wheel should be stopped, to ensure the wheel does not come into contact with the operator's hands. Once the workpiece is in place, the operator should manually position the grinding wheel over the work piece using the x-axis and y-axis control wheels. The head is brought down under power feed until just above the workpiece. For machines without a power feed function the fine feed adjustment can normally be lifted up allowing the coarse feed handle to freely rotate, and manually bring the head down to the workpiece. Figure 5.9 shows the fine feed adjustment control lifted to allow free rotation of the hand wheel, and in the example used (which shows a Jones & Shipman 540 surface grinder), the power feed controls can be seen in the background.

It is crucial that the wheel does not come into contact with the workpiece when being rapidly feed down. When this occurs the wheel ploughs into the workpiece damaging it, and normally stops the wheel. The wheel will then

Figure 5.9 Raised fine control wheel on coarse feed hand wheel

have a flat spot on it and will be unbalanced. This then requires the grinding wheel to be re-dressed.

Once the head is close to the work piece the operator uses the fine feed to bring the head until it appears at the surface of the workpiece. The operator should then move the table backwards and forwards on the x-axis and traverse the table to begin to identify the high spot. This can be undertaken quite rapidly and after the first pass with no high spot identified the table traverse feed can be engaged providing a reciprocating action. The operator will then lower the head a small amount and traverse the table on the y-axis until spark down occurs. Spark down is the point where the grinding wheel connects with the highest point of the workpiece, and it is this point that the grinding process commences.

Surface grinding is one of the slowest metal removal processes, however, provides one of the highest surface finishes. This requires small depths of cut, employed repetitively. The depth of cut will vary, depending on the grade of wheel being used and the material being ground, however, it should be in the region of 0.025mm – 0.01mm (0.001″ – 0.0004″) for roughing cuts and 0.005mm for finishing.

Once spark down has been achieved and the depth of cut set the operator then ensures that the grinding wheel passes over the entire surface being ground, by leaving the x-axis table on its reciprocating feed and rotating the y-axis control hand wheel. The amount that is traversed should never be more than the width of the grinding wheel. For most standard situations where a 19mm (3/4″) wheel is being used this could be an entire rotation. Many operators use a half turn to give a degree of over cutting, however, if a narrow wheel is being used or the machine has a high geared traverse screw this may be too much and the operator should establish what the traverse distance of one rotation is prior to use. Many grinding wheel manufacturers and engineering data books give some guidance on traverse distances, and in general terms it is suggested that for roughing grinding operations for unhardened steels that the traverse distance but up to 75% of the wheel width, where for stainless or hardened steels this be no more than 50% of the wheel width. For finish grinding operations, traverse distance will be approximately 30–40% of the wheel width, with it being on 25% of the wheel width for harder materials.

Once spark down has been achieved, and the initial cut commenced, the operator can engage a cross feed. On machines that have this function, it merely involves the rotation of a switch to the right or the left to change the direction of feed. An example of this type of control can be seen in Figure 5.10 on a Jones & Shipman 540 grinder. The silver switch is the feed direction control, which is rotated right to drive the table away from the operator and to the left to advance it towards the operator. The central position is when the feed is disengaged. The plastic thumb wheel is to lock the table cross feed in position.

Figure 5.10 Cross feed mechanism

The operator merely then needs to watch the cross feed until the wheel begins to leave the workpiece, at which the cross feed is returned to the off position, the next cut applied and it rotated to the opposite side, beginning the process again. The length of feed is normally set by a cam, which can be adjusted by the operator. Each model of grinder is different and the operator needs to consult the user manual for information on how to adjust the stroke length.

Initial cuts often do not seem to provide much gain with few sparks being seen other than at the high spot. As sequential cuts are made, if the surface being ground has been well machined the area being ground rapidly expands until the wheel will have a continuous sparking action throughout each pass. For general surface grinding to provide a flat smooth surface grinding only continues until the operator can see no un-ground areas on the surface. While cast irons and many types of steel do not tend to clog the grinding wheel in use, if a lot of material is to be removed, or materials that do tend to clog the wheel are being ground the wheel will require re-dressing. It is fairly obvious when a wheel requires this, as the sound changes, the appearance of the surface changes, the surface of the wheel becomes discoloured (normally black) the sparking changes and there may be some burning to the surface. Taking a cut that is too heavy accelerates the requirement for dressing. If the operator has positioned his workpiece towards an end of the magnetic vice or it is narrow, it is common for the workpiece to not be removed and the dressing diamond positioned adjacent to the workpiece. If removal of the workpiece is required this rarely causes a problem when replaced as long as the vice is wiped down. It is normal for the workpiece to be wiped down after re-dressing if it has not been removed, and coolant is not being used to remove any abrasive debris. Care should be taken to avoid contact with the grinding wheel and the use of a soft brush recommended. Dressing does not remove a significant amount of the wheel's periphery, but the operator will need to spark down again.

Where, a high surface finish is required, it is normal for the wheel to be dressed after roughing cuts and a series of very fine cuts of approximately 0.005 mm applied. These cuts continue until the workpiece has been ground to size, or the standard of finish achieved. The final cut is continued back, and forth until 'spark out' is achieved. This is where no more sparks can be seen to be coming from the between the grinding wheel and the workpiece, and the process is complete.

5.8 SETTING FEED STOPS FOR LONGITUDINAL FEED

Most grinding machines that are set up for grinding under power feed will have a fixed stroke length, and an adjustable feed rate. The feed rate on many workshop machines provides only basic adjustment and while optimal feed rates linked to material removal rates can be calculated, it is not common for machines to offer more than a basic adjustment. However the basic stroke length of the table is set for the maximum stroke of the machine. This is not a particular problem, but as surface grinding is not an especially quick process, it is useful if adjustable trips can be set to reverse the table direction as soon as the grinding wheel has passed over the workpiece to reduce any non-contact time.

Again differing machines will have different approaches to the way that reversing mechanisms are adjusted. One common method is the setting of mechanical trips. These trips are simple blocks of metal that are fastened to a track and retained in position by a screw. Loosening the screw allows the block to be moved up and down the track. These blocks then trip a lever as the table moves down the machine reversing the direction of the table. There are two blocks one at each end of the track to provide a reversing action on each stroke. Figure 5.11 shows one such trip block mounted on a Jones & Shipman 540 grinder adjacent to its feed reversing lever.

To set the block positions the block is loosened and the table moved by hand to a position where the grinding wheel is just off the workpiece. The reversing lever is set to the trip position and the block moved into contact with the trip face and the locking screw tightened as shown in Figure 5.12.

Figure 5.11 Feed trip block and reversing lever

Figure 5.12 Feed trip block being set in position

The other end is the adjusted in a similar manner. When in normal operation the stroke to the table feed will then reverse as the wheel exits the workpiece, substantially reducing the time for each stroke. Surface grinding by its nature is a relatively slow process compared to other machining operations, therefore, any reduction of non-material contact time will substantially reduce the time taken to grind a surface.

Adjustment of the cross feed stroke mechanism is more complex and rarely undertaken, as the incremental step is normally less than the width of a grinding wheel. Again the type of mechanism varies with respect to machine manufacturer, however, a lever drive mechanism with an adjustable pin fitted into a slot is not uncommon. To adjust the length of stroke a locking screw is loosened and the lever moved closer to the centre of a cam to shorten the stroke, or out to the cam periphery to length the stroke. Once the set length is adjusted the screw is re-tightened and feed can be applied. Figure 5.13 shows one type of mechanism on a Jones & Shipman 540 grinder, and it can be seen that adjustment is more difficult than for longitudinal feed.

5.9 WORKPIECE SUPPORT

Workpiece support, and ensuring workpiece stability is crucial to successful grinding. While the emplacement of a steel workpiece onto a magnetic vice requires little consideration when a plane surface is to be ground on a workpiece with a substantial surface area, if a workpiece is to have a feature ground that is of an awkward shape, or has a small area in contact with the surface of the magnetic device, it will require some support. The key object of supporting a workpiece is to prevent lateral movement or rocking while being subjected to a grinding operation.

One approach is to utilise any skirt that may be fitted to any vice to prevent a workpiece from sliding off the vice during grinding. This requires a proportion of the workpiece to be over a part of the vice that is magnetised and while some models of may have a magnetic portion that goes to the edge of the vice it is common for models to have an un-magnetised portion between the edge of the vice and the central area. If the component is small it may not be possible to use a skirt.

Where workpiece configuration or geometry conspire to prevent firm work holding, support structures can be used. This is where blocks of steel, such as machine cubes, angle plates, vee blocks and sheet materials are used to support a block. Figure 5.14 shows and example of a set-up for grinding the end surface of a vee block. The workpiece is stood on its opposing end which does not have enough surface area to resist the overturning moment which will be applied by the grinding surface. The block is therefore being supported by the emplacement of another vee block behind it, and the emplacement of a round bar which engages in the vee of the workpiece. The round bar has been turned and faced on a lathe to ensure that the bar end is exactly square to the base to ensure good contact between the workpiece vee faces and the magnetic base. The bar itself has also been supported with a small vee block as the bar diameter is not significant compared to its height.

Figure 5.13 Cross feed adjustment mechanism

Where non-ferrous materials or complex shapes need to be secured prior to grinding, it is not uncommon for a plain or universal vice to be clamped onto the magnetic vice. These configurations generally require care in setting up, which can be a lengthy operation. If the vice has been specially manufactured

Figure 5.14 Vee block supported for grinding

for grinding then it will have a ground edge that is parallel to the fixed jaw, and ground sides. These sides can be pushed up against the skirt of the magnetic vice to ensure squareness. Where a standard machine vice, or universal vice is to be used there will be no ground datum edge and the workpiece will need to be clocked in using a DTI to ensure squareness with the x-axis or y-axis. However, vices have a tendency to bow or lift a workpiece from a flat plane and errors of flatness are especially prevalent when a universal vice is being utilised. The scale of a universal vice is rarely calibrated in anything lower than one degree increments, therefore, it is normal to have to clock in the upper surface of the workpiece to ensure it is as flat as required. Figure 5.15 shows a universal vice being used to mount a component.

Where possible it is useful if the operator can position and work holding device in a position on the magnetic vice that will still allow them to mount a dressing diamond directly to the magnetic vice. This will allow for dressing the wheel as required, without it being necessary to move the workpiece. While moving the workpiece off the machine when it is only a plane surface being ground is not a problem and it can be returned to continue grinding, if the set-up involves grinding to a shoulder then the alignment will have to be re-established.

Figure 5.15 Universal vice used for work holding on surface grinder

5.10 HEAT AND WORKPIECE WARPING

The process of cutting material in any machining operation results in the conversion of power used in cutting into heat. This is no different for grinding, as each grain of abrasive is itself a cutting tool. In conventional machining the heat generated passes into the tool as a well as the workpiece, and it is this mechanism and friction that dulls a cutting edge. In the case of a grinding wheel the cutting edge is maintained by the friability of the grain and the strength of the vitreous bond. Grinding wheels are also not good conductors of heat, and there is also frictional resistance from the workpiece when grinding leading to significant heat build-up. Unless coolant is being used, almost all the heat is transferred to the workpiece, although the breeze generated by the rotation of the grinding wheel does create some cooling effect.

Warping or bending due to frictional heat may not be immediately apparent, and tends only to be noticed when a component is removed from the vice, left to cool for a short period of time while the magnetic vice, or other work holding device is cleaned and the work piece remounted. There is an expectation when grinding that grinding an opposite surface, will be quicker that the first side as the ground face will not have a high spot, so the reverse side previously used to locate the workpiece should be more planar. However, if warping has occurred it will be an immediate and clear to the operator as spark down will occur at either both ends of the workpiece or in the centre. In extreme cases the cut depth can be significant and burning of the surface observed. If this occurs the grinding operation needs to be suspended the magnet released and the workpiece allowed to cool, if thin ends or a reduced central thickness is to be avoided.

For larger components heat transfer is not a significant problem as the surface area and mass of the workpiece allows for the dissipation of heat, however, thinner components can suffer from significant amounts of warping. Workpieces that have a significant surface area in comparison to their thickness are especially vulnerable, and this can continue for workpieces of substantial thickness, where they are of a large surface area. The use of substantial amounts of coolant is the most effective method for removing this problem, however, it is not a complete cure. The use of finer cuts is one method as larger cuts depths significantly builds up frictional heat. Another method is to pause to allow the workpiece to cool between each, or a series, of cuts. This significantly increases the time to grind a component, but where flatness or absence of warping is critical it may be a requirement.

5.11 GRINDING SLOTS

In many respects grinding a slot is little different than grinding a surface, although there is often a considerable degree of confinement and the added complexity of grinding the sides of a slot.

The type of slot to be ground and its width is the first aspect an operator will need to consider. While it may be tempting to grind a narrow slot to depth in one pass, this relies of there being zero wear on the wheel throughout the operation, and the slot generated will be the width of the wheel. If the depth of the slot is significant then the area being ground at any one time will be substantially more than for surface grinding, and the frictional resistance may be enough to stop the wheel. A more controlled approach is to use a wheel that is slightly narrower than the slot and relieved on both sides. Figure 5.16 shows the basic configuration of a relieved grinding wheel being used to grind the bottom and sides of a slot.

The width of the slot will drive the width of side dressing and relief. The length of the side dressing will be driven by the depth of the slot to be ground (i.e. the wheel needs to be relieved for the full depth of the slot plus some clearance). The grade of wheel selected for slot grinding tends to be harder than might ordinarily be chosen, as they have reduced wear characteristics compared to softer grades. While this can affect surface finish, side grinding will always produce a different surface finish than edge grinding, and the internal appearance of slots its generally less of an issue than plane surfaces.

The operator should select a grinding wheel to the nearest width to the slot. Operators will have a personal preference as to whether this should be larger or narrower than the slot. Substantially larger will result in a lot of edge grinding to bring the working section of the wheel to the correct width but will leave a more substantial centre. Narrower wheels will only require edge relief to be dressed into the wheel, however, very thin slots will often be narrower than the thinnest commercially available wheel. It is theoretically possible to dress a wheel down to three grains wide as grinding wheels are remarkably rigid, however, this type of grinding is specialised and the wheel will be incredibly fragile at this width.

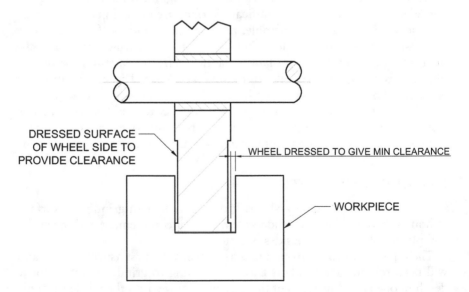

Figure 5.16 Schematic of side relieved grinding wheel for side and bottom grinding of slot

To grind a slot the wheel is mounted trued, or dressed as for surface grinding. Then both sides of the wheel need to be dressed and relieved. Dressing the sides of the wheel performs two functions. The first is to ensure that the sides of the wheel are 'trued' and are absolutely parallel, very much in the way that new grinding wheels are trued on their outside edge prior to use. To do this a standard dressing tool can be used, or better, a side dressing tool purchased specifically for side dressing of grinding wheels used.

Initial side dressing is intended to bring the wheel down to a thickness slightly less than the width of slot to be produced. The wheel is relieved on both sides to provide a depth of cut that is slightly more than the depth of cut required. The clearance is there to ensure the operator can clearly see the grinding operation and to prevent any inadvertent grinding of the outer plane surface by the remaining grinding wheel shoulder. It is quite normal for operators to dress the sides of the wheel such there is a graduated slope from the width required to the original width.

Following side dressing to width the grinding wheel needs to have a further dressing operation to provide a relief diameter. This is where more material is removed from the side of the wheel but leaves a ring of material at the dressed side at the periphery of the wheel. The width of the ring of material, and the depth of relief should be as small as possible, but enough to allow for re-dressing of the wheel if required during extended grinding operations. Figure 5.17 shows a sectional view of the desired wheel profile.

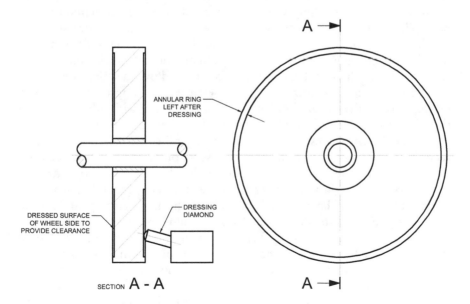

Figure 5.17 Sectional view of grinding wheel showing relieved sides

The relieved section performs two functions. The first is that it substantially reduces frictional drag by reducing the surface area in contact with the workpiece. In surface grinding the area in contact with the workpiece is tiny and limited to the depth of cut × the width of the wheel. In slot grinding there is this area in contact plus the area of the side of the wheel in contact. By relieving the sides of the wheel most of this contact area is removed, reducing the frictional drag. The second function is to maintain the effective peripheral cutting speed. Most materials are cut at a rate expressed as surface metres/min (SMM) or surface feet/min (SFM), as with any rotating tooling it is for a peripheral velocity that a spindle speed is set at. As the radial distance from a surface moves towards the centre the effective rotational velocity decreases until the centre is reached at which the velocity is zero. Therefore, the rotational velocity of the grinding wheel decreases towards the centre of the wheel. This is not normally a problem with surface grinding, as it is only the periphery of the wheel being used, however, when slot grinding the sides of the wheel are being used as well. By relieving the sides of the wheel the rotational velocity can only reduce to the effective inner diameter of the working rim. This preserves the surface finish produced and allows the grains to cut correctly rather than being dragged through the workpiece. The narrower the rim width is set to the closer the peripheral velocity and inner velocity will be. However, the width of rim needs to allow for the amount of dressing that is likely to be required, as if the rim is removed through over dressing the wheel periphery, then the wheel sides will need to be relieved again to create a new rim.

To side dress a grinding wheel and to create the relief a diamond must be applied to both sides of the wheel. When using a standard dressing tool the diamond is placed at 10–15° to the machine y-axis to preserve the diamond, and the magnetic vice locked. The diamond is brought into contact with the side of the wheel and material is removed over a series of passes to a required depth. Operators will have their preferred approach with some taking most of the material off one side of the wheel, with others reducing the wheel to a width equally from both sides. There are arguments for and against both approaches, but in general an equal reduction from both sides seems to be common, but at least one dressing operation needs to occur on each side of the wheel to 'true' it up. The width that this needs to be dressed to need only be slightly narrower than the slot being ground as it is only intended to remove the frictional forces from the area being ground and to give more control of the slot sizing. While it could be dressed to substantially narrower, this merely takes more time, provides a wheel with lower lateral strength, and reduces the amount of re-dressing, or relieving the wheel sides.

Following dressing of the sides the relief section needs to be generated. This is undertaken by plunging the diamond into the wheel and using the x-axis hand feed to carefully remove material until a rim is created. While some operators like to do this by hand, the use of a diamond fastened to the table makes it a more consistent relief, especially when a deep slot is being formed. The width of the rim being left should be as narrow as possible, but allows for re-dressing, while retaining peripheral speed. This process should be undertaken on both sides of the wheel.

Once the grinding wheel has been prepared the workpiece is mounted onto the machine. It is particularly important to ensure that the slot to be ground is parallel to the machine table axis to avoid damage to the wheel, given its reduced lateral strength, and to ensure accurate grinding of a pre-machined slot.

Grinding a narrow slot has four considerations which are:

- slot width;
- slot depth;
- squareness of the slot bottom corners; and
- surface finish of the slot sides.

It is normal for the to pass the wheel manually through the slot by hand to check that the wheel is clear of the sides before engaging any automatic feed. The operator then proceeds to spark down in the normal manner. To preserve the bottom corner it is usual for the operator to then traverse the table until one side of the slot incurs spark down. The operator should apply very small incremental cuts to avoid any deflection of the wheel, and while the relief of the wheel means that it only cuts at the periphery the grinding wheel does not enjoy excessive side force applied. Where fitted, utilising a fine feed mechanism as shown in Figure 5.18 (on a Jones & Shipman 540 surface grinder) allows good control.

The slot side is then ground until the required surface is attained, and/or the distance of the slot wall from a datum edge is achieved. The table is then traversed until the other side of the grinding wheel sparks down on the slot wall and the incremental grinding process is applied until the desired slot width, or distance from a datum edge is generated. The relief of the wheel and the narrow width of the side cutting edge reduces scratching of the slot wall and provides a good surface finish.

The operator has a choice of whether to grind to depth before traversing to grind the slot side walls, or to grind a side wall and then progress to depth before traversing to grind the opposite wall, or to grind the side walls and then go to depth. There is no one specific method, however, the operator is looking to maintain a square or unworn corner profile of the wheel. The slot being ground will normally have a max corner radius stated on the component drawing, and this can be a determinant of the approach taken. Dressing a wheel using translational movements will produce an almost perfectly square corner, however, square corners are susceptible to wear. While wear can be accommodated in the corner radius tolerance, where this is a very small allowance it is likely to be a better approach to spark down in

Figure 5.18 Fine traverse control wheel

the bottom of the slot and to then traverse to grind the slot wall without significant edge cutting at the corner. The slot is then ground to depth with the majority of any wear being radial, protecting the corner and only requiring minor side movement to provide a plane bottom to the slot.

Measurement of these slots is most easily, and accurately achieved using gauge blocks (slips). The table can be stopped reciprocating, and the width checked during grinding operations, with a high degree of accuracy using slips, and the remaining curt width identified. Slot depths can be determined using a depth micrometer, or slip gauges and a DTI used as a comparator. However, unless ceramic slip gauges are being used and a non-magnetic DTI base any vice magnet will need to be released during measurement.

The use of coolant assists in reducing heat and washing away grinding products which in some circumstances substantially improve surface finishes achieved on slot side walls. Where coolant provision is available operators are advised to use it, however, the lack of provision does not prevent a closely ground slot being achieved.

5.12 GRINDING TO A SHOULDER

Grinding to a shoulder has much similarity with slot grinding. It differs from slot grinding in that it is normal for only on side of the wheel to be relieved, and it is quite common for the workpiece to be held at an angle of 45° and the periphery of the grinding wheel to be ground at 45°. The advantage of this is that the operator has more control of the depth of cut allowing progression into a corner. This is due to the effective depth progression being 0.7 of the depth set on the head height adjustment due to the angle of the workpiece. The operator also good visibility. The disadvantage of this approach is that the workpiece set-up is more complex and requires the use of a sine table, the use of an angle wheel dressing tool, and a very fragile grinding wheel corner. However, when grinding a shoulder that has an undercut there are clear advantages of using this approach.

Dressing a wheel using an angled dressing tool requires a specialist dressing tool to be purchased, setting it to a precise angle and applying it to the periphery of the grinding wheel until the required angle is produced. On the that is to be used to grind the shoulder side the wheel is relieved in the same manner as for slot grinding, but only on the side to be used. Figure 5.19 shows a diagram of the set-up for grinding of a shoulder at an 45° angle.

Where a shoulder is not undercut a more conventional approach can be taken. Again the grinding wheel preparation is undertaken in a similar

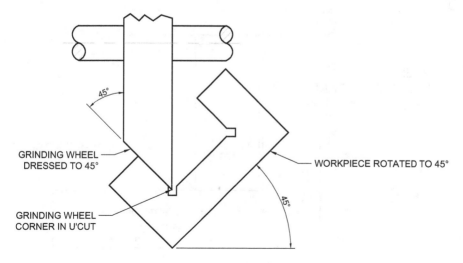

GRINDING WHEEL DRESSED TO 45°

WORKPIECE ROTATED TO 45°

GRINDING WHEEL CORNER IN U'CUT

45°

45°

Figure 5.19 Set-up for precise grinding of undercut shoulders

manner to a slot, although only the face engaging with the workpiece shoulder requires relief, and the width of the wheel is generally unimportant.

The workpiece is carefully aligned with the machine *x*-axis. It is normal for the wheel to be placed very close to the shoulder and spark down of the wheel periphery undertaken. The table is then carefully traversed to grind the shoulder. Where the workpiece shoulder has previously cut it may be possible to grind to depth and the traverse the wheel out of the work in a series of cuts. This approach preserves the corner of the profile of the wheel and erosion of the corner of the open side is less important as the flat periphery of the wheel removes any radius during successive cuts. Where a number of passes are required to bring the shoulder down to depth the operator is advised to return the table to adjacent to the shoulder before applying the next incremental cut to avoid cutting on the corner edge of the wheel, and continue to work out of the shoulder as shown in Figure 5.20.

5.13 CUTTING DEPTH

Grinding is not different from any other metal cutting process, in that depth of cut feed rate, tool life and surface finish are all considerations. Surface grinding is a process that is considerably slower to remove material than milling, turning or drilling, however, it provides substantially better surface

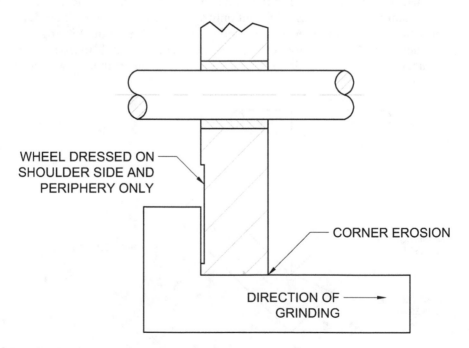

WHEEL DRESSED ON SHOULDER SIDE AND PERIPHERY ONLY

CORNER EROSION

DIRECTION OF GRINDING

Figure 5.20 Approach to 'working out' of a corner to preserve wheel corner profile

finishes, and the precision of the machine allows its cutting depth accuracy often to be an order of magnitude better than could be achieved from conventional workshop machines.

There is a considerable amount of data available concerning the relationship between material removal rates and grinding wheel life, and a series of indices come from this relationship. Differing materials will affect the grinding wheel life in varying ways, such as friability of the grains, breaking of the vitreous bond, or wheel clogging, and how often the wheel is dressed. The cutting action is not in any real way different for grinding than for milling, as a chip is formed, and the size of the chip is related to the depth of cut. The depth of cut will vary, depending on the grade of wheel being used and the material being ground, however, it should be in the region of 0.025–0.01mm (0.001–0.0004″) for roughing cuts and 0.005mm for finishing.

Specialist literature on grinding identifies grindability groupings and provides calculations and indices for depth of cut and feed rates. In a manufacturing environment where significant amounts of grinding are being undertaken this is an important factor in establishing a lean environment and one where tool life is maximised. This is often less important in a workshop environment where the cost of time spent calculating the most efficient rate may outweigh any saving, and many workshop machines have limited adjustment of feed rates preventing implementation of optimum settings.

It is often tempting for operators to use a heavy cut to break through a surface, or to reduce the number of grinding passes. This should be avoided as it results in clogging of the wheel and increases the requirement for dressing of the wheel. While this is a relatively simple process if the grinder is fitted with an auto-dresser, it takes time to have a vice mounted diamond to dress the wheel. Operators need to assess if their depth of cut is too deep by observing the sparks that come from the wheel. In general a solid orange 'flame' of sparks coming from the wheel as the workpiece is fed under it indicates a cut is too deep. Sparks that can be seen, but the operator can see them as individual sparks and see through the area of sparking will have the correct depth of cut. Operators need to apply a degree of common sense and if a poor or burnt surface finish is observed, and the machine sounds like it is struggling then the depth of cut needs to be backed off until the correct surface finish, and wheel life is established.

5.14 COMMON GRINDING FAULTS

Poor surface finish

Surface grinding is no different than any other cutting process and the causes of poor surface finish have similar origins. The depth of cut and the grade of wheel used are often primary factors in the generation of a poor finish. Coarser wheels create a larger chip creating a poorer surface finish, which may be good for material removal rates, however, will not provide a high standard surface finish. A solution to this is to reduce the depth of cut or mount a finer grade of wheel, however, care must be taken to ensure that the grade of wheel is suitable for the material being ground and will not immediately clog as this would introduce a different problem such as burning. If the operation is using coolant it could be that the flow is not enough to remove chips fast enough and the operator should increase the flow where possible. Over dressing the wheel can also be a factor as it can leave grains standing proud which may also be a factor.

Burning

This is a common fault and is identified as clearly visible dark discolouration of the workpiece. It indicates that there has been too much frictional heat. In extreme cases, and when grinding sensitive materials, burning can result in heat treatment of the workpiece surface i.e. hardening, micro cracking, or the creation of internal tensile stresses. It is most commonly found in dry grinding, but can occur when coolant used if flow is not enough to remove the frictional heat. The most common cause is using a cut that is too heavy for the wheel and material being ground. It can also occur if the wheel is glazed or loaded as more frictional heat is built up due to rubbing, rather than cutting, by trying to push the workpiece under the wheel too quickly. Where this is the case, the solution is to dress wheel more frequently or reduce the feed rate.

Warping

Warping is often not noticed until the workpiece is removed from the grinder and placed back onto a magnetic vice to grind the opposite side, with a significant high spot in the centre or at the ends identified. Thin workpiece sections are especially vulnerable and the operator needs to recognise this prior to starting grinding. In extreme cases the workpiece bows while being ground and the grinding wheel encounters a section of heavy cutting. The causes of warping are almost identical to burning in that it is caused by excessive heat generated by friction. The solution is again the use of lighter cuts and where possible the use of coolant. Where very thin sections are being ground that are susceptible to warping, the operator may need to make a series of light cuts, then leave the workpiece to cool for a while before returning for the next series.

Chatter/rippling

Chatter or rippling is recognised by the generation of a distinct pattern or series of ripples on the surface of the workpiece, and a chattering grinding process can often be heard. The causes of rippling or chatter can be vibration in the machine, which is rare, but a solution is to increase the speed of the wheel or reduce it. A more common cause is the use of an out of true wheel. This is either because a newly mounted wheel has not been trued and dressed correctly, or more commonly, after the wheel has suffered an impact or the wheel has been suddenly plunged into workpiece. The most common reason for this is an operator bringing the head down to the workpiece under power feed and fails to stop just above the workpiece surface. Another, albeit less common cause of rippling or chatter, is the grinding wheel slipping on its spindle due to an insufficiently tightened spindle flange.

Low wheel life

A correctly selected, dressed grinding wheel being given an appropriate depth of cut for the material being ground at a feed rate that is similarly appropriate, should have a substantial life. Grinding wheels that are judged to have a low life have normally suffered from a number of causes: insufficient coolant, wheel speeds set too low, over-dressing a grinding wheel (i.e. dressing it before it requires it), or taking overly large and too many dressing passes. More importantly the cause of low life spans of grinding wheels is likely to be the use of a wheel that is too hard or too soft for the material being ground. This means that the wheel 'loads up' requiring more frequent dressing, or wears away too quickly.

Scratching

Scratching of the workpiece can be seen in two areas. The first is where side grinding has taken place such as where the side of a grinding wheel has come up against a shoulder. While the pattern left on a workpiece feature will always be different where the side of a wheel has been used compared to the peripheral surface, the pattern produced is normally attractive and regular. Where wheel side has not been trued and dressed will result in a series of irregular and undesirable scratches. The solution is simply to dress the wheel side that will be brought into contact with the feature. The second cause of scratching is where coolant filtration system is not fully effective and swarf is not completely removed from re-circulating coolant and gets fed back under the wheel. Replacement of coolant filters or membranes normally cures this issue.

Slow cutting

While surface grinding is by its nature never going to be a fast material removal rate, it should be efficient. Slow cutting rates are almost always caused by feed rate and wheel speeds not fast enough. The solution is to increase wheel speeds and feed rates to a max point before burning occurs. However, most surface grinders will have a fixed speed, and only a limited ability to adjust feed rates. These settings will have been provided by the manufacturer of the machine to provide the optimum settings for a range of grinding operations. If the operator considers that the cutting rate is slow, investigation of other criteria such as wheel type and the amount of dressing required should be undertaken.

Wheel not cutting

This normally happens when the grinding wheel has been glazed by true-ing. If initial truing of a wheel has been undertaken using excessively heavy passes of the diamond the surface of the wheel can become glazed due to the heat generated. The solution is to re-dress the wheel using lighter slower cuts until the glaze is removed and a clean periphery is achieved.

5.15 WHEEL CHANGING

Wheel changing is relatively straightforward on most surface grinders, although in commercial workshops there is normally a requirement to have undergone training in grinding wheel safety and changing wheels, in order for the operator to demonstrate competency. This training is not onerous, but in the event of an incident regulatory bodies would investigate training and competency levels.

To change a wheel the machine should be electrically isolated to prevent inadvertent initiation of spindle start. Given that the power switch is a button on the front of the machine this can be easy to do. The guarding around the wheel and wheel flange is then dismantled exposing the wheel assembly. It is unusual for a surface grinder to have a spindle lock, therefore the operator grips the wheel with one hand and uses a peg spanner or other similar tool to loosen the spindle flange. It is unlikely that a hammer will be required and the flange should only be hand tight.

The grinding wheel is then pulled form the spindle, and the area around the wheel cleaned of any grinding deposits, using a small brush or lint free cloth, leaving a clean spindle, and wheel housing.

The grinding wheel selected for use first needs to be checked that its maximum rated speed is equal to, or more than the operating speed of the surface grinder spindle. It then needs to be checked for for cracks, chips or any other defect. This is primarily visual, however, grinding wheels should also have a process called 'ringing' undertaken before mounting onto the machine spindle. A ring test is where the operator holds a plastic or wooden shaft (often the handle of a hammer or mallet) and then gently tapped on the side using another wooden or plastic implement. The process is very gentle and should not damage the integrity of the wheel in any way. This tapping should result in a clear 'ringing' note being heard. This ringing note identifies that the wheel is free from cracks or internal defects that may leave to wheel failure or bursting.

A wheel that does not have a clear ringing sound and sounds a dull note is probably cracked, should be broken or smashed, and disposed of.

The wheel should also be checked to ensure that paper discs know as 'blotters' which have the grinding wheel nomenclature and manufacturers details printed onto them, are on both sides of the wheel. The paper discs are an important feature as they form the interface between the grinder driving flanges on the spindle and the grit of the wheel. The driving flanges on the spindle are the mechanism for transmission the spindle rotary motion to the grinding wheel. As the flanges are generally steel, and the grinding wheel is made from an extremely hard and incompressible surface the blotters on each side of the wheel allow the grit to bite into the paper and provide a surface for the flanges to engage with providing an even frictional lock, that does not require a high compressive load, prevents slippage that would otherwise damage the flange and/or the grinding wheel, or provide an unsafe condition.

Once all of the checks have been undertaken the wheel is mounted on the machine spindle. It is normal for grinding wheels to be purchased with a centre hole matched to the diameter of the machine spindle. However, for some grinding wheel types and grades this may not be possible and the wheels are manufactured with a larger central hole. In these situations the manufacturer will normally supply a series of plastic inserts that reduce a larger hole to each standard spindle size and provide a close fit, and do not affect performance once the wheel has been installed and trued. Grinding wheels are very sensitive to imbalance, and any vibration that is caused will transmit into a poor surface finish. While the spindle does not drive the grinding wheel it does require a close fit to remove any radial movement in use. Figure 5.21 shows a wheel fitted onto the spindle prior to the spindle flange being re-installed.

Once the wheel is mounted the spindle flange is screwed back on, in a reverse operation to its removal. The flange is tightened using the spanner hand tight. It does not need, and should not experience any heavy torque loads to be applied. It only requires to be tight enough that is does not slip when experiencing the frictional forces of grinding, and this is the function of the paper discs on the grinding wheel. The wheel should then be spun by hand to check that the wheel has free rotation and the wheel guard reassembled.

At this point the power can be switched on to the machine, the dressing diamond mounted and the and the wheel started. Truing a grinding wheel is essential to remove any imbalance and ensure that vibration due to any wheel eccentricity is removed. It can take a little time as it needs to be undertaken gently enough that the wheel does not 'glaze' from the heat being

Figure 5.21 Grinding wheel mounted on surface grinder spindle

produced, however, the wheel needs to be precisely round. Coloured wheels, often provide clear evidence of the wheel being precisely round as parts of the periphery will be discoloured where the diamond has not reached, and the truing process continues until a uniform periphery is seen. White wheels in pristine condition present more of a challenge, and listening to the sound made by truing is a clear indication to the operator whether the diamond has a continuous cut or interrupted. once the wheel has been trued and then finely dressed. the wheel is stopped and the machine cleaned of any grinding debris, ready for use.

5.16 GRINDING DIFFERING MATERIALS

The vast majority of surface grinding operations are undertaken on steels or cast irons, however, surface grinding is not exclusive to these types of material. Practically all materials can be ground, given that the cutting action of grinding is not different to any other cutting operation, however, the nature of grinding is a little less forgiving than single point or multi-point tool cutting.

Extremely ductile materials or those that have a tendency to tear, will require a different approach and grinding wheel to extremely hard materials. Ductile or soft materials can be more susceptible to burning and wheel clogging than other types of wheel. Silicon carbide tends to be the abrasive material of choice for grinding of non-metallic materials and non-ferrous metals, with diamond being used for grinding of carbides, very hard tool steels, ceramics, and glass.

Softer materials may clog up a wheel faster than hard etc. however, the selection of the correct wheel, and correct grinding technique will reduce the requirement for constant dressing. The websites and product catalogues of abrasive wheel manufacturers are the first place to identify the correct grinding wheel type for materials being ground. While quality workshop handbooks and manufacturers product data provide solutions for most situations and have good advice; for more esoteric applications the 'contact us' section of most wheel manufacturers' websites is a good mechanism for identifying the specific wheel type for more unusual applications.

5.17 SAFETY ISSUES

Surface grinding is one of the less complex machining operations given its limited number of degrees of freedom. However, the rotational speeds of the cutter, its construction, the rate of feed and the proximity of the operator make this a process that has more risk than many others and consequently requires more focus. While the outcome of any incident is probably no more or less severe than entanglement with any other machine, there is often little or no notice of a problem when surface grinding. It is important that operators undertaking surface grinding operations do not have distractions and can focus on the operation.

Of considerable concern is the risk of having a wheel burst. This is where the grinding wheel fragments, and given the proximity of the operator to the process and the speed in which it happens, it is unlikely that the operator will have time to move out of the way of debris. Grinding wheels in general are remarkably robust cutting tools when used in their designed condition. Wheel bursts are very rare, and almost all of them have been caused by either cracked or damaged wheels, operating them out of their intended performance parameters (i.e. material type, depths of cut and feed rate, or improper mounting). To ensure the integrity of a grinding wheel the operator should check for cracks by 'ringing' a wheel before mounting, when not being used store them where they are not subject to damage, and selecting the correct wheel for the task. It is also good practice for the operator to check any already installed wheel for defects prior to use.

Grinding wheels are highly abrasive and normally rotating at speeds of 2500–3500 rpm. Contact with the operator's bare flesh will result in the generation of immediate and deep burns. While loose clothing is never a good idea in a workshop, and heavy gloves often impractical to wear. The operator should consider always wearing clothing or a dust coat with long sleeves. It is quite common to leave a grinding wheel running while checking a workpiece during grinding or setting up the dressing diamond, to ensure the wheel does not become unsettled. The safest condition is to turn the wheel off during these operations, however, where this is not desired the operator should move the table to an extreme such that the wheel is a far as possible away from the operator while handling workpieces.

The rotational speed of grinding wheels and the friction that is generated means that any unsecured workpiece will be ejected from the machine with considerable velocity. Modern workshop surface grinders tend to have a completely enclosed guard to separate the operator from the workpiece. However, there remain a substantial number of workshop grinders that are open. Almost all of these will have a raised end guard to prevent the workpiece from being ejected from the machine. Any operator who looks at a surface grinder will almost always find that the raised guard is dented and the depth of the dents indicates the force involved in an ejection. This is normally due to a workpiece being ejected from a magnetic vice because it was not energised. The second most common reason is a workpiece not being sufficiently supported. Where a magnetic vice has not been switched on the workpiece is thrown out on spark down and little damage is done. Where taller workpieces that have not been properly supported get ejected, they often dig into the grinding wheel as the they rotate. When this has occurred the operator will need to carefully check the wheel for any cracks, and will in any event need to re-dress and true the wheel.

Excessive cut depths are not in themselves especially dangerous and one of the reasons that surface grinders are belt driven is that if a depth of cut is so great it just causes the wheel to stop, with the drive belt typically becoming displaced. Most of the damage is caused to the workpiece, however, as with all rotating machinery, the machine should be switched off and isolated before re-establishing the belt onto its drive pulleys.

Where a number of people have access to a machine, it cannot be assumed that all operators of a machine leave it in a safe condition. In almost all cases the first task undertaken in surface grinding is to dress the grinding wheel. Most machines will have a power switch for the wheel rotation as this is driven off an electric motor, with the feed power having a separate switch as this is normally driven off a hydraulic system. While it is almost unknown to dress a grinding wheel using any sort of feed if the feed is engaged when the power is turned on and the diamond is positioned for dressing it is normal for the dressing tool to be driven into the grinding wheel. This normally causes chipping and a deep gouge to be made in the periphery of the wheel, and can cause ejection of the dressing tool. The damage to the wheel is normally substantial and best practice is to replace them. Damaged wheels should always be broken into pieces to prevent inadvertent use.

When grinding into shoulders using a cross feed function operators should be circumspect as the wheel approaches the shoulder. The sudden impact of the grinding wheel into a shoulder where cross feed may be half the width of the grinding wheel, will result in impact damage commensurate with feed into a dressing tool. Where feed is not being used, but an operator inadvertently introduces the side of a wheel to a shoulder it may result in side loading of the wheel, and more probably twisting of the workpiece. While grinding wheels are extremely robust, operators should use a degree of circumspection when approaching a vertical surface as the area being ground increases massively.

Chapter 6

Cylindrical grinding

Cylindrical grinding is a complementary process to surface grinding. Where surface grinding is concerned with providing plane flat surfaces and slots by passing a workpiece translationally underneath a rotating wheel, cylindrical grinding is entirely focussed on producing a fine ground circular profile along a shaft. This shaft may be entirely cylindrical or can also be tapered. Cylindrical grinding is characterised by having a rotating grinding wheel and rotating workpiece. Almost all cylindrical grinding machines in workshops are for external grinding, and while the grinding of internal cylindrical profiles is common, this is often undertaken using an attachment fitted to the cylindrical, or more commonly a lathe to overcome the requirement for the workpiece to be mounted between centres on a standard workshop cylindrical grinder.

While it is possible to cylindrically grind curvilinear profiles, in many workshops this is limited to a width formed by profile generated on the surface of the grinding wheel, rather than extended elliptical sections, as the most workshop machine dynamics do not allow for continuous path movement.

Typical workshop cylindrical grinders are used for producing precision ground components, to close tolerances, and generally of smaller sized components. The size of machines found in many workshops will only be able to accommodate maximum workpiece diameters in the region of $\varnothing100$mm, and lengths of approximately 400mm. While there are machines available that can take workpieces much larger and longer than this, and those which offer an off-centre facility, they are comparatively rare in a general engineering workshop.

Cylindrical grinders mount the workpiece between centres and this allows the grinding of a continuous cylindrical surface, or series of differing diameters or tapers as long as they share the same axis. However, it is possible to cylindrically grind diameters that do not share a common axis as long as pairs of offset centres are provided. Once the workpiece has been mounted between centres, it is continuously rotated around its axis and the rotating grinding wheel is introduced to the workpiece generating a ground surface.

DOI: 10.1201/9780429298196-6

6.1 THE CYLINDRICAL GRINDER

The traditional cylindrical grinder consists of a grinding wheel, normally somewhat larger in diameter than for a surface grinder, positioned perpendicular to the machine table which is also offset from the wheel axis. On the machine table is a tailstock centre which can be moved longitudinally along the table to allow for varying lengths of workpiece. At the opposite end of the table is a powered headstock and centre which is used to provide workpiece rotation. For external grinding the workpiece rotates in the same direction as the wheel. For internal grinding the opposite approach is taken with the workpiece rotating in the opposite direction than the wheel.

Many machines have the capability of swinging the table off is axis to allow the grinding of tapers. A typical example of a traditional cylindrical grinder is shown in Figure 6.1. While the general principles of cylindrical grinding are identical as with those for surface grinding the approach differs. While the table traverses along an Z-axis to allow grinding of a workpiece surface there is no Y-axis movement and axial movement of the grinding wheel and setting of cutting depth is undertaken by moving the head into the workpiece on the X-axis.

There are two approaches to cylindrical grinding where the grinding wheel is traversed along the diameter of a workpiece until a required diameter is achieved, or alternatively the table locked in position and the grinding wheel plunged into the workpiece until the component dimension is achieved. Plunge grinding is often referred to as in-feed grinding, and while can be used to grind a plain diameter it can also be used to grind a profile that has been dressed into the grinding wheel. The limitation of the plunge grinding approach is the width of the grinding wheel, which is commonly approximately 20mm in width, therefore longer cylindrical sections need to be traverse ground. The limitation for grinding a plain length of workpiece is the limit of the table traverse length, although for smaller diameter workpieces there will be a minimum diameter/length ratio where deflection will occur irrespective of the depth of cut applied.

6.2 GRINDING WHEEL TYPES

There is little difference between wheels that are used for surface grinding or off hand grinding and cylindrical grinding. A grinding wheel is one group of a set of materials and products know as abrasives. In general workshops wheels used for cylindrical grinding have an abrasive substance held within a vitrified bond. While it is not unusual for wheels with a rubberised organic bond to be used for cutting off on a surface grinder this would not take place on a cylindrical grinder as the workpiece is mounted between centres, and would break into two halves when cut through damaging the wheel, workpiece and potentially, the operator. As with surface grinding the wheels

Figure 6.1 Workshop cylindrical grinder

require dressing with a diamond to form a plane flat surface prior to use but hold their shape during use.

Again there is a high degree of similarity between wheels used for cylindrical grinding and surface grinding, with the vast majority being made from silicon carbide or aluminium oxide abrasives locked into a vitreous bond. This is logical given that the materials being ground in cylindrical grinding will be from the same group of materials as for surface grinding. The advantage of silicon carbide and aluminium oxide grinding wheels is their hardness and brittleness which gives them a friability. It is this property that causes grains to break off presenting a new sharp edge to the workpiece, and they are the type generally used in cylindrical grinding.

Aluminium oxide wheels are likely to be the most common type selected by an operator as they are best suited to grinding of a wide range of steels and can accommodate material conditions from fully annealed to hardened. Silicon carbide wheels are more likely to be used when grinding non-metallic materials, non-ferrous alloys, non-magnetic stainless steels, and cast irons. Silicon carbide is also known to chemically react with ferrous metals.

Grinding wheels tend to be a significantly larger diameter than those used for surface grinding, and as such their maximum speed in RPM will be lower than for a surface grinding wheel as the peripheral speed and effective cutting rate in surface metre/min (SMM) or surface feet/min (SFM) will be the same which is in the region of 1980m/min for grinding wheels. As with surface grinding they are supplied in a number of grades, each of which have differing properties. There are five main properties that affect the performance of the wheel when in use:

- **Abrasive** – the type of abrasive material used (i.e. silicon carbide/aluminium oxide etc.).
- **Grain** – the particle size of the abrasive grain on a scale of 4 (very coarse) to 1200 (very fine). The grain size relates to the application it is to employed in. Coarse grains tend to be used for rapid material removal grinding operations such as plough grinding. The surface finish is proportionately rougher than that for surface grinding operations, where the grain size will normally be much smaller giving a slower material removal rate but a very good surface finish.
- **Grade** – graded soft – hard representing the tenacity of the bonding material to retain the grain. The grade is an indication of the strength of the bond that is retaining the abrasive particles, with an A grade indicating the softest bond and a Z grade indicating the hardest bond. Where the abrasive material is less likely to fracture presenting a sharp cutting edge, a weaker bond is useful as it allows the grain which may have a dull cutting edge due to friction with the workpiece surface to break away presenting a new cutting edge. Where the abrasive easily fractures a harder bond is appropriate, to retain the abrasive and improve the wheel life.
- **Structure** – The porosity of the wheel, with higher numbers denoting higher porosity. The significance of porosity is that the grains that are held in place by the bond do not fill all of the space and there are gaps

between each grain and its supporting bond. These gaps allow the heat to dissipate, allow for coolant to circulate and for the temporary storage of swarf chips. When selecting a coarse grade of wheel which will cut thicker chips, the porosity of the wheel will need to be greater.

- **Bond** – the bonding material used in the wheel construction. There are seven generic types, and while a resinoid bond cutting wheel is sometimes used on a surface grinder for cutting off very hard materials the normal bond for surface grinding is vitrified.

One limitation of all grinding wheels is their peripheral speed. All wheels are manufactured to a maximum speed which is expressed either as a peripheral surface speed in surface metres per min or surface feet per minute. Alternatively it will clearly state a value for its maximum RPM. Operators when selecting a wheel should ensure that the machine speed setting does not exceed the maximum RPM for the wheel. It is more common for cylindrical grinders to have a variable speed setting, and operators need to check that the machine speed set is not above the maximum rated speed of the wheel being used.

There are a number of national and international standards such as BS EN 12413, ISO 525, or ANSI B7.1. These cover a number of aspects relating to grinding wheel type, and size. They also relate to the marking of abrasive wheels, and the information that is carried on them is universal. All wheels will have the following information marked on them:

- company or brand name;
- lot no. or traceability code;
- dimensions of the wheel;
- maximum operating speed, stated as RPM or peripheral speed in surface m/min; and
- wheel specification data.

The wheel specification marking gives the operator exact information about the wheel. All wheels have a series of letters and numbers to inform the operator of the abrasive type, grain size, grade (hardness), structure (porosity), and bond type (i.e. vitreous). Figure 6.2 shows a typical grinding wheel.

The label on the grinding wheel shows the manufacturer's coding and from this wheel it can be seen that A36P5V (printed on the label) are the specific properties of the wheel. In this case they mean the following:

- A = aluminium oxide.
- 36 = grain size.
- P = hardness on a scale of A (soft) to Z (hard).
- 5 refers to the structure going from 1 (dense) to 15 (open). This value is optional and not on all wheels.

Figure 6.2 Typical grinding wheel for cylindrical grinding

- V refers to the bond type (in this case vitreous).
- 10 × 1 × 3 refers to the physical size of this wheel (ie Ø10″, 1″ wide, and Ø3″ spindle hole).

Grinding wheels are also supplied in a series of colours. Aluminium oxide wheels are found in white, pink, ruby red, brown, or grey colours. Each of the wheel colours indicates grinding characteristics, with grey and brown wheels tending to be used in heavier production grinding or surface grinders on low to high carbon steels. Pink and white wheels are generally used on harder steels which need cooling and friable properties to avoid burns to the surface of the workpiece. Wheels with a ruby red grit are usually reserved for grinding of tool steels. Silicon carbide wheels are normally found in black or green colours. Black silicon carbide has a very sharp grit and is normally used to grind non-ferrous metals as well as plastic and some rubbers. Green silicon carbide abrasive is sharper than black and is commonly used for the grinding of carbides, titanium and materials that have been deposited via a plasma spraying process.

When an operator is considering selecting a wheel type for surface grinding, detailed information is available on manufacturers websites and in good quality workshop handbooks. However, in broad terms, aluminium oxide is used for grinding high tensile strength steels and materials, with silicon carbide being used for lower tensile strength steels, cast irons and non-ferrous metals. Coarser grain sizes are used for rapid removal of materials and finer grits are used for improved surface finishes. Softer grades of bond are used when the machine power is low or good surface finish is important and harder grades where maximum wheel life is important and finish requirements are less crucial.

6.3 WHEEL DRESSING

Prior to use all wheels require either truing, following mounting of a new wheel, or dressing. Dressing of a grinding wheel on a surface grinder refers to the process of moving an industrial diamond across the periphery of the grinding wheel to provide a plane surface that is concentric with the machine spindle axis. It can also be used on the side of the wheel if grinding into a shoulder is to be undertaken. When a wheel has first been mounted on a machine it will requires dressing, effectively a precise 'truing' process to ensure that it is perfectly concentric with the machine spindle it has been mounted on. After this the wheel will require periodic dressing.

In a perfect grinding environment a wheel would never require dressing, as the combination of workpiece material behaviour and grinding wheel selection where the natural friability of the abrasive, or the loosening of the bond, constantly provides a sharp fresh cutting edge to the workpiece. However, the properties of the material being ground, the depth of cut, the use or failure to use coolant, the grinding wheel type being used, and the type of grinding being undertaken often means that the wheel requires periodic dressing.

In particular using the wrong type of wheel for the material being ground has a significant effect on the regularity of dressing, as does operations where surface scale or heat treatment debris needs to be penetrated, and in this situation it is not uncommon for the operator to have to undertake frequent dressing of the wheel until clean material is exposed.

Dressing, and truing, is undertaken using a single point diamond, the size of the diamond. A diamond, normally purchased with it retained in a mild steel bar is itself mounted in a steel housing. This housing should be angled approximately 10–15° down in the direction of the rotation of the wheel to maximise its life, and ideally, also 10–15° to the axis of the machine spindle. For cylindrical grinding this requires a diamond holder to be bolted down into a slot in the machine table to rigidly mount the diamond. The position the diamond is bolted to the bed and the angle the diamond point is held in the mounting ensures the 10–15° down angle and cross angle is achieved. Severe down angles in relation to the curved surface of the wheel and diamond run the risk of the wheel surface interfering with the diamond retaining bar rather than the diamond. Figure 6.3 shows a diamond dressing bar presented to the wheel for dressing.

Prior to starting the machine it is essential for the operator to check that any automatic table feed has been left engaged. The result of this being engaged is to immediately drive the diamond into the grinding wheel leading to either a wheel that requires significant dressing to return to condition, damages the wheel requiring its replacement or ejection of the dressing device. Either of the last two outcomes represent an unsafe condition, and may result in the operator experiencing a wheel burst at some future point, or being impacted by the ejected dressing device.

Having started the machine the diamond is advanced into the grinding wheel. For newly mounted wheels it is normal for the operator to hear that

Figure 6.3 Cylindrical grinder wheel dressing

the wheel is on being dressed for a portion of the periphery, and as the dressing process advances it will become apparent that the diamond is in contact with the wheel all around its periphery.

The process of dressing a grinding wheel takes care, and should be given time. Effective dressing will ensure that the wheel is in an optimum condition, for material removal and production of a high-quality surface finish. The depth of cut for each pass should be no more than 0.025mm (0.001″) and for fine grain wheels or those being used for high finish grinding operations limited to no more than 0.010mm (0.0004″). The cross feed rate is similarly controlled as it is important to avoid having a spiral pattern dressed into the wheel. In broad terms coarser grain grinding wheels can cope with faster cross feed, and a finer grain wheel will require a slower cross feed. What limits the speed of cross feed when dressing the wheel is the grain size. When wheel dressing the operator is trying to ensure that there is a plain flat debris free surface across the diameter of the wheel. The diamond needs to engage with all of the grains around the periphery of the wheel across the whole surface to ensure it is ready for grinding a workpiece. This means that the cross feed velocity when dressing is going to be a function of the wheel diameter, machine RPM and grain size. It is important that the grinding wheel does not have a series of spiral grooves dressed into it and that the entire surface needs to be dressed until it is a plane flat surface. The diamond is slowly fed across the face of the wheel and given a generally large peripheral length this takes time. It is important that the operator gives the wheel time to fully rotate and the important aspect of wheel dressing is being gentle. The diamond should be carefully introduced to the wheel using the table hand wheel until it is just at the surface of the grinding wheel. The cut is applied by bringing the head towards the operator using the y-axis adjustment wheel. When the diamond can be heard to touch the wheel the diamond is passed across the surface of the wheel using the x-axis hand wheel. A series of very light cuts are undertaken with each one conducted in one smooth pass across the face of the wheel.

Sequential cuts are taken until the wheel is clean, and any grinding debris is removed. This is quite easy for the operator to see and as the diamond passes over the surface of the wheel the colour lightens until a clean, new surface is presented. On initial use, or after any impact it is likely that the wheel will not be true. In this case dressing should continue until the diamond can be heard to be cutting continuously around the periphery, and the surface is clean.

Where the grinding is going to result in the wheel being brought into contact with a shoulder, the side of the wheel coming into contact with the shoulder should be dressed. This often requires a separate holder that allows for the diamond to reach from the periphery of the wheel to at least the depth of the shoulder if not to the blotter/drive flange. Dressing the side of the wheel, will ensure that there is no scratching of the workpiece and a high standard of surface finish is achieved on the shoulder sides.

6.4 MOUNTING OF WORKPIECES BETWEEN CENTRES

One of the advantages of mounting a workpiece between centres is that it provides accurate axial location. This means that the workpiece can be removed for measurement if required, although this tends to be only undertaken for measurement of axial lengths, which can be difficult to establish when a workpiece is on the machine, whereas diameters can easily be determined safely without dismounting the workpiece.

The workpiece must be pre-drilled with a suitable sized morse taper centre hole to allow engagement with the machine workpiece centres. It is normal for the workpiece to be driven by a peg rotating a driving dog attached to the workpiece. The operator needs to determine if a short plain boss of minimal diameter drilled with a morse centre should be pre-machined on the workpiece to allow for attachment of the driving dog. This can be applied at one, or both ends. The requirement for this approach will depend on whether the finished component is allowed to have a centre hole remaining in one or both ends, whether the workpiece driving dog would interfere with any aspect of grinding, or whether attaching the driving dog to the workpiece in an are requiring grinding would result in the workpiece be rotated 'end for end' to allow the section originally having a driving dog attached ground. Where short ends for centres are required, these would need to removed post grinding, but once removed it would not be possible to re grind the component, without significant difficulty in accurately re-establishing a centre into the workpiece, which would then not be able to be removed.

Most driving dogs involve the workpiece being inserted into a frame and locked in place with a clamping screw which normally provides a point load. It is the friction generated between the clamping screw point and the two tangential faces touching the surface of the workpiece that provides the workpiece rotation. If the workpiece does not have a sacrificial boss to attach the driving dog to, then it will need to be attached to the workpiece itself. It is normal for a soft packing shim, such a brass or a thin soft plastic to be placed between the frame of the drive dog and workpiece to avoid marking of the workpiece surface. Differing sizes of driving dog are available to accommodate workpiece diameters, and an example is shown in Figure 6.4.

Where a workpiece diameter exceeds the maximum size of driving dog, an offset hole can be drilled into the workpiece to allow a headstock drive

Figure 6.4 Set of cylindrical grinding workpiece driving dogs

peg to be inserted directly into the workpiece or a sacrificial boss on one or both ends will need to be utilised.

Once the workpiece has been fitted with its driving device it is ready to be mounted onto the grinding machine. The tailstock clamping screw should be loosened to allow the tailstock to slide up and down the machine table. The operator then has two choices, either to measure the distance between the centres, allowing for any sprung loaded centre, or centre adjustment, and then tightening the clamping screw or mechanism, or alternatively holding the component engaged in the headstock centre and sliding the tailstock until engaged in the rear centre and then tightened. The choice of approach tends to depend on how heavy the workpiece is; however, it is important that the workpiece is fully engaged between centres and that the workpiece driving dog is fully engaged with the driving peg to ensure the workpiece has no residual movement either laterally or off axis, and will rotate smoothly while being ground. Figure 6.5 shows a workpiece mounted and ready for grinding.

6.5 WORKPIECE ROTATION SPEED SELECTION

The workpiece rotates to continually present a new part of the periphery to be cut by the grinding wheel to provide a round ground surface. Most common workshop machines will have a number of speed settings for workpiece

Figure 6.5 Workpiece mounted between centres and ready for cylindrical grinding

rotation. The speed that is selected is determined by the workpiece diameter and the type of material being ground.

The range of peripheral speeds will vary from approximately 15 surface m/min for hardened tool steels to 45 surface m/min for aluminium alloys. Rotational speeds for standard carbon and alloy steels will be in the region of 20–30 surface m/min depending on whether they are in a hardened or fully annealed condition.

Common workshop machines do not always have a numeric value for speed setting on the machine but may, instead, have a series of speed setting. Figure 6.6 shows the workpiece rotational speed control, which has six positions rather than rotational speeds marked in RPM.

It is likely that the machine handbook will identify the rpm values for the speed settings where they are not marked on the machine, however, where this is not available they can be identified by measuring the rotational speed using a rev counter, applied to a workpiece of known diameter.

It is relatively simple for the operator to calculate the rotational speed setting required by dividing the peripheral length of the workpiece in metres into the recommended work surface speed for the material as shown in the example below.

Figure 6.6 Cylindrical grinder workpiece rotational speed selector

---

Example 1 (metric)

To identify the rotational speed for a ∅50mm workpiece:

∅50mm = 0.05m

Using $\pi \times D$ = Circumference

$\qquad$ = $\pi \times 0.05$m = 0.157m

Using 30m / min as surface cutting speed

Rotational speed $= \dfrac{30\text{m / min}}{0.157\text{m}}$

$\qquad$ = 191rpm

Calculation will identify a speed for each type of material, and for machines that do not have an infinitely adjustable workpiece speed control the nearest speed setting will need to be used.

6.6 USE OF COOLANT

Almost all cylindrical grinders have a facility for, or actively use, coolants supplied through a pipe system to the face of the grinding wheel The primary role of this flow of coolant is to reduce the temperature of the grinding wheel and the workpiece. The intention of this process is to remove the frictional heat build-up, prevent any heat treatment of the workpiece as a result of heat inadvertently introduced as a by-product of the grinding process and most importantly prevent any bowing or distortion of the workpiece while being ground. The degree of frictional heat generated when grinding can be substantial as witnessed by the 'blueing' of the workpiece. The energy used to cut metal is efficiently converted into heat and the frictional resistance between the workpiece and the wheel, coupled with the lower thermal conductivity of the wheel results in heat remaining with the workpiece.

Most cylindrical grinding operations can be undertaken without using coolants and some wheel types and wheel/material combinations recommend dry grinding, however, the use of coolants where recommended allows for increased in-feed depths and traverse speeds. It also significantly increases the period between wheel dressing.

The cutting action of a grinding wheel is no different than any other cutting tool in that it involves a shearing process where the material is deformed and converted to a chip, either continuous or as a series of small swarf chips. The majority of energy involved in this process is converted into heat which is focussed at the tool point and along the surface of the tool where swarf rubs over the surface of the tool. Heat build-up and wear causes the abrasive to shear off exposing a new sharp edge, and the vitreous bond of most wheels also gives way due to frictional forces. The addition of relatively small amounts of coolant rapidly quenches this heat build-up and substantially extends the period before re-dressing is required.

The selection of coolant can be quite involved as they vary from specific cutting oils, solids, and dry powders to gas cooling. The selection of a cutting fluid or coolant can depend on the machinability of the material being cut, the cost and the effect that it may have on the health of the operator, especially long term. However, most machines found in a general machine shop use essentially one type for logistical ease and low cost. These tend to be from a group of water-miscible fluids. These fluids are an emulsion of oil, water and an emulsifying agent. The proportion of mixture varies, but a mix of 1 part oil/emulsifying agent to 20–30 parts water is quite common. Machining stainless steels may again involve a richer mix of approximately 1:5, however, differing manufacturers products will have recommended mix values, and operators should consult product data sheets for the optimum mix. Most coolants available coolants are fully synthetic, and it is unusual to get coolants that use a fat-based emulsifier, however, operators should ensure that coolants used for grinding do not use a fat-based emulsifier as this tends to clog the grinding wheel.

The advantage of this emulsion of oil and water is that water has excellent cooling properties, and the oil content provides good lubrication, with a secondary advantage of providing some corrosion protection to the machine. Some materials that are vulnerable to a chemical reaction such as copper, brasses and bronze etc. may be subject to staining from chemicals within the coolant especially some sulphur or chlorine compounds. Where components that have aesthetic requirements, or high surface finish requirements, adjusting the in-feed depth and traverse rates to reduce cutting temperatures and not using coolant, or changing to a passive type are the options open to the machine operator.

Delivery of coolant to the workpiece and tool cutting edge is normally via a sectional/flexible pipe, or a rigid pipe to the top of the wheel which is then taken down onto the workpiece. There are some disadvantages to this system, as while it provides excellent cooling and lubrication to the periphery of the wheel, if grinding to a shoulder or wide flange the amount of coolant can be restricted as it requires a splashing effect from the peripheral delivery to provide cooling. While this can be addressed by increasing the flow, there can be a problem with coolant being sprayed out onto the floor below any guard creating a safety hazard.

While coolant manufacturers and literature suggests that coolants should be delivered at 10lt/min – 20lt/min, this represents a best case. In general it is not a requirement to deliver large quantities of coolant to the tool point which can often be a rather wet process, and depending on the type of coolant being used can be a long term hazard to the operator. However, the delivery of a stream to the tool such that there is little or no steam witnessed rising from the tool/cutting area is often a good indicator. Workpieces that are more than hand hot when checked at the end of each cut are an indicator of insufficient cooling. Poor surface finish is also an indicator of insufficient cooling.

It is important that excessive heat is not built up during grinding as the workpiece is being held between centres. Heat will induce thermal creep into the workpiece and as it is rigidly held at each end, it can only bend. This results in an increase in the depth of grinding at the maximum point of distortion and when cooled, the workpiece will not have a parallel profile and will have one that is slightly concave.

When machining more exotic materials such as magnesium or metals that may react with water no coolant would be used. In general this should not be a problem for the operator, as the cutting speed and feed rate should not be so high that heat does not dissipate fast enough, and the operator needs to have an awareness of these marginal conditions.

6.7 CUTTING DEPTH

Cylindrical grinding is not different from any other type of grinding or indeed metal cutting process, in that depth of cut feed rate, tool life and surface finish are all considerations. Cylindrical grinding is a process that is a little slower to remove material than surface grinding given its reduced feed rates, and the degree of eccentricity it is trying to remove from the workpiece. There is however, a significant degree of commonality with surface grinding when it comes to establishing depth of cut.

There is a considerable amount of data available concerning the relationship between material removal rates and grinding wheel life, and series of indices come from this relationship. Differing materials will affect the grinding wheel life in varying ways, such as friability of the grains, breaking of the vitreous bond, or wheel clogging, and how often the wheel is dressed. The cutting action is not in any real way different for grinding than for milling, as a chip is formed, and the size of the chip is related to the depth of cut. The depth of cut will vary, depending on the grade of wheel being used, the material being ground and whether the operation is to undertake a roughing cut or be for a finishing pass. However, it should be in the region of 0.05mm (0.002") for roughing cuts across most material types and 0.005–0.01mm for finishing cuts on most steels and alloys.

It is often tempting for operators to use a heavy cut to break through a surface, or to quickly eliminate any eccentricity of the machine workpiece diameter. This should be avoided as it results in clogging of the wheel and increases the requirement for dressing of the wheel. This takes time as it is common to have to remove the workpiece to install the dressing diamond and re-establish the spark down position. Operators need to assess if their depth of cut is too deep by observing the sparks that come from the wheel. In general a solid orange 'flame' of sparks coming from the wheel as the workpiece is fed under it indicates a cut is too deep. Sparks that can be seen, but the operator can see them as individual sparks and see through the area of sparking will have the correct depth of cut. Operators need to apply a degree of common sense and if a poor or burnt surface finish is observed, and the machine sounds like it is struggling then the depth of cut needs to be backed off until the correct surface finish, and wheel life is established.

6.8 TRAVERSE CYLINDRICAL GRINDING

Once the grinding wheel has been dressed, the workpiece inserted between centres and the workpiece rotation speed selected, grinding can commence. The operator starts the wheel and workpiece rotation and by rotating the y-axis control wheel, should draw the grinding wheel towards the workpiece until apparently just above the workpiece. The longitudinal position that is selected for spark down depends on whether the operator is undertaking a plunge grinding operation, or will be traversing along the length of the workpiece as it rotates. Spark down is the point where the grinding wheel connects with the highest point of the workpiece, and it is this point that the grinding process commences. If the workpiece is seen to be noticeably eccentric as it rotates at one end, it is wise to start there, similarly if the workpiece is bowed a more central location for spark down is advised. The traverse wheel is rotated until the wheel comes into contact with the surface of the workpiece and sparking is witnessed. It is usual to traverse the workpiece using the x-axis control wheel to identify any high spots. Where a high spot is found it is normal to back off the cut slightly until an appropriate depth of cut for a roughing cut is achieved.

Cylindrical grinding, as with surface grinding is one of the slowest metal removal processes, but, is a process that provides one of the best surface finishes. This requires small depths of cut, employed repetitively. The depth of cut will vary, depending on the grade of wheel being used and the material being ground, however, it should be in the region of 0.025–0.01mm (0.001–0.0004″) for roughing cuts and 0.005mm for finishing.

Once spark down has been achieved the table is traversed such that the grinding wheel is off the surface of the workpiece. When doing this care must be taken to ensure that the grinding wheel is not inadvertently traversed into the tailstock centre or into the driving dog assembly. The depth of cut is then set by the operator and then the grinding wheel is passed over the entire periphery of the surface being ground, by rotating the Z-axis control hand wheel. The depth of cut normally has two methods. The main X-axis handwheel provides relatively quick traverse of the wheel head, but is often calibrated in increments of 0.0025mm (0.0001″), and it is also common to see a micrometer adjustment for extremely precise cutting depth adjustment. The amount that is traversed for each workpiece should never be more than the width of the grinding wheel. Many operators use a slow rate of traverse to give a degree of over cutting and allow the material to be removed, however, if a narrow wheel is being used or the machine has a high geared traverse screw this may be quite a slow operation and the operator should establish what the traverse distance of one rotation is prior to use. Many grinding wheel manufacturers and engineering data books give some guidance on traverse distances, and in general terms it is suggested that for roughing grinding operations for unhardened steels that the traverse distance but up to 75% of the wheel width, where for stainless or hardened steels this be no more than 50% of the wheel width. For finish grinding operations, traverse distance will be approximately 30–40% of the wheel width, with it being on 25% of the wheel width for harder materials. Figure 6.7 shows a plain bar being cylindrically ground.

The operator merely then needs to pass the grinding wheel along the axis of the workpiece until significant sparking ceases. With roughing cuts this may mean that the wheel needs to traverse to the opposite end of the workpiece, and back to the starting position. Alternatively a slower traverse or lighter cut taken. A cut can be applied at either end of the workpiece, and a series of lighter cuts is often better than heavy cuts as wheel life is maintained and heat removed more effectively.

While there are recommended cut depths for roughing cuts, where a workpiece is long or it has a high length to diameter ratio, the workpiece can easily deflect in the centre. This is not always apparent to the operator

Figure 6.7 Plain cylindrical grinding

and can result in a considerably smaller diameter in the centre of the work-piece compared to that at the ends. The way to address this issue is through use of extremely light cuts, or the use of a different process such as centreless grinding, where that option is available.

Initial cuts often do not seem to provide much gain with few sparks being seen other than at the high spot. As sequential cuts are made, if the surface being ground has been well machined the area being ground rapidly expands until the wheel will have a continuous sparking action throughout each pass. For general cylindrical grinding to provide a smooth surface, grinding only continues until the operator can see no un-ground areas on the surface, and the lack of an interrupted sound during grinding provides to operator with another indicator that the surface is being continuously ground. While cast irons and many types of steel do not tend to clog the grinding wheel in use, if a lot of material is to be removed, or materials that do tend to clog the wheel are being ground the wheel will require re-dressing. One of the advantages of mounting the workpiece between centres is that for most situations the workpiece can be removed from the centres for dressing to be undertaken and replaced, on the same axis. It is fairly obvious when a wheel requires re-dressing, as the sound of the grinding changes, the appearance of the surface changes, the surface of the wheel becomes discoloured (normally black) the sparking changes and there may be some burning to the surface. Taking a cut that is too heavy accelerates the requirement for dressing. That said given the generally substantially larger diameter of cylindrical grinder grinding wheels compared to those used on a surface grinder, the requirement for dressing is much less frequent than that for surface grinding. Dressing does not remove a significant amount of the wheel's periphery, but the operator will need to spark down again.

Where, a high surface finish is required, it is normal for the wheel to be dressed after roughing cuts and a series of very fine cuts of approximately 0.005mm applied. These cuts continue until the workpiece has been ground to size, or the standard of finish achieved. The final cut is continued back, and forth until 'spark out' is achieved. This is where no more sparks can be seen to be coming from the between the grinding wheel and the workpiece, and the process is complete.

When a workpiece is being ground to a tolerance diameter the operator has the opportunity to remove the workpiece in order to measure the diameter or leave set up between centres. While it is normal to leave a workpiece unmoved to eliminate and possibility of eccentricity, by keeping the centres clean this is not normally a problem. Irrespective of either approach the wheel should be switched off and have ceased rotating before any measurement takes place. Once the remaining depth of cut is established the operator can decide whether to continue with roughing cuts or finish cut depths, which are applied using the scale of the y-axis handwheel.

6.9 CYLINDRICAL GRINDING TO A SHOULDER

Cylindrical grinding to a shoulder is in many ways exactly the same as for traverse grinding of a plain diameter. The difference between the two is that there is a restriction on the traverse length on one or both sides of the workpiece. The concern for the operator when grinding to a shoulder is whether it is acceptable to allow the side of the wheel to come into contact with the shoulder, or whether this has to be protected. While it is unusual for the wheel not to be allowed to come into contact with the shoulder, the workpiece will either require an undercut at the root of the shoulder or will have a small unground section.

There are a number of considerations to be taken into account when grinding into a shoulder. The first of these is the depth of the shoulder, as if the depth that can be ground is limited to the distance between the periphery of the grinding wheel, and the central boss of the wheel flange, or any part of the machine head assembly that may prevent the wheel being advanced into the workpiece. While this is normally only an issue found when grinding off-centre features, it is a limitation. Another consideration is whether the sides of the wheel require dressing. In surface grinding the wheel is often relieved, however, in cylindrical grinding, while entirely feasible side relieving tends to be undertaken less often, and the driver for whether to do this is often related to the amount of side grinding, or if side grinding of a shoulder is the primary objective. By dressing the sides of the wheel the operator ensures that the abrasive surface is correctly prepared, and that the sides of the wheel a perpendicular to the peripheral face ensuring a square corner. when the grinding wheel comes into contact with the shoulder it will grind that surface, and leaves behind a series of machining marks that look like tangential radial spokes with a fine ground finish. Where the sides of the wheel have not been dressed, these spokes are often less even and give a duller slightly scratched appearance.

Prior to starting grinding the end stops are set. On many machines there are a pair of opposing stops mounted onto a rail that allow their position to be moved up and down the table of the machine. These stops will come up against a fixed post on the bed of the machine providing a limitation to the left and right traverse distance of the machine table. Figure 6.8 shows an example of this arrangement.

To set the end stops first any adjusting screw, where fitted, is adjusted such that there is enough protruding from the stop to allow adjustment (i.e. retraction during grinding). If the traverse fixed stop is of the type that is

Figure 6.8 Typical cylindrical grinder table traverse stop

retractable, the operator will need to move this into its raised position. Then the retaining screw fitted to each clamp is loosened and the table traversed until the (non-rotating) grinding wheel gently touches the shoulder and the moveable stop is pushed up against the fixed stop and the clamping screw tightened. Where grinding is taking place between two shoulders the same process is undertaken for the opposite side. Once the stops have been set, it is usual for any fine adjustment screw to be advanced a small distance to bring the grinding wheel out of contact with the workpiece.

The wheel and workpiece rotation is then started and a spark down process commenced as for plain traverse cylindrical grinding. The table is traversed until it reaches the end stop, at which point the operator will back off the fine adjustment screw until the side of the wheel comes into contact with the shoulder and begins to grind. For operations where grinding is taking place between two shoulders, the table is traversed and the fine adjustment screw rotated until contact is made on the opposite side. The operator can then begin to grind in a normal manner. When grinding shoulders the operator should be aware of the significantly increased surface area being ground, and light cuts are advised. Heavily distorted shoulders may take a significant amount of grinding, and operators may need to consider machining any gross distortions off on a lathe rather than try and grind away large amounts of material.

Where grinding to a specific width, or lateral position is required the fine adjustment screw allows controlled movement, with a high degree of precision. Grinding into a shoulder is also entirely feasible without use of traverse stops, however the operator has to be careful to ensure that a heavy cut is not applied and that the side pressure applied is consistent throughout the cutting phase. Excess side force applied in the direction of the tailstock has the potential to overcome the pressure of any sprung loaded centre causing ejection of the workpiece, and damage to the grinding wheel, workpiece and potentially, the operator. It is always better to use the traverse stops where fitted as this gives a fixed stop to prevent excess pressure being applied and gives the operator something to 'lean into'.

Careful wheel dressing provides a very 'square corner' to the grinding wheel, and as such, when grinding into a shoulder corner, the resulting form will be similarly square. This corner will wear and some form of radius will become apparent over time. However, where the component drawing dimension give a MIN RAD dimension this is rarely a problem. Where the root corner has a dimensioned and toleranced radius this will require the wheel to be dressed with a radius. This is relatively straightforward to achieve using a radius dressing tool, however, as this is a process that removes material and can only increase in radius it is suggested that the wheel to be dressed to the radius at maximum material condition (i.e. the smallest radius), allowing subsequent re-dressing to progress towards minimum material condition (i.e. maximum radius). After this point has been reached the wheel will need to be re-dressed to remove the radius, often requiring a considerable amount of material to be removed from the wheel and it re-dressed with a radius, or the wheel changed for a new one and the process repeated.

6.10 PLUNGE/IN-FEED CYLINDRICAL GRINDING

Plunge, or in-feed grinding, is a process where following dressing the grinding wheel is aligned with the section to be ground and the table locked to prevent lateral movement. The diameter to be ground can be either plain and smooth or profiled. Where the diameter to be profiled, this requires the wheel either to be dressed to the profile by the operator, or more commonly buy mounting a pre-profiled and trued wheel from a specialist supplier. The wheel is then slowly introduced to the workpiece and the root diameter ground to the size required. The method used is not especially different from plain grinding other than there is no traversing of the machine table. Plunge grinding is normally only undertaken on workpiece sections where the width of the grinding wheel is at least as wide as the length of workpiece section to be ground. The limiting factor is the maximum width of the grinding wheel for the machine being used. In addition to this, the distance between the periphery of the grinding wheel and the centre diameter, or component of the machine head that interferes with the workpiece and limits plunge depth is also a limiting factor.

Any grinding that has to take place between two shoulders will require the grinding wheel to be the correct width for the gap required. While it is possible for a narrower wheel to be used, or one where the grinding wheel has been dressed to a correct width, the operator needs to consider if it is advisable to grind a diameter and two shoulder faces simultaneously. The amount of material to be removed from the shoulder sides, the depth of the shoulder and the amount of material to be removed from the diameter all factors in this. Where there is little material to be removed from the shoulders and the depth of shoulder is not significant, plunging in to grind a plain diameter, is normally a fairly straightforward operation. However, where the depth of shoulder is greater than the width being ground, and there is a significant amount of material to be removed from the shoulders, the task is unlikely to be completed in one operation. As the grinding wheel enters the slot, all of the cutting is being undertaken by the corners of the grinding wheel. Where heavier cuts are being undertaken, there will be erosion of the corners creating a rounded root fillet once the wheel has plunged to depth and the plain diameter ground. This may not be a problem where a root fillet radius is permissible, however, the wheel may require re-dressing to re-establish a square corner as a finishing operation.

Operators should also consider the machine power requirement when plunge grinding into a cavity between two shoulders. As before shallow plunge depths rarely provide any real challenge to a machine, however, where a heavier cut is being taken, and particularly, the depth significant the area of material being cut can become such that the machine slows. This is therefore another limiting factor. In such situations the operator is advised to use a narrower wheel, and use a plunge grinding approach to remove any significant amount of material from one shoulder, and then the other. Traverse grinding can then be used to provide the plain diameter between the shoulders. Alternatively, two plunge grinding operations can be used to rough out the shoulders, and then a finishing plunge grinding operation using a wider wheel undertaken to finish cut the shoulder sides and grind the bottom plain diameter.

6.11 LIMITATIONS OF GRINDING CYLINDERS

Grinding of cylinders requires a number of issues to be addressed if the external and/or internal diameters are to be perfectly cylindrical with respect to each other and aligned with the nominal cylinder centre axis.

Internal cylindrical grinding differs from external cylindrical grinding in that the grinding wheel operates in the opposite direction to the rotation of the workpiece, whereas for external cylindrical grinding the workpiece rotates in the same direction. The reason for this is that internal grinding is undertaken on the face towards the operator, and external grinding is undertaken on the side of the workpiece opposite the operator, and the direction of rotation ensures the relative movement between the grinding wheel and the workpiece is always in opposition.

Internal cylindrical grinding requires access to the workpiece cavity, therefore, preventing the workpiece from being mounted between centres, and must be mounted in a chuck, or chuck and fixed steady for longer workpieces. Where this is required, the cylindrical grinder must have an interchangeable head between the normal dead centre and driving head, and a powered chuck arrangement. A powered grinding head is also required and is fitted where the grinder tailstock is normally mounted. This interchangeability is not common for standard workshop cylindrical grinders, and is more commonly undertaken using specialist sub-contract grinding companies, where the highest levels of precision are required and undertaken on bespoke internal grinding machines, or using an attachment which can be mounted to the saddle of a workshop lathe, where the workpiece is held in the lathe chuck and rotated, and a mounted point on a powered spindle with a fine adjustable screw to give x-axis movement and allow the mounted point to be fed into the work piece. Traverse is provided by movement of saddle along the z-axis of the machine.

The grinding of the external surface of plain cylinders requires the manufacture of end plugs to allow the cylinder to be mounted between centres. A key consideration when manufacturing the end plugs and the size of the grinding allowance is any eccentricity between the bore of the cylinder and the external surface.

As the workpiece is mounted between centres using end plugs that locate in the bore, the axial alignment of the end plugs and the concentricity of the internal bore to the external surface is a variable. If the internal surface of the bore has been machined at the same time as the external surface there will be a high degree of concentricity between the inner bore and external diameter. However, if the workpiece has been hot formed tube, which has subsequently been externally clamped in lathe chuck, prior to machining the external surface then there may be significant eccentricity between the two diameters. Where this is the case, the centre hole for the end plug can be adjusted, or a significant grinding allowance left on the external diameter of the workpiece.

To create an end plug, it is normal for a pair of plugs to be machined on a lathe with one having an extended spigot to allow the workpiece drive dog to be attached. It is also normal for the end plugs to have a close-fitting shoulder to provide good concentricity, and enough friction to transmit rotational drive. This often necessitates the use of an H7 fit, or better, clearance or transition fit. The use of an interference fit may be attractive, as it provides a high degree of frictional loading and if made from aluminium rarely distresses the workpiece, but can provide significant difficulty in removal unless holes have been drilled into the ends to allow a drift to be inserted and the plug hammered out. Figure 6.9 shows details of a typical end plug, with a stepped shoulder recess. It is also normal for the external diameter of the end plug to be less than the finished external diameter of the cylinder being ground to ensure that there is no contamination of the grinding wheel, where softer materials such as aluminium, have been used

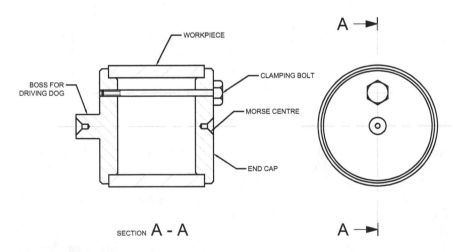

Figure 6.9 Typical cylinder end plug configuration

to manufacture the plugs, and wheels have been matched to harder materials such as cast irons, or high carbon steels. It also ensures that the cylinder corners are consistently generated.

It would be normal to drill the morse centre holes used to mount the workpiece between centres in one operation when manufacturing them on a lathe. However, where significant eccentricity exists between the inner bore and external surface, it makes sense for the operator to drill the holes when a common centre position can be established.

Following manufacture (minus the morse centres) of the end plugs they are fitted to the cylinder bores, and the workpiece placed on a surface plate. A digital, or vernier height gauge is then set to the nominal centre height of the workpiece. The centre of the cylinder end plug is then scribed in at least three planes by rotating the workpiece over the surface plate. This produces a series of scribed lines which leaves a plain 'cocked hat' in the centre. The end plug and workpiece are then marked to identify rotational alignment, and allow refitting at the same place, and then removed from the workpiece.

The workpiece plug is then centre drilled on a pedestal drill in the centre of the cocked hat with a morse centre drill sized to match the cylindrical grinder centres. The plugs are then re-inserted into the workpiece ensuring the marks are in alignment. This ensures that the plugs remain in the same radial orientation they were marked in. This method then ensures that the centres are aligned closely with the external surface to be ground and not with the internal bore, effectively eradicating any gross concentricity issues between the workpiece bore and its external diameter.

Once the end plugs have been inserted the workpiece can be mounted in a similar manner to any other workpiece, and traverse grinding undertaken. The operator does need to ensure that the axial loading and the fit of the end caps is good enough to transmit drive, and prevent any off axis movement while grinding is being undertaken. The use of an H7 type fit is usually enough to ensure this, however, where an error has been made in manufacture of the caps or the internal surface of the caps precludes a good fit being provided alternatives made need to be employed. This can be either use of a light structural adhesive, or employing lighter cuts. The choice will be driven by the type of material being used, its sensitivity to adhesives and the end load provided by the machine centres.

6.12 RADIAL LIMITATIONS

There are two aspects to the radial limitation of a cylindrical grinder. The first is the distance between the centre of the workpiece and the machine table. This distance forms the maximum radius of a plain workpiece that can be accommodated on the machine, and double this distance identifies the maximum workpiece diameter which can be accommodated. Figure 6.10 shows the maximum radius that can be fitted to a typical workshop machine.

Figure 6.10 Maximum radial distance available on typical cylindrical grinder

Any workpiece capable of being mounted above the machine table will be capable of being mounted such that the distance between the workpiece centre axis and the face of the grinding wheel allows for some clearance. However, when plunge grinding or grinding to a shoulder, another limiting factor is the distance between the periphery of the grinding wheel and its central boss. This distance between the periphery and any wheel flange or central boss is often greater on the left-hand side of the wheel than on the right hand side. Figure 6.11 identifies the difference that can be encountered.

This is normally because the drive shaft and wheel drive head assembly is on the right-hand side. Normally this is of little consequence, however,

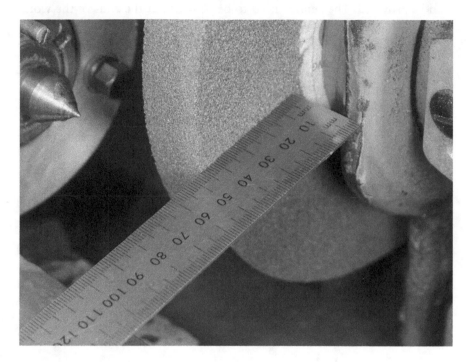

Figure 6.11 Maximum radial distance available for grinding between shoulders

if grinding to a shoulder it is likely that a greater depth of shoulder can be ground on one side than the other. When attempting to grind a workpiece that is close to the limits of the machine the operator may wish to consider the workpiece orientation to avoid a workpiece clash.

6.13 OFF-CENTRE GRINDING

Off-centre grinding is no more complex than ordinary traverse grinding, and grinding between shoulders. Probably the most common form seen is grinding crankshaft journals where a cylindrical feature requires grinding off the main workpiece axis. While the mass of these components does not generally factor in general workshop cylindrical grinding, and therefore does not need balancing, the 'throw' of the workpiece can be considerable, and it is entirely possible that the workpiece diameter for off-centre grinding may be considerably smaller than for plain cylindrical grinding.

While most crank shafts are ground on specialist machines, off-centre grinding is entirely feasible in the general workshop if any off-centre journal is ground during manufacture or a precision end cap with off axis centres can be provided. The centres need to be provided at the ends of the workpiece as any provision of centres at the journal being ground would result the overhanging sections impacting the powered workpiece head and tailstock.

The radial limitation of the machine also needs to be established and compared to the maximum 'throw' of the workpiece. This throw is identified by determining from the workpiece drawing, the maximum distance from the offset centre to the maximum extremity of the workpiece on the opposite side of the workpiece, as shown in Figure 6.12.

To grind an offset diameter, the workpiece is mounted between centres on the centreline of the diameter being ground. The workpiece is checked for free rotation, and the feature is ground as for normal traverse, or plunge grinding between shoulders.

Where end caps are not being used it is common for a workpiece to have flanges carrying the offset centres located at the ends of the workpiece. These are commonly machined off following cylindrical grinding, therefore the operator should ensure that any offset journal, is finished to size and protected, as once the centre is removed and the end journals ground, it is difficult to return to grinding any off-centre journal, as it would require new off-centre caps to be manufactured, which potentially introduces alignment problems, and any retaining mechanism will potentially damage the surface finish of the workpiece end journals.

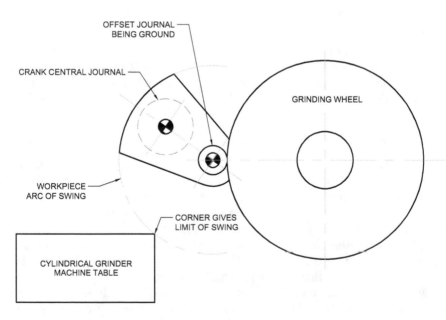

Figure 6.12 Maximum throw distance for offset journal

6.14 TAPER GRINDING

Grinding a taper is entirely feasible on many standard workshop cylindrical grinders. The normal approach is to pivot the machine table to provide the correct angle. Each end of the grinder table commonly has a set of clamping bolts and clamping flange to prevent the table from pivoting when not required. At one or both ends of this table it is normal to find a scale to indicate the amount of pivot. The scales may be the same at both ends or could be two differing types. One that measures taper in inches/foot, or in degrees. Older machines tend to have the imperial inches/foot scale with metric machines tending to have degree scales, given that increments of mm/m are less useful. Figure 6.13 shows a typical table taper angle scale inscribed in units of 'taper ⅛"/foot', although the opposite end of the table has its scale marked in degrees.

Most machines that have a taper grinding facility will have a displacement angle of approximately 10° and while this will vary between machine types, only specialist machines will have significantly larger angles. This may not seem large but it is important to remember that an included angle will be generated, which is twice the angle set. Some machines especially those graduated in a taper/ft are almost certainly graduated for the included angle, however machines with scales may be calibrated in a simple angle and the machine operator should refer to the machine manual to determine if this is plain angle or covers the included angle. Alternatively check, by measurement from the workpiece centre or manufacture a test piece, if they are unaware of the machine configuration.

Figure 6.13 Taper angle scale

To convert an angle to a value for inches/ft, or to work back and convert inches/ft into a known angle simple trigonometry is used. The worked examples below show the methodology for this.

Example 1 – Convert inches/foot to degrees

Using tan angle $= \dfrac{\text{opposite}}{\text{adjacent}}$

Assuming taper required is 2″/ft then:

Tan angle $= \dfrac{2''}{12''}$

Tan angle $= 0.1666$

$\text{Tan}^{-1}\ 0.1666 = 9.46°$ (derived from tables or calculator)

To set the table swing to produce a taper of 9.46° the offset would be half of this and set at 4.73° as the overall taper is an included angle and the angle set on the machine is half of the angle required, unless calibrated for included angles.

Example 2 – Convert degrees to inches/foot

If the workpiece length is 6″ and intended to have an included angle of 8° ground, then the following calculation is undertaken:

Using tan angle $= \dfrac{\text{opposite}}{\text{adjacent}}$

Tan 8° $= 0.1405$ (derived from tables or calculator)

$0.1405 = \dfrac{\text{opposite}}{\text{adjacent}}$

$0.1405 = \dfrac{\text{opposite}}{6''}$

By transformation: Opposite $= 0.1405 \times 6''$

Opposite $= 0.843''$

To resolve into a setting/ft: 12″ is divided by the workpiece length:

$$\dfrac{12^2}{6^2} = 2$$

Therefore the offset distance which in this case is 0.843″ is multiplied by 2 to give the taper per foot value, to be set on the machine.

$$0.843″ \times 2 = 1.686″/\text{ft}$$

Prior to swinging the table it is normal to first dress the grinding wheel. This can be accomplished after the table has been swung, however, if the dressing tool is set to angle that uses any slot in the table, it is worth undertaking any dressing to ensure the angle presented by the diamond to the grinding wheel is at its optimum angle. To swing the table, any clamping arrangement at each end needs to be loosened such that the table can swing relatively freely. It is then set to the angle required using the scale and it is unlikely that any machine with the capability to taper grind would not have a scale somewhere on the machine to indicate the offset. Once the offset has been set the clamps are tightened and the workpiece can be mounted between the centres.

Where a workpiece has been turned at the correct angle the starting point for spark down is not critical as the workpiece will effectively be parallel. However, where a slow taper is being ground there may have been no opportunity, or requirement, for pre-turning to the taper angle and the entire taper will be generated by grinding. Where a workpiece is a plain diameter prior to grinding, it is important that spark down and initial grinding starts at the offset end of the workpiece that is closest to the grinding wheel as can be seen in Figure 6.14.

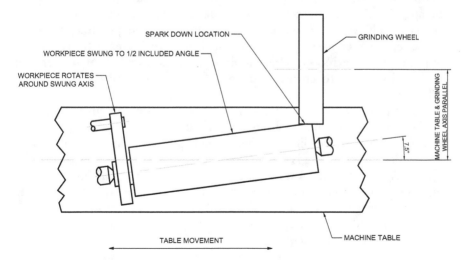

Figure 6.14 Spark down position for plain diameter workpiece being taper ground

Failure to do this results in a heavy cut as the grinding wheel is traversed down the workpiece. This leads to burning of the workpiece and potentially heat treatment. While on very slow tapers it can be tempting to continue by slowing the rate of traverse, this is rarely successful, as clogging of the grinding wheel, and a radius can be formed on the lead corner of the wheel.

Once workpiece grinding is complete the table needs to be returned to a position for parallel grinding. The accuracy of the scale, and any parallax issues that may be encountered means that the table position needs to be checked. There are two approaches to checking the table is correctly returned to a datum position. The first method is to mount a plain ground bar, that has previously been ground parallel to its centres. The length of this bar should be the maximum that can be fitted in between the workpiece centres, but should be evenly mounted either side of the table pivot. A Dial Test Indicator (DTI) is mounted on the bed of the machine or attached to the grinding head and positioned at one end of the test bar. The table is then traversed and the dial observed to indicate if any extension or deflection is identified by the DTI. Where a deflection is detected the table clamps are slackened and the table traversed by half the deflection measured. The table is then re-clamped, the DTI scale reset to zero, and the table once again traversed. Any deflection should have been substantially reduced or removed. This process is repeated until any deflection is removed and the table is then firmly clamped.

A second method to ensure that the table is set such that parallel grinding is achieved is to provide physical proof. The usual approach is to use a piece of ground stock bar, drill morse centre holes in each end on a lathe and mount between centres in the cylindrical grinder. Having set the angle of the table to zero the workpiece is ground until a fully ground surface is achieved. This workpiece is again as long as possible, however, as the length increases the diameter will need to increase to avoid any deflection while grinding. This should also be mounted evenly either side of the table pivot to ensure that any offset will effectively be equal at either workpiece end.

Once the bar has been ground such that a continuous ground surface has been provided, then each end is carefully measured with a micrometer. Where the end readings are the same the table is demonstrated to be set parallel. Where there is a difference in readings the bar is shown to be tapered. As before a dial test indicator located on the machine bed, or grinding head, and applied to one end of the workpiece is used to measure either a positive or negative displacement, where half of the difference between the micrometer readings are applied to remove the taper. Once this adjustment has been made the surface of the test piece is then re-ground and further measurements taken. This process is continued until the test piece is found to be completely parallel.

6.15 MEASUREMENT

One of the advantages of any machining process that is undertaken between centres is that the workpiece can be removed form the machine and reinserted without losing its centre. This means that when cylindrically grinding measurement can be undertaken on machine or off machine. When removing workpieces from the machine the only key operation is to ensure that the centre points remain clean and that the centre cavities in the workpiece are free from swarf of debris prior to re-mounting between the centres.

To measure the workpiece on machine, the grinding wheel should be moved off the workpiece, the workpiece rotation stopped and the grinding wheel rotation stopped to ensure that it does not come into contact with the operator when rotating. The workpiece is then measured using a micrometer, and good practice suggests this should be done in at least two positions to check if there has been any taper generated or a slightly convex workpiece profile where deflection in the centre has resulted in lower amounts of material removal.

Grinding is a precision process, and micrometers tend to be a better measuring instrument for establishing workpiece sizes than using vernier or digital callipers. Vernier callipers in particular, tend to be limited to an accuracy of 0.02mm, but also are subject to handling errors and can easily mark the workpiece surface with any sharp chisel points machined onto the jaws of many types of calliper. Figure 6.15 shows a gross example of a handling error (on a rectangular bar) that can introduction error into the value obtained from a measurement.

Irrespective of the measuring instrument used the component drawing tolerance will denote what instrument is used. It would be impossible for valid measurement to be obtained for a dimensional tolerance of ±0.01mm using a measuring instrument with a stated accuracy of 0.02mm. A good general engineering approach suggests that the measuring instrument should have a resolution of roughly four times that of the workpiece tolerance. Standard micrometers have a resolution of 0.01mm, and relatively inexpensive versions fitted with a vernier scale allow measurements to within 0.001mm. Some expensive variants are available, with significantly smaller resolution, however, when this level of accuracy is required, the workshop environment often needs temperature and humidity control to give gauge room conditions, where thermal expansion issues are addressed. The advantage of using external micrometers is that they have a flat face, and handling errors which can commonly be experienced with callipers are removed, and accurate measurement of workpiece diameters undertaken.

To measure a plain diameter measurement on machine once the workpiece and grinding wheel rotation has ceased, and the coolant switched off any guarding can be removed to give full access to the workpiece. It is usual, but necessarily essential to wipe the workpiece to remove any coolant remaining. The workpiece is then measured ideally in three places to

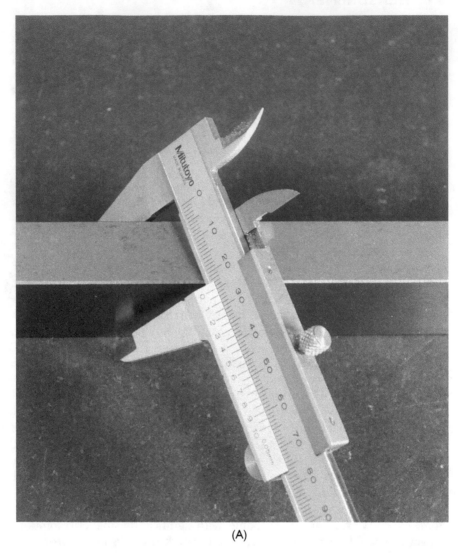

(A)

Figure 6.15 Operator handling issue introducing measurement reading error

confirm the uniformity of the workpiece profile and assess its size. Following measurement, any guarding is replaced and either another roughing cut is set using the head *x*-axis handwheel, or a finishing cut to depth set, and the grinding operation completed.

There is no essential requirement to leave the workpiece on the machine and as the workpiece is mounted between centres it can be removed, wiped clean and measured in a metrology area away from the machine. However, it is essential to ensure that the workpiece centres are clean and free from

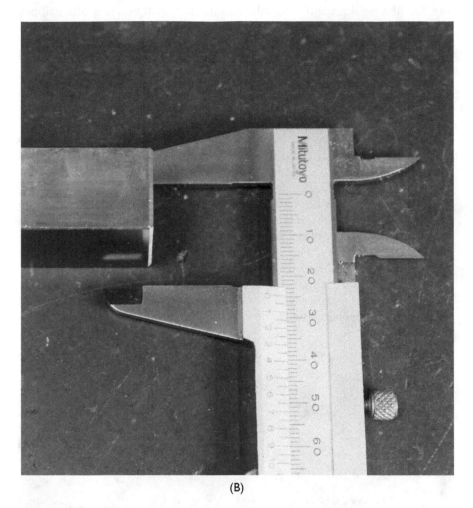

(B)

Figure 6.15 (Continued)

any debris or the workpiece will not return to its previous axis. To remove this potential for error introduction the operator could choose to leave the workpiece in situ, and the only concern will be where to place the measuring instruments so that they will not become covered in coolant or grinding chips, or in a place where they can be dropped.

Measurement of the distance from the workpiece end face to a shoulder can be a little more difficult to determine on machine, and the workpiece often has to be removed as tailstock centres and the workpiece driving head often prevents access for measuring instruments. The approach to measuring these depths is again driven by the resolution required and the resolution of any measuring instrument. Simple measurements can be obtained from the end of a workpiece to a shoulder using the depth function of a calliper, or by using a depth micrometer. In this situation, despite the depth micrometers better resolution, the surface area of the feature used as a datum, may not be sufficient to make a stable platform for the depth micrometer anvil leading to error. Where a measurement of high accuracy is required or a measurement with difficult access is required a stack of gauge blocks (slips), and 'finger clock' (DTI) can be used as a comparator. When the stack height of the gauge blocks is shown to be the same by moving the stylus of the DTI from the gauge stack to the datum face of the workpiece and getting a zero deflection the length is equal to the stack height of the gauge blocks, and a valid measurement obtained. Figure 6.16 shows a typical example of this method being undertaken.

Measuring distance between shoulders is easier, and often undertaken on the machine. While there are a number of approaches that can be taken to measuring distance between shoulders, it is easy to determine a width to a high level of accuracy. The simplest and quickest approach is to use the internal jaws on vernier or digital callipers. These types of measuring equipment are quick and easy to use, however, as with measuring external features they are prone to measurement inaccuracy due to handling error, scratching of the workpiece faces, and in the case of vernier callipers, often only have

Figure 6.16 Distance to shoulder being established using gauge blocks and comparator

a resolution to 0.02mm and therefore cannot give a valid measurement to a greater accuracy than this.

Internal micrometers are another option, however, these have similar handling error issues that callipers are prone to, and have a minimum width, and while accurate to 0.01mm or for models with a vernier scale 0.002mm often cannot be used to measure gaps narrower than 25mm, given the minimum length of the micrometer body that has to be inserted in the gap. It is possible to use a telescopic gap gauge between the faces and then measure using and external micrometer, however, given the precision that grinding is often undertaken to, there is the potential for a multitude of errors to be introduced.

One of the simplest and most accurate methods is to use a pack of gauge blocks (slips) to establish the gap between two faces. Given the rectangular nature of the blocks any measurement error from handling is removed. Gauge blocks are calibrated to varying accuracies depending on the grade of the set, however, a standard workshop grade set (grade 2) are going to be manufactured to an accuracy of 0.0005mm. Inspection grade and calibration grade gauge blocks are manufactured to closer standards than this. Any box of gauge blocks will have a calibration table identifying each blocks error, and as such it is relatively simple to undertake precise measurements simply, and remove any manufacturing error from the gauge block.

To measure the gap between two shoulders a pack of slips are built up until they can just be inserted between the shoulder faces. This pack should be able to be pushed in the gap without using force, however the pack should have no easy movement or rocking when inserted in the gap. Once a pack has been built that just fits the gap, addition of all of the block sizes gives a precise value for the gap between shoulders. Figure 6.17 shows an example of a slip pack inserted between two shoulders.

To determine sizes for a taper requires the workpiece to be removed from the machine, and taken to a surface plate. Depending on the method used to dimension the taper on the component drawing, generally drives the approach. If the taper has been dimensioned using taper value/inch, or a dimensioned included angle then the simple method used in Example 1 can be used. If the drawing has been dimensioned with a taper/inch or included angle, and must be ground to a known diameter established at a reference point as shown in Figure 6.18, then the approach shown in Example 2 also needs to be utilised. It is almost always a requirement to check both the angle of taper, and the diameter of a taper at a known point.

Taper measurement using rollers, gauge blocks and a micrometer is a simple accurate method for measuring tapers using either type of dimensioning system. However, mounting the workpiece precisely vertical or horizontal is important, and if an inspection fixture with centres is available this can simplify mounting, however, the use of vee blocks and face plates which are generally available in workshops provide a good substitute. A simple set up for measuring a taper can be seen in Figure 6.19.

Figure 6.17 Gauge block pack used to determine distance between workpiece shoulders

Example 1 – Measurement of taper expressed in inches/foot

Firstly the workpiece needs to mounted vertically or horizontally. Often vertically, is the easiest method, as retaining rollers in place can be more straightforward, however, using inspection mounts between centres becomes more difficult. two sets of gauge blocks of identical height are then placed adjacent to the taper as shown in Figure 6.19. The height of these blocks is not especially important, although operators should ensure that there is enough remaining length of taper to be able to take a second measurement, which should be at

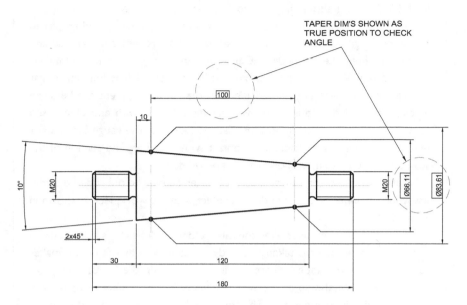

Figure 6.18 Taper dimensions on workpiece component drawing

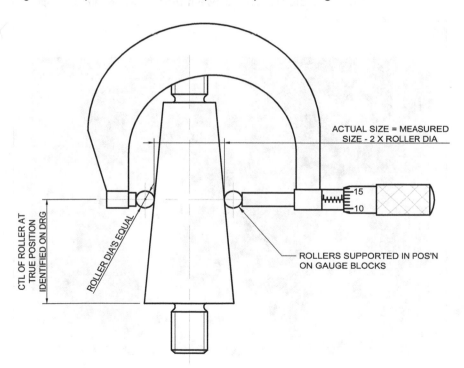

Figure 6.19 Example showing set-up for measurement of taper

least 1″ (25.4mm) above the first. For slower tapers it may be advantageous to have longer a length, to ensure accuracy. A pair of rollers should be placed on top of the gauge block stacks. The diameter of the rollers is not critical, and operators often take the pragmatic approach and use drill shanks, end milling cutters, dowel pins or sections of ground bar as rollers. It is important that the diameter of the rollers is sufficient that that a measurement can be taken across the outside of the rollers without interference with and block used to mount them. It is also important that the height of the gauge block stack ensures that the rollers come into contact with a fully ground section of the workpiece and that any end distortions are avoided.

To establish the taper a measurement is taken across the outside of the rollers when the inner faces are in direct contact with the workpiece as shown in Figure 6.20.

Once this has been undertaken the gauge block stack height is increased and a second measurement is taken in exactly the same way as the first. It makes sense when deriving tapers expressed in inches/ft then the increase in gauge block stack height should be increased in whole inches, preferably of an even number to simply multiplication. The smaller value is then subtracted from the first to give a value for the difference between the two widths and the taper calculated.

Figure 6.20 Measurement across rollers being undertaken

Borrowing values from an earlier example if a workpiece is intended to have taper of 1.686"/ft then the following measurements and calculations need to be undertaken. It can be seen that the difference between the two diameters is 0.1405" when the increase in gauge block stack height is 1". This value is multiplied by 12 (1"= 1/12 ft) to give the value for inches/ft.

$$0.1405'' \times 12 = 1.686''/ft$$

This approach does not require the roller diameters or the axial position to be known, and only confirms the taper angle. The approach is also used to establish the taper angle when expressed in degrees. In this case the difference between measured diameters is divided by two to give a value for a right-angled triangle.

Using a metric example the difference between the two measurements was 3.569mm, and the increase in height was 25.4mm. The value of 3.569mm was divided by 2 to give 1.784mm. This allowed the angle for aa right-angled triangle to be calculated.

$$\text{Using}: \text{Tan angle} = \frac{\text{opposite}}{\text{adjacent}}$$

Distance opposite angle = 1.784mm
Distance adjacent to angle = 25.4mm

$$\text{Therefore}: \quad \text{Tan angle} = \frac{1.784\text{mm}}{25.4\text{mm}} = 0.07025$$

$Tan^{-1}\ 0.07025 = 4.018°$
$4.018° \times 2$ (to give included angle) $= 8.04°$

When generating tapers, it is normal to undertake the above calculation to establish that the correct angle of taper is being ground, and then to undertake a second measurement to establish that the taper has been ground to a required diameter at a known position. To determine a value for the diameter of a taper at a datum point, or identified dimensional check position, the approach shown in Example 2 needs to be used.

Example 2 – Establishing component drawing reference size at designated position

The approach taken to determine a known diameter requires a pair of rollers should be placed on top of the gauge block stacks. The diameter of the rollers is again not critical but must be of an equal size and of a known diameter. The diameter can be established by accurately measuring, however, operators often take the pragmatic approach and use drill shanks, end milling cutters, dowel pins or sections of ground bar as rollers. The height that these rollers are set at, must be such that the centre line of the rollers be mounted at the datum check point as designated by the component drawing.

Establishing the datum height, can require some ingenuity as they use the end that the workpiece is resting on. It is normal when mounting the workpiece to use a gauge block stack to support the workpiece on any datum face. This gives a known distance from the datum face to a known base such as a surface plate. From this known surface two gauge block stacks are built to the height of the datum position MINUS half the diameter of the measurement rollers. This ensures that when the measurement rollers are placed on top of the gauge block stack the centre line of the rollers is position on the axis of the component datum line, as shown in the diagram illustrated in Figure 6.19.

Once the rollers have been mounted in position the distance across the rollers is measured, and both the diameters of the rollers subtracted from the value obtained to give the size across the taper at the point where the inner diameter is in contact with the workpiece.

6.16 COMMON CYLINDRICAL GRINDING FAULTS

Poor surface finish

Surface grinding is no different than any other cutting process and the causes of poor surface finish have similar origins. The depth of cut and the grade of wheel used are often primary factors in the generation of a poor finish. Coarser wheels create a larger chip creating a poorer surface finish, which may be good for material removal rates, however, will not provide a high standard surface finish. A solution to this is to reduce the depth of cut or mount a finer grade of wheel, however, care must be taken to ensure that the grade of wheel is suitable for the material being ground and will not immediately clog as this would introduce a different problem such as burning. If the operation is using coolant it could be that the flow is not enough to remove chips fast enough and the operator should increase the flow where possible. Over dressing the wheel can also be a factor as it can leave grains standing proud which may also be a factor. That said, one other factor to check, is whether the grinder is working correctly. The operator should check if the rotation of the workpiece smooth and continuous, are the centres correctly inserted, and is there enough end load applied to the workpiece to retain it firmly between the centres and preventing any axial, or trans axial movement of the workpiece. Spiral marking of the workpiece can often relate to poor alignment, overly fast traversing rates, or the grinding wheel being out of true.

Burning

This is a common fault and is identified as clearly visible dark discolouration of the workpiece. It indicates that there has been too much frictional heat. In extreme cases, and when grinding sensitive materials, burning can result in heat treatment of the workpiece surface (i.e. hardening, micro cracking, or the creation of internal tensile stresses). It is most commonly found in dry grinding, but can occur when coolant used if flow is not enough to remove the frictional heat. The most common cause is using a cut that is too heavy for the wheel and material being ground. It can also occur if the wheel is glazed or loaded as more frictional heat is built up due to rubbing, rather than cutting, by trying to push the workpiece under the wheel too quickly. Where this is the case, the solution is to dress wheel more frequently or reduce the feed rate.

Tapering

Tapering of the workpiece intended to be parallel, is common when extreme care has not been taken when re-aligning the table to a zero position following deliberate taper grinding. The tolerances usually applying in cylindrical grinding often require the machine to be set exactly parallel. Good engineering practice suggests that this should always be the case, but returning the workpiece axis to match the machine axis can take time. The use of a long test piece mounted equally astride the table pivot point to allow grinding and measuring of diameters at each end. Once the difference between diameters produced at each end is eliminated the table has been proven to be parallel, and any error in form removed.

Low wheel life

A correctly selected, dressed grinding wheel being give an appropriate depth of cut for the material being ground at a feed rate that is similarly appropriate, should have a substantial life. Grinding wheels that are judged to have a low life have normally suffered from a number of causes. Insufficient coolant, wheel speeds set too low, over-dressing a grinding wheel (i.e. dressing it before it requires it), or taking overly large and too many dressing passes. More importantly the cause of low life spans of grinding wheels is likely to be the use of a wheel that is too hard or too soft for the material being ground. This means that the wheel 'loads up' requiring more frequent dressing, or wears away too quickly.

Scratching

Scratching of the workpiece can be seen in two areas. The first is where side grinding has taken place such as where the side of a grinding wheel has come up against a shoulder. While the pattern left on a workpiece feature will always be different where the side of a wheel has been used compared to the peripheral surface, the pattern produced is normally attractive and regular. Where wheel side has not been trued and dressed will result in a series of irregular and undesirable scratches. The solution is simply to dress the wheel side that will be brought into contact with the feature. The second cause of scratching is where coolant filtration system is not fully effective and swarf is not completely removed from re-circulating coolant and gets fed back under the wheel. Replacement of coolant filters or membranes normally cures this issue.

Slow cutting

While surface grinding is by its nature never going to be a fast material removal rate, it should be efficient. Slow cutting rates are almost always caused by feed rate and wheel speeds not fast enough. The solution is to increase wheel speeds and feed rates to a max point before burning occurs. However, most surface grinders will have a fixed speed, and only a limited ability to adjust feed rates. These settings will have been provided by the manufacturer of the machine to provide the optimum settings for a range of grinding operations. If the operator considers that the cutting rate is slow, investigation of other criteria such as wheel type and the amount of dressing required should be undertaken.

Wheel not cutting

This normally happens when the grinding wheel has been glazed by truing. If initial truing of a wheel has been undertaken using excessively heavy passes of the diamond the surface of the wheel can become glazed due to the heat generated. The solution is to re-dress the wheel using lighter slower cuts until the glaze is removed and a clean periphery is achieved.

6.17 WHEEL CHANGING

Wheel changing is relatively straightforward on most cylindrical grinders, although in commercial workshops there is normally a legal or approved code of practice requirement for personnel to have undergone training in grinding wheel safety and changing wheels, in order for the operator to demonstrate competency. This training is not onerous, but in the event of an incident regulatory bodies would investigate training and competency levels.

To change a wheel the machine should be electrically isolated to prevent inadvertent initiation of spindle start. Given that the power switch is a button on the front of the machine this can be easy to do. The guarding around the wheel and wheel flange is then dismantled exposing the wheel assembly.

This guarding may be of a type that comes around both sides of the wheel, but it is more common for a cover plate to be bolted onto one side of the wheel housing, and removing this cover exposes the wheel and hub assembly. It is also unusual for a cylindrical grinder to have a spindle lock, therefore the operator grips the wheel with one hand and uses a peg spanner or other similar tool to loosen the spindle flange. It is unlikely that a hammer will be required and the flange should only be hand tight.

The grinding wheel is then pulled from the spindle, and the area around the wheel cleaned of any grinding deposits, using a small brush or lint free cloth, leaving a clean spindle, and wheel housing. Significant deposits that have been in place for some period of time may require more aggressive methods to remove any encrusted grinding deposits and cleaning with a wire brush may be required.

The grinding wheel selected for use first needs to be checked that its maximum rated speed is equal to, or more than the operating speed of the surface grinder spindle. It then needs to be checked for cracks, chips, or any other defect. This is primarily visual, however, grinding wheels should also have a process called 'ringing' undertaken before mounting onto the machine spindle. Ringing is where the operator holds a plastic or wooden shaft (often the handle of a hammer or mallet) and then gently tapped on the side using another wooden or plastic implement. The process is very gentle and should not damage the integrity of the wheel in any way. This tapping should result in a clear 'ringing' note being heard. This ringing note identifies that the wheel is free from cracks or internal defects that may leave to wheel failure or bursting.

A wheel that does not have a clear ringing sound and sounds a dull note is probably cracked, and should be broken or smashed, and disposed of.

The wheel should also be checked to ensure that paper discs, or Blotters, with the grinding wheel nomenclature and manufacturers details are on both sides of the wheel. Blotters are an important feature as they from the interface between the grinder driving flanges on the spindle and the grit of the wheel. The driving flanges on the spindle are the mechanism for transmission the spindle rotary motion to the grinding wheel. As the flanges are generally steel, and the grinding wheel is made from an extremely hard and incompressible surface the papers on each side of the wheel allow the grit to bite into the blotter and provide a surface for the flanges to engage with providing a frictional lock, that does not require a compressive load. This would otherwise damage the flange and/or the grinding wheel, or provide an unsafe condition.

Once all of the checks have been undertaken the wheel is mounted on the machine spindle. It is normal for grinding wheels to be purchased with a centre hole matched to the diameter of the machine spindle, and often cylindrical grinding wheels have a substantial sized mounting hole to allow fitment to larger diameter machine spindles. To accommodate multiple machine spindle diameters the grinding wheel manufacturer will normally supply a series of plastic inserts that reduce a larger hole to each standard spindle size and provide a close fit, and do not affect performance once the wheel has been installed and trued. Grinding wheels are very sensitive to imbalance, and any vibration that is caused will transmit into a poor surface finish. While the spindle does not drive the grinding wheel it does require a close fit to remove any radial movement in use. Figure 6.21 shows a wheel fitted onto the spindle and the spindle flange re-installed.

Once the wheel is mounted the spindle flange is screwed back on, in a reverse operation to its removal. The flange is tightened using the spanner hand tight. It does not need, and should not experience any heavy torque loads to be applied. It only requires to be tight enough that is does not slip when experiencing the frictional forces of grinding, and this is the function of the paper discs on the grinding wheel. The wheel should then be spun by hand to check that the wheel has free rotation and the wheel guard reassembled.

At this point the power can be switched on to the machine, the dressing diamond mounted and the and the wheel started. Truing a grinding wheel is essential to remove any imbalance and ensure that vibration due to any wheel eccentricity is removed. It can take a little time as it needs to

Figure 6.21 Cylindrical grinding wheel mounted on spindle

be undertaken gently enough that the wheel does not 'glaze' from the heat being produced, however, the wheel needs to be precisely round. Coloured wheels, often provide clear evidence of the wheel being precisely round as parts of the periphery will be discoloured where the diamond has not reached, and the truing process continues until a uniform periphery is see. White wheels in pristine condition present more of a challenge, and listening to the sound made by truing is a clear indication to the operator whether the diamond has a continuous cut or interrupted. once the wheel has been trued and then finely dressed. the wheel is stopped and the machine cleaned of any grinding debris, ready for use.

6.18 SAFETY ISSUES

Cylindrical grinding is one of the less complex machining operations given its limited number of degrees of freedom. However, the peripheral speed of the grinding wheel, its construction, the risk of entanglement with the rotating workpiece, and the proximity of the operator make this a process that has more risk than many others and consequently requires more focus. While the outcome of any incident is probably no more or less severe than entanglement with any other machine, there is often little or no notice of a problem occurring. It is important that operators undertaking cylindrical grinding operations do not have distractions and can focus on the operation be performed.

Of considerable concern is the risk of having a wheel burst. This is where the grinding wheel fragments, and given the proximity of the operator to the process and the speed in which it happens, it is unlikely that the operator will have time to move out of the way of debris. In cylindrical grinding the wheel is normally directly opposite the operator, and facing them. The wheel guarding, and machine head, provide much protection from debris, however, it is likely that sections of a burst wheel will be ejected from the front of the wheel. While this is probably moving downwards, fragments are also likely to ricochet into the operator. Grinding wheels in general are remarkably robust cutting tools when used in their designed condition. Wheel bursts are very rare, and almost all of them have been caused by either cracked or damaged wheels, operating them out of their intended performance parameters (i.e. material type), depths of cut and feed rate, or from improper mounting. To ensure the integrity of a grinding wheel the operator should check for cracks by 'ringing' a wheel before mounting, when not being used store them where they are not subject to damage, and selecting the correct wheel for the task. It is also good practice for the operator to check any already installed wheel for defects prior to use.

Grinding wheels are highly abrasive and normally while generally rotate at lower speeds than surface grinding wheels due to a larger diameter they will often still be rotating at speeds of 2500rpm with a high peripheral velocity. Contact with the operator's bare flesh will result in the generation of immediate and deep burns. While loose clothing is never a good idea in a workshop, and heavy gloves often impractical to wear the operator should consider always wearing clothing or a dust coat with long sleeves. It is less common to leave a cylindrical grinding wheel running while checking a workpiece during grinding or setting up the dressing diamond, to ensure the wheel does not become unsettled. However, the safest condition is to turn the wheel off during these operations and certainly stop the workpiece rotation before reaching into the machine.

The rotational speed of grinding wheels and the friction that is generated means that any unsecured workpiece will be ejected from the machine with considerable velocity. Modern workshop cylindrical grinders tend to have a completely enclosed guard to separate the operator from the workpiece. However, there remain a substantial number of workshop grinders that are open. Where no built-in guard is fitted to the machine, it is recommended that a separate vision screen be placed between the operator and the work-piece/wheel area. This needs to be firmly mounted to ensure it does not fall into the working area, and also needs to be robust enough to remain in place if impacted. The most common reason for workpiece ejection when cylin-drically grinding is a workpiece not being sufficiently supported between its centres. When any workpiece has been ejected there has almost certainly been some damage to the grinding wheel. When this has occurred the opera-tor will need to carefully check the wheel for any cracks, and will in any event need to re-dress and true the wheel.

Excessive cut depths are not in themselves especially dangerous and one of the reasons that surface grinders are belt driven is that if a depth of cut is so great it just causes the wheel to stop, with the drive belt typically becom-ing displaced. Most of the damage is caused to the workpiece, however, as with all rotating machinery, the machine should be switched off and isolated before re-establishing the belt onto its drive pulleys.

When grinding into shoulders operators should be circumspect as the wheel approaches the shoulder. The sudden impact of the grinding wheel into a shoulder where cross feed may be half the width of the grinding wheel, will result in impact damage commensurate with feed into a dressing tool. Where an operator inadvertently introduces the side of a wheel to a shoulder it may result in side loading of the wheel, and more probably twist-ing of the workpiece. While grinding wheels are extremely robust, operators should use a degree of circumspection when approaching a vertical surface as the area being ground increases massively.

Chapter 7

Drilling

Drilling is one of the most common operations undertaken in mechanical manufacturing, and is one of the oldest manufacturing processes. Holes are provided for a number of reasons from providing a basic hole to be tapped with a thread, clearance holes, holes to allow for corners to be formed in fabrications, and perforations to lighten structures. While holes can be produced through thermal processes, such as oxy/propane, plasma, laser profiling and water jet cutting, by far the most common workshop drilling process is the use of a standard twist drill. Drilling is characterised as a process that involves a tool being held firmly in a chuck that rotates at speed and is pressed down into a workpiece to cut away material leaving a cylindrical cavity. There are some variations such as deep hole drilling, however, these generally require special machinery, and the following sections do not go into these in any detail, but instead focus on drilling operations undertaken in general machining workshops using common types of tooling found within them.

7.1 THE WORKSHOP DRILL

Drills come in a wide range of sizes and configurations from large multi-spindle drills, down to a simple hand drill. By far the most common types of drilling machine found in a general engineering workshop are the pillar drill, bench drill, and radial arm drill. All three types perform essentially the same function, and as a broad principle, all use the same tooling and all have only two degrees of freedom. Drills are mounted either into a chuck such as a Jacobs chuck, which requires parallel shanked drills, or drills mounted directly into the spindle using morse taper drills. It is the size of the morse taper that controls the largest size of hole which can be drilled on the machine.

Pillar drill

The pillar drill is normally floor standing and comprises a base, column, table, and drilling head. The size of the base and column can vary greatly

DOI: 10.1201/9780429298196-7

from some simple arrangements where the column is made from a relatively thin walled steel tube to substantial castings. What differentiates the machines is the size of hole that can be drilled. Lighter models of pillar drill are unlikely to be able to drill holes greater than Ø10–12mm into mild steel, whereas more substantial models will cope with drilling holes of approximately Ø32–50mm. In the modern era almost all Pillar drills have their motor located on the top of the supporting pillar. Drive can be direct, and is a feature of large, or more expensive models, but more commonly, belt drive systems are used to power rotating spindles. Figures 7.1 and 7.2 show examples of a large pillar drill manufactured from castings, with an integral gearbox, and a more modern, but less capable belt-driven pillar drill.

Belt-driven machines are simpler in design and cheaper to manufacture as a gearbox is not required, and pulley systems allow, for simple speed changing, however, belt drive systems also have a power restriction in that the size of belt, and angle of engagement limit the power that can be transferred through a belt drive system, and larger machines require a geared drive.

All drills have the ability to manually feed the drill into a workpiece normally by use of a simple lever, which is commonly a capstan mounted on the side of the drilling head, however, more substantial machines have a feed mechanism, which can be required to ensure large holes a successfully drilled without significant work hardening, due to the inability of the operator to maintain enough downward pressure.

On all drills the workpiece is either directly mounted on table of held in some type of work holding device such as a hand vice. The height of the table almost certainly is adjustable, which may be through release and tightening of a simple friction clamp, with the height adjustment made by the operator lifting or lowering the table by hand, or through use of a geared mechanism where a handle is rotated to raise and lower the table. The table requires adjustment to accommodate varying workpiece heights and configuration, and also to accommodate the limited stroke of the quill.

It is less common for drills to be fitted with any coolant system in general workshops, although production machinery has a higher incidence of coolant systems. It is more common for operators to lubricate the workpiece cutting edge with cutting oils, than use high flow emulsified oil coolants.

Bench drill

The bench drill has a high degree of similarity with almost all of the features of the pillar drill. Again it has a base from which a short pillar emerges. An adjustable table is mounted to the pillar with a motor and spindle assembly are mounted on top of the pillar. The drive system and speed adjustment is almost certainly a simple belt drive system. Figure 7.3 shows a typical example of a simple belt-driven bench drill.

Figure 7.1 Pillar drill with direct drive, feed mechanism and manufactured from castings

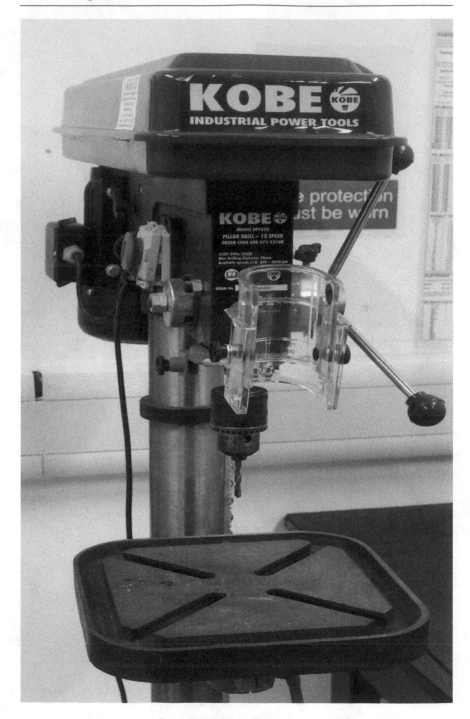

Figure 7.2 Common workshop belt-driven pillar drill

Figure 7.3 Typical belt-driven bench drill

The capacity of bench drills to drill large holes is very limited in work-pieces made of materials that are much harder than mild steels, although softer materials such as aluminium, and many engineering plastics can be successfully drilled on them. The capacity of these machines is often only Ø10mm when drilling into mild steels, and the size of chuck fitted to the spindle, and the spindle morse taper limits the size of drill which can be fitted into the machine.

The bench drill, however, can give the operator much greater feedback than larger types of drilling machine. As the workpiece tends to be closer to eye level there is normally good visibility of the hole being drilled. Given the relatively light construction of this type of machine there is also a lot of feedback felt during the drilling operation, which allows the operator to be more sensitive to drilling conditions.

Radial arm drill

The radial arm drill is the third type of drill which is commonly found in workshops. While they tend to be confined more to fabrication workshops, than toolroom type workshops they are still very common. Radial arm drills vary considerably from bench or pillar drills in that they do not have a table that moves up and down a column. Radial arm drilling machines are normally used for heavy work involving high downward loads, either as a result of the mass of the workpiece, or the pressure exerted by the drilling operation. It is normal for a radial arm drill to have a large base or machining cube to mount the workpiece on. This means that height adjustment is achieved by moving at head and arm assembly up a substantial column. Radial arm drills will normally have two motors: one to power the drilling spindle, and one to move the head up and down the column.

The head of a radial arm drill is mounted on an arm which pivots on the machine column. The head can also move transversely along this arm to allow its positioning at differing radial distances from the column. The configuration of an arm pivoting on the machine pillar and the ability of the head to move along this arm gives a significant amount of positional freedom.

The power is transmitted through a geared system and it would be unusual to find a radial arm drill that has a belt-driven spindle. Almost all machines have power feed mechanisms, and advancement of the machine quill into the workpiece may have a manual function but any lever or capstan is commonly an actuator for an electric feed.

7.2 MACHINE CONSTRAINTS

All machines have constraints, however, machinery used for drilling can have more issues to deal with than drilling on a lathe or on a milling machine. Principal issues are as follows:

- machine power required to drill size of hole through the required material;
- size of component able to be drilled such that it is securely mounted on the drill table;
- radial distance between the drill support pillar and centre line of the spindle;
- size of chuck, or spindle morse taper;
- stability of the drill; and
- ability of the table to be adjusted to a point where the correct depth of hole can be achieved.

All drilling machines have a limit to the distance the quill can move and as a result there is a limit that a hole can be drilled to. While this is rarely a problem as it is the drill bit that provides the limit that a hole can be drilled to given the length of flute, however, it can be an issue if holes are to be drilled in the bottom of a pocket, or have a feature that introduces a clash between part of the machinery and the workpiece. Where this is an issue, operators should give consideration as to whether the hole could be drilled from the opposite side, and while this is not always a possibility, it is the first consideration to make that may avoid having to create special purpose tooling.

Table size and mass, are often an issue. Some machines have tables of substantial mass, and it is not unusual for the height adjustment to consist of a simple friction clamp, which provides easy and highly flexible adjustment, but also requires the operator to move the table up and down the pillar support by hand, and to tighten the friction clamp also requires the operator to support the table with one hand. In extremis, especially where the table needs to be precisely rotated to allow any workpiece that is already clamped to the table to be moved into position, assistance may be required. However moving the table to a height where there is only a small clearance between the drill point and the workpiece is advantageous, as it ensures that the distance the quill will need to be displaced is within its maximum stroke, and it removes the requirement for multiple rotations of the quill feed arm or capstan. That said, if multiple drilling operations are to be undertaken, such as centre drilling, followed by a pilot hole and desired drill size, then the table should be set to a height such that the longest drill can be inserted into the chuck, unless the machine has a cranked height adjustment mechanism.

The distance between the rear pillar and the centreline of the machine spindle is a key constraint on the size of workpiece that can be mounted on a drill. While it may be possible for a workpiece to be reversed, care must be taken to avoid a significant overhanging mass, and trying to drill holes into a large workpiece on an undersized drill should be avoided.

While rarely a constraint for floor mounted pillar drills and radial arm drills, bench drills can experience limitations with the maximum height that can be achieved between the end point of the drill and the table. The real flexibility of pillar drills which are floor mounted and have a table with a friction clamp are demonstrated when long workpieces need to be drilled, as the table can normally be dropped to the base or swung out of the way allowing the workpiece to be mounted onto the floor pedestal.

7.3 SIZE OF DRILL AND POWER

There is a link between the size of a drill and the maximum size of hole that can be drilled. In broad terms the size of chuck fitted is a clear indicator of the maximum size of drill, as is the morse taper within the spindle. Taper shank drills are supplied with shank tapers that increase in standard increments (i.e. no. 1, no. 2, no., 3 morse, etc.). The largest size drill that fits a spindle taper is a broad indication of the maximum size of hole that could be drilled on that machine. The tendency for larger drills to have the soft shank turned down on a lathe to allow it to be fitted into a chuck, is not good practice and can either damage the machine, or lead to the drill becoming wedged in the workpiece and the machine stalled. However, even when the physical machine constraints are taken into consideration, the material being drilled and its thickness become an aspect of the ability of the machine to drill a hole.

Drilling mild steels with a high speed twist drill is a good benchmark for the performance of many machines. Many bench drills are fitted with a chuck that will accept $\varnothing$12mm or ½"dia drill bits, yet struggle to drill mild steel at this diameter, however it is common for these machines to be perfectly able to drill $\varnothing$10mm holes into mild steel. This identifies a constraint for the machine and the material. Harder materials such as alloy steels, and some softer materials that can be tricky to drill such as brass may cause the machine to stall at smaller diameters. Therefore, whist drill bits may fit in the chuck of the machine, the operator needs to have an awareness of the capability of the machine to be able to drill a hole in the workpiece of differing materials, and that for some materials a larger machine with more power may be required. Being able to continuously drill, without causing work hardening, and damage to tooling, is in part down to the power of the machine.

7.4 TYPES OF DRILL BIT

A HSS twist drill is one of the most commonly found cutting tools in an engineering workshop. The most common type of drill found in the workshop is a drill comprising two flutes and a point angle of 118°. Cutting takes place across two lips which have a 10–12° clearance angle behind them. As the lips cut into the workpiece, the swarf chips are forced up into the flutes of the drill and exit at the surface of the hole, either as two long continuous chips or as a series of small chips. The material type, rotational speed, feed rate and continuity of feed all affect the form of swarf chips produced. The shank of the drill has two types. Parallel shank drills are designed to be held in a chuck, and taper shank drills are designed to be inserted directly into the drilling machine spindle, or a morse taper sleeve, to correctly size it for the machine being used. Parallel shank drills are commonly found in sizes up to ∅16mm (⅝″) which tends to be the largest size Jacobs type chuck commonly found in workshops, although manufacturers will commonly be able to supply diameters up to ∅20mm. Thereafter taper shank drills tend to be used. Taper shank drills are also available for smaller sizes, however, these are less common, due to their increased cost compared to a parallel shanked Jobber equivalent.

These drills are also supplied in a number of lengths with, stub drills, jobber drills, and long series drills being the standard formats. There is little difference in the configuration of the cutting edges and flute profile, however, shorter stub drills are more rigid and can often be used at higher feed rates, and rarely require support, however, are limited in the depth they can drill to. Jobber drills are the general purpose drill that has extra length but still retains a degree of rigidity. Long series drills are for drilling deeper holes, however, are less rigid and require pecking every diameter of depth drilled to ensure that swarf chips can progress up the flutes. Extra length drills are also available from manufacturers and these have substantial length and can be used where otherwise specialist gun drilling would be required. Extra length drills, and in many cases long series drills need to be used with a process of stage drilling where stub and jobber drills are first used. The longer the drill the more potential for oversize drilling is experienced. However, extra length drills do provide an opportunity for drilling deep holes, without resorting to specialist gun drilling, albeit with less accuracy and the problems of getting coolant to the bottom of the hole being drilled.

Twist drills are manufactured to a number of standards including BS, DIN, and ASME standards, and within these standards there are a number of sub types. All have essentially the same configuration, and the configuration of a drill manufactured to an ASME standard, is not going to be significantly different to a DIN or British Standard. The standard point angle is 118°, which gives a balance of centre finding, good concentric cutting and relatively flat-bottomed holes. The main cutting edge or lip of the drill point removes the material therefore longer main cutting edges remove material faster and this is one of the reasons why the common point angle is 118° as it gives the balance of length of cutting edge, and centring ability. It is important for the cutting edges to be the same length to prevent oversize hole cutting.

However, some materials benefit from having significantly different angles, with brittle materials such as some engineering polymers benefitting from a point angle down to 60° which allows material to break away from the lower surface of the workpiece as the drill pushes through. Acute point angles also affect the shape of the cutting edge which due to the flute configuration make the lips slightly convex, and while in many cases will not be a problem, in situations where a high standard of cutting is crucial the face of the flute along the cutting lip needs to be ground away to reduce the helix angle and improve the cutting performance in brittle materials. Flatter points up to 140° can assist with drilling difficult materials such as those like stainless steels and titanium which are prone to work hardening, but again changing the point angle affects the profile of the cutting lips which become in this case become concave. When drilling difficult materials in any amount it is advised that the operator procure a drill designed for that material, which will have good cutting properties.

There are a considerable number of specialist drills for difficult materials or specialist applications where the point angle, point configuration, flute helix angle and coatings all vary. The standard drill point angle is useful for most operations. Smaller angles are better at centring, especially on curved surfaces, and do not wander as much. Larger angles have a reduced 'tapping time' which is the time taken to bite into full diameter, however, higher contact pressure is required and they are less good a centring hence the use of a spot drill or morse centre to identify a location for the drill point to grip when entering the workpiece.

For larger drills the use of a very thin radiused grinding wheel can be used to produce thin grooves across the cutting face, creating a contoured profile across the edge of the lips. This functions very much in the same manner as a the cutting edge of a ripper cutter used in milling, and breaks long continuous swarf chips into small sections. While this is advantageous when drilling ductile materials that tend to produce long chips, it is limited to larger sizes of drill.

Drills can also be ground with a double angle. Where abrasive materials are being drilled such as some types of reinforced plastic and hard materials such as cast iron, the outside corners of the drill are subject to rapid wear, and as erosion of the corners increases, the drill cutting performance rapidly declines. To avoid this the drill is left with its standard 118° point angle for 50% of the lip length and then ground at an angle of 90° to the outside corner. This retains the centring capability, and cutting action where the rotational velocity is low, but reinforces the corners of the drill, so they do not rapidly erode. This type of drill point is also useful when drilling cast iron as it prevents 'frittering' of the iron as the drill breaks though the lower surface of the workpiece leaving a clean circular hole.

Slow spiral drills have flutes with a much lower angle twist drills with a lengthy helix of approximately 10–19° than a standard jobber drill which have a normal spiral have a helix angle of 19–40°. They are used for drilling short chipping materials such as brass, bronzes, some harder plastics and magnesium alloys. Whereas high helix drills with a (rammed) short spiral angle of 27–45° have a spiral are used for drilling long chipping materials such as copper, aluminium, aluminium alloys and soft plastics. Worm pattern drills have an even greater helix. They are often coated, and used for drilling deep holes in steels and materials that produce a continuous chip up to strengths of $1300N/mm^2$. The profile of flute improves swarf removal, different widths are available. The wider the groove the easier swarf is removed from the cutting face and ejected from the workpiece cavity. Congestion in the flutes leads to heat build-up which can be enough to anneal the drill and cause it to break. Therefore selecting a drill bit with the correct flute profile and helix, can prevent drill breakage or excessive wear.

While HSS drills can be used on most materials, the behaviour of some hard materials when being drilled such as titanium, titanium alloys and nickel based workpiece materials means that drill bits with 5% cobalt content may be a better option as use of HSS requires the spindle speed and feed rate to be about 66% of normal to be able to produce a correctly formed hole, and not have excessive wear on the drill point. Similarly coated drills assist with drilling of very hard materials, or those that are heat and acid resistant. While the drilling of non-ferrous materials such as brasses copper and aluminium benefits from selecting a high or low helix flute, the use of bright finish drills rather than coated or plain HSS drills provides a good alternative.

The selection of twist drill for an application can seem like a complex area, and in many respects it is. However, manufacturers data sheets for products provide clear advice on their application, and the combination of flute helix, flute profile, cutting point angles, and coating types are only normally only crucial in a volume production environment, where materials with difficult machining properties are being drilled, or high accuracy is required. Otherwise a standard two-flute HSS twist drill is suitable for most situations.

Other types of drill are available for more specialist applications. DIN1412 shows the form of a spot weld point drill, which consists of a small pointed spur at the centre to penetrate sheet steel and a flat cutting edge to remove the area of spot weld. Three- and four-flute drills are also freely available, and these are used for opening out holes to a larger diameter. They cannot be used for drilling a plain hole as they require an hole to already have been drilled. They are useful where an existing hole has to be opened up by a small amount and do not suffer the chatter that a two flute drill bit would have. Consideration of a suitable pilot hole size and main drill size can often avoid holes having to be opened out, however, modifications to a workpiece can lead to a hole having to be opened up, and the use of a three or four flute drill will prevent damage at the hole mouth.

7.5 USE OF CHUCKS AND CHANGING CHUCKS

For the drilling holes of any diameter under Ø12mm, and potentially Ø16mm on larger machines a three jaw 'Jacobs' type chuck is almost certainly used. These require a parallel shank drill bit which is placed within the jaws, which are widened, or tightened, by inserting the peg of a chuck key into one of several holes of the chuck body, allowing the gear teeth on the body and the key to mesh. This is rotated by the operator until the drill bit is held tight. While no mechanical assistance is required to ensure the drill is tight, holding the chuck key such that the operators thumb can push onto the top of the key tommy bar provides the greatest turning force. Figure 7.4 shows a drill being tightened in a typical 3 jaw chuck. failure to properly tighten a drill bit into a chuck, normally results in the drill bit

Figure 7.4 Tightening drill bit into Jacobs chuck

getting stuck in the workpiece and the chuck continuing to rotate scoring the soft shank of the drill. This also often creates a significant burr on the drill shank which requires careful removal to ensure correct insertion in the chuck on successive operations. In extremis this scoring can be so severe as to require disposal of the drill.

All chucks have a maximum size of drill that can be fitted, however, there tends to be a lower limit as well. While high quality, and expensive, chucks have close fitting jaws many are manufactured with less precision. The jaws also often have a small flat which acts as the frictional surface. This small flat removes the corner of each jaw and depending on the size of it prevents the very smallest drills being fitted. Operators who wish to drill small holes such as $\varnothing$0.5–1mm can find it easier to mount minute drills in a pin vice and insert the pin vice into the chuck of the drill. Figure 7.5 shows this arrangement.

The majority of chucks are retained within the machine spindle using a morse taper system. To insert a chuck into the machine spindle only requires the chuck to be gripped by the operator around its body and then inserted into the spindle using a sharp upward motion. If the chuck does not fall out it is retained in the spindle. To remove a chuck, a tapered drill drift inserted into an oval shaped hole in the spindle is required. The taper of the drift passes through the spindle and will engage the tang of the chuck morse taper. The arrangement is shown in Figure 7.6.

To eject the chuck the end of the drill drift is given a sharp tap with a hammer, which breaks the morse taper and the chuck drops out. To protect the chuck from damage, it is wise for the operator to either catch the chuck, or provide a soft medium for the chuck to drop onto, in order to prevent damage to the bottom face of the chuck, and machine table.

Where larger drill sizes are required it is normal to fit a tapered shank drill bit directly into the spindle. The method for inserting and ejecting these type of drill bits is identical to the process of inserting and ejecting a chuck. However, protecting the drill point when ejecting is more important than for a drill than for a chuck, as damage to the drill point and machine is more likely.

7.6 USE OF TREPANNING TOOLS

Trepanning is a method of cutting larger hole diameters on smaller machines or where the power of the machine is not enough to allow the use of large size drill bits. It has its advantages when cutting thin materials as the point angle of a large drill bit can 'grab' thin materials as it penetrates, and use of a trepanning tool can avoid that problem. In general trepanning is only undertaken on thin materials and any heavy cutting can be considered a milling operation, and may be better undertaken using a slot drill and rotary table. That said, where a workpiece is too large to fit on a milling machine, or being undertaken in a fabrication shop, it can be utilised on relatively

Figure 7.5 Mounting arrangement for ∅1mm drill

heavy sections. When considering whether to trepan, the operator should consider the thickness of the material, as often heavy steel workpiece thicknesses may be better perforated by burning or plasma cutting and dressed using a grinder than trepanning.

Figure 7.6 Chuck ejection using drill drift

Trepanning tools come in a variety of designs with some modern designs intended for CNC use having carbide inserts and two or three cutting points. However, the traditional workshop type has one or two cutting points, mounted on an adjustable bar to allow discrete diameters to be set often over a very wide range.

The design of the majority of cutters, has a tool not unlike a lathe roughing tool outed vertically on a arbor. The body of the tool has a plain central shaft which is inserted in to the drill chuck and commonly has either a twist drill, or plain diameter protruding to a level just below that of the cutting tool point. This central drill or peg acts as an anchor during cutting and prevents the tool from wandering on its radius and ensures a round hole is cut about a central datum. An example of a typical workshop trepanning tool can be seen in Figure 7.7.

Where tools with two cutting points are provided it is essential that these are set to the correct axis, and it is normal for the depth of one point to be set a small distance such as 0.2mm lower than the other to provide progressive cutting.

To use a trepanning tool the workpiece needs to be firmly retained in place as the trepanning process can generate high turning forces. The tool is set to the correct radius and inserted into the chuck of the drilling machine.

Figure 7.7 Typical workshop trepanning tool

The speed that the machine is set to will relate to the type of cutter (i.e. HSS or carbide), and the material removal rate in surface m/min or surface ft/min. In order to calculate the correct speed the operator needs to know the following information:

- hole diameter;
- cutting tool material (i.e. HSS); and
- surface machining rate for the material being cut.

To identify the cutting speed for a ∅100mm hole using a HSS trepanning tool bit in mild steel, the following calculation is undertaken:

Where S = spindle speed
$\qquad D$ = hole diameter

V_c = surface machining rate (= 27 surface m/min for mild steel)

Using $S = \dfrac{V_c}{\pi D}$

$$S = \dfrac{27}{\pi \times 0.1} = 86 \text{ rpm}$$

Having centre punched, or centre drilled a centre point, the workpiece is either drilled to a size to accept the trepanning tool spigot, or drilled using the trepanning tool centre drill/support, and the quill advanced until the cutting tool starts to score the surface of the workpiece. Where the drilling machine has a coolant supply this should be directed over the area being cut. However, this is less common on workshop machines and it is more normal for the operator to periodically apply cutting oil over the circumference being cut to prolong tool life, especially if cutting metals. Downward pressure on the quill feed capstan is applied until the material begins to cut, and this pressure is applied until the workpiece is perforated. It is important for the operator to keep the pressure applied evenly through the cut, but not to overload the tool, to ensure a successful and efficient cutting action.

7.7 WORKPIECE HOLDING

Workpiece holding when drilling is very subjective, as many operators never do more that hold a workpiece in a hand vice, while others use complex clamping arrangements or fixtures to mount the workpiece prior to drilling. The three key factors to be considered are as follows:

- size of hole being drilled;
- positional accuracy required; and
- stability of the workpiece on the machine table.

The size of the hole being drilled is usually a good indicator of the amount of load that is going to be transmitted to the workpiece. While large holes can be drilled using hole saws, or trepanning tools the use of a large diameter twist drill bit is going to apply a high torque loading on the workpiece, which if not restrained will try and rotate, instead of being cut.

Drilling, commonly involves the operator holding the workpiece by hand, either while mounted in a hand vice or just holding onto the workpiece in an area that is free from swarf sweep. This is generally no problem as the loads imparted by the drilling process are low enough that the strength of the operator is enough to allow a correctly formed hole to be drilled. Where problems arise is when the hand vice is large, or the hole being drilled crosses the threshold of an operator being able to prevent rotation, and this is not immediately apparent. Drilling thin materials, or holes into materials such as brass, can generate snatch loads with pull the workpiece from the operators grasp. One method to control this is to position the workpiece such that one edge of the workpiece, or handle of a hand vice rests against the left hand side of the machine pillar as shown in Figure 7.8.

The advantage of positioning any work holding device or workpiece is that any snatch loads are absorbed by the pillar of the machine which provides a physical stop to prevent rotation. As the load from the drill is pressing downwards the operator in many situations does not need to hold the workpiece, or vice, as it is unable to move. That said holding the workpiece can provide additional stability. Providing a mechanical stop is especially important when drilling thin workpieces which are hand held. It is common for sheet materials not to be held in a hand vice due to their size, or the likelihood of them bending when tightly clamped or drilled. Where these types of workpiece are not clamped down to the table by other means such as the use of clamps, or other locking tools, it is common for the drill to grab the workpiece as the drill point breaks through if the operator has become distracted from the operation. This materials are prone to this and hand held snatch loads are likely to cause a serious cut to the operators hands.

Figure 7.8 Method of positioning workpiece to prevent rotation

Drilling can be a quickly performed operation, but an approach to work holding that ensures that the workpiece is stable, squarely mounted and does not present a potential hazard to the operator is essential.

7.8 USE OF VICES

There are two types of vice commonly used when drilling. These are the hand vice and the machine vice. The type used tends to self-select as larger machine vices tend to be of a size that only fit on larger pillar drills, and are used when drilling larger holes, and smaller operations tend to use a hand vice which will fit on any machine.

Hand vices come in a wide variety of sizes and designs, but often have many common features. All will have a cast frame, fixed jaw and sliding jaw, with a rotating handle used to provide clamping pressure, and to act as a handle for the operator to hold. Many, however, also have stepped edges, and both vertical and horizontal grooves. The stepped edge is to allow the fixture of workpieces above the frame of the vice to allow the drill to pass through a workpiece without drilling into the vice frame. The vertical and horizontal grooves may be in one or both faces of the vice jaws and are intended for holding round components. Figure 7.9 shows an example of a typical hand vice.

Figure 7.9 Typical hand vice used on workshop drilling machines

While most hand vices have a hole in the handle which allows the insertion of a tommy bar to provide greater clamping force this should not be a pre-requisite and a vice firmly tightened by hand should be enough. In drilling all the loads are downwards into the workpiece, and rotational. Therefore, comparatively little force is required to hold the component, as long as it is correctly mounted in the vice. Most hand vices have machined slots in the base to allow them to be bolted directly to the table of the machine. It is unusual for bolting down to be undertaken, but can be useful as a form of fixture if repetitive drilling operations are to be undertaken.

When drilling all the way through a workpiece the drill will make a hole in the frame of the vice if no provision has been made for this. For components that have a square lower edge this can easily be mounted using the stepped edge as can be seen in Figure 7.10.

Where this is not possible, and the drill point cannot be guaranteed to pass though workpiece into any gap in the hand vice frame, sacrificial packing needs to be placed underneath the workpiece using either wood, engineering plastics or a differently coloured metal. This change of colour will become apparent to the operator when the swarf is lifted up the flutes of the drill and deposited onto the upper surface of the workpiece, indicating the drill has passed through. The workpiece can also be lifted using a machine parallel or pair of parallels. This can be more problematic as it is probable that the parallel will be damaged by any drill passing through and use of

Figure 7.10 Workpiece mounted in stepped edge of hand vice

two parallels employed. This however, can be a tedious process to achieve as ensuring the workpiece is correctly position square in the vice and parallels can be found that are short enough to avoid excessive overhang can present problems.

One method that is often seen is to overhang the workpiece from the hand vice. This approach needs to be used with caution. Hand vices are designed to have a load applied within the width of the jaws. Where loads are applied outside of this an overturning moment is generated. When the overhang is significant, this either requires the operator to hold the vice down. More commonly, the frictional force applied by the jaws is overcome by the downwards force of the drill, and the workpiece pivots about the outside edge of the hand vice jaws. This can lead to ejection of the workpiece, or the hole being drilled at an angle. An example of this displacement is shown in Figure 7.11.

The use of a machine vice can often provide substantial stability, and often facilitates the drilling of larger sized holes that would otherwise require clamping to the machine table. They are used in exactly the same manner as they would be on a milling machine in that machine parallels are often used to support the workpiece and ensure it is square. Machine vices used in drilling are often found to have a degree of movement in their jaws and as such may require packing to get pairs of parallels inserted under the workpiece to 'lock up'. This type of vice is not designed to be held by hand, although often are. The rotational positioning of the vice such that one part

Figure 7.11 Workpiece displaced by force of drill point

of it is placed against the machine pillar is advised. When this is not possible, and larger holes being drilled it is important to bolt the vice down using tee bolts, or plain through bolting, as the rotational forces involved will be significant.

7.9 USE OF FIXTURES

From time to time holes are required to be drilled into a workpiece whose shape does not lend itself to being clamped in a vice, or its overhang is such that the distance from the vice jaws would not provide enough stability. The solution to this problem is to use angle plates, machine cubes, or a bespoke fixture. Fixtures tend to be used in a production environment, or where multiple operation requiring precision are used, and the size of machine cubes, tends to restrict their use to larger machines such as radial arm drills, however, angle plates are commonly employed to retain a workpiece in place during drilling operations. The advantage of using angle plates in particular, as well as machine cubes and fixtures, is that the workpiece can be correctly mounted off the machine and it is common to see the workpiece get fastened to an angle plate on a workshop surface plate. This allows the operator to ensure that the workpiece is securely retained, without having access problems on machine, and also is at a precise orientation.

Once the workpiece has been fastened the entire assembly can then be brought to the drilling machine and bolted down in position. On larger machines this is achieved by use of tee bolts inserted into slots in the table, with the shafts of the bolt passing through slots in the angle plate and secured with nuts and washers. The remaining problem is aligning the centre of the drill point, or centre drill with the required position on the workpiece. This can require some movement and rotation of the assembly to achieve the required position, but once bolted down, the workpiece is rigidly mounted and will not move. Figure 7.12 shows an example of a workpiece bolted to an angle plate.

7.10 TABLE HEIGHT

Almost all drilling machines have an adjustable table height. This is to ensure that the machine quill does not run out of travel when using shorter drill bits, while still accommodating the maximum size drill for the machine, and taller workpieces. The adjustment is normally achieved either by rotation of a handwheel operating a rack and pinion system or by hand where a friction clamp adjacent to the machine pillar is loosened and the table raised and lowered by hand, with the friction clamp tightened once the height has been set.

The parameters for setting the height are relatively simple. The table height should be set with the drill bit relatively close to the surface of the workpiece such that the operator does not have to make multiple rotations

Figure 7.12 Workpiece mounted on angle plate

of the capstan, and that the remaining quill extension once the tip of the drill has touched the surface of the workpiece is sufficient to reach the bottom of the hole to be drilled. However, if the workpiece is large and either, heavy or bolted to the machine table, if pilot holes, or centre holes are to be drilled prior to the finished size, then a gap needs to be provided to allow any drill bit to be extracted or inserted into the drill chuck. This is sometimes not apparent and the operator needs to be aware of the longest drill that needs to be inserted. Where an adjustment needs to be made, it is generally no problem where the table can be wound up or down using a handle, however, where friction devices are the clamping mechanism and the workpiece relatively heavy this can be problematic as the table is prone to drop, and the radial position will almost certainly move. Where this type of problem is likely to be encountered the operator needs to have an awareness of this issue.

For taper shank drills that are directly inserted into the machine spindle, the drill needs to be inserted prior to the workpiece being oriented underneath it or the table dropped to change drills. This is a common problem, especially where a workpiece is to be drilled with a centre drill, and possibly a pilot hole before the chuck is removed and a much larger and longer taper shank drill is inserted. The ability of the machine to have an adjustable and controllable table height mechanism is a factor in machine selection when the operator is considering which machine to use to drill any hole.

7.11 SAFETY AND GUARDING

Drilling in workshops, especially when using a bench drill or small pillar drill brings the operator into closer proximity to the cutting face of the workpiece than the majority of other workshop processes. It is not uncommon for the operator to have the workpiece at chest height and close to an operators face. One hand is normally operating the spindle capstan, and the other holding a workpiece-holding device such as a hand vice. This means that there is little protection to the operator from swarf chips ejected from the workpiece, long continuous chips raking the hand holding any hand vice, and in extremis fragments of broken drill bits, or workpiece ejection.

To protect against theses hazards a number of simple devices and items of personal protective equipment can be worn. The simplest and easiest device to fit to a drill is a guard around the chuck. These are cheap and easy to fit and provide protection in an arc in excess of 180°. They are able to be rotated either, upwards or to one side, and normally have an extending sleeve to allow for shielding of long series drills. An example is shown in Figure 7.13.

Some versions of this type of guard include an inter-lock preventing the machine from being operated without it, however, the majority are simple manual devices as shown in the figure. The problem with many of this type of guards is that small diameter or short stub drills, prevent the guard from being dropped down as the guard can be longer than the drill, requiring the multiple rotations of the drill capstan or handwheel until the drill can touch the workpiece. An alternative to this type of guard is a DC brake.

A DC brake type of device is a safety brake that stops the machine almost instantaneously. They provide an effective means of stopping the machine, but do not provide any protection from swarf ejection. The brake is activated by the operator hitting a flexible bar hanging down from the machine head causing a brake to be applied to the spindle and the drive power shut off. They are effective as they immediately stop the spindle, but are perhaps best used in conjunction with a Perspex drop-down guard. The required size and location for these devices means that they are often only able to be fitted to larger machines, and therefore, are more common on larger pillar drills and radial arm drills than bench drills.

When drilling, along with most machining processes, wearing of safety glasses or goggles, is always recommended. They are cheap, readily available, and manufactured to nominated standards. A more contentious area is the wearing of gloves. Many organisations and educational establishments require the wearing of heavy leather work gloves when drilling, while others

Figure 7.13 Pillar drill chuck guard

forbid it. Handling of workpiece materials that may have sharp edges, and be hot suggest that gloves could be a good idea. However, wearing gloves when in close proximity to rotating machinery, especially in a pressured production environment, risks entanglement. While the glove provides protection from hands being raked by continuous chips, correct drilling technique and guarding can remove this hazard and the added dexterity of not wearing gloves can be provide more control and reduce the chance of entanglement. If a workpiece is well bolted down, or being drilled slowly on a larger machine gloves are not really adding anything, when handling workpieces with sharp edges or brushing away swarf they should be worn. Common sense needs to be applied by the operator, and if in doubt use gloves, but ensure that hands cannot come anywhere near rotating parts.

7.12 FEEDS AND SPEEDS

Feeds and speeds as for milling, and turning are an important factor in successful drilling. The use of the correct speed and feed rate generates correctly drilled holes but also significantly prolongs the life of the drill bit.

The vast majority of drills used in any engineering workshop remain of the plain twist drill type, with jobber drills probably the most common, although stub drills and long series are also present. While carbide drills are freely available, there is a tendency for these to be used on CNC machines and are less common as a standard drill bit, however, coated drills are regularly seen in typical workshops, and have their advantages.

To adjust the speed on the drilling machine two approaches are taken. These are either by setting gear levers to provide a range of spindle speeds, rotating a speed setting handle, or moving a drive belt up and down a series of conical pulley's. The easiest speed changing mechanism is the rotating selector, which also often has a high range and a low range. An example of this can be seen in Figure 7.14.

To change the spindle speed in a belt-driven machine, the power should be isolated before the cover is taken off the belt drive mechanism. There is normally an overlocking tensioning device which is then released which slackens the drive belt. The drive belt is then moved up or down a pair of grooved cones which change the relationship between the diameter of the pulley at the motor and the diameter of the pulley on the spindle to select the closest speed to that required. Figure 7.15 shows this configuration. A plate is normally attached to the machine indicating which pulley combination produces with RPM value. Once the correct combination is et the tension device is re-engaged, and the cover closed ready for use.

Table 7.1 shows values for RPM covering a range of drill sizes and surface machining rates rates. Machining rates can vary quite considerably with some stainless steels having a surface machining rate of 9 surface m/min, mild steels at 27 surface m/min and soft aluminium at 55 surface m/min.

Spindle speeds can be quite easily calculated using the formula below:

$$\text{RPM} = \frac{\text{surface material removal rate} \left(\text{m/min} \right)}{\pi \times \text{drill diameter} \left(\text{mm} \right)}$$

Figure 7.14 Pillar drill spindle speed change mechanism

Figure 7.15 Belt-driven machine speed adjustment system

The type of twist drill being used can allow significant adjustment of the spindle speed. In broad terms the use of stub drills, which are less likely to wander than long series drills can have a spindle speed calculated at a greater material removal rate. The values identified in the table below, are based on use of a standard HSS uncoated jobber drill, and are the normal starting point for drilling operations. If the operator considers that the material is not drilling well then speed adjustment is the normal approach. One method of identifying a new spindle speed value is to consider that in general stub drills can be used at a surface material removal rate approximately 2–5m/min faster than that for a jobber drill, and long series drills can have a wide variation where the spindle speeds should be calculated at 3–10m/min slower than that for a jobber drill. Drill manufacturers websites are a good source of information for the correct drill/material combination and materials suppliers are often able to advise on suitable machining rates for discrete materials.

Effective drilling requires the speed of the machine to be set to the optimum setting for the drill bit, and material surface removal rate. As machines generally do not have infinitely adjustable speed control this often means selection of the nearest value for the material and size of hole being drilled. The performance of the drill is a combination of the workpiece material type, drill bit configuration, any coolant or cutting oil used, the downward force into the workpiece, and the spindle speed setting. Therefore it is not uncommon to adjust speed settings if satisfactory results are not being generated and speed calculations provide the operator with a good starting point, but do not replace experience, where differing speeds or drill types may be advantageous. Good-quality workshop handbooks (such as *Machinery's Handbook*, Industrial Press, 30th edition 2016) had a number of tables and formulae to enable an operator to calculate the exact machine power, rotational speed and feed values for most materials, however, this is often more precise than required and standard tables can be used as a guide to feeds and speeds for varying materials, which the operator can adjust as experience suggests. Table 7.1 below shows spindle speeds for a common range of diameters, against a set of material surface machining rates.

Machines that have a feed mechanism built in tend to be the larger, pillar drills and almost all radial arm drills. Some metals, such as stainless steel, manganese, and high-tensile steels, have a tendency to work harden to a point where drilling with HSS drills becomes unachievable. To overcome this problem the use of drilling machines with a feed mechanism is advised to ensure that the drill does not dwell at the bottom of the hole causing work hardening. When machining the these materials operators should consider reducing the spindle speed and using a heavier feed/rev than on more easily machined materials.

Of all the tools used in metal cutting twist drills can be the most stressed. They have to cut material using only two cutting edges, have an applied axial load and also have to resist the torque loading applied. When manual feeding into the workpiece this can be felt, and allowed for by the operator, but when power feeds are being used, consideration must be given to the feed rate applied to prevent breakage. Feed rates are again influenced with

Table 7.1 Drill spindle speeds for range of drill diameters and material removal rates

Surf rem'l rate m/min	Drill diameter (mm)									
	Ø3	Ø5	Ø6	Ø8	Ø10	Ø12	Ø16	Ø19	Ø25	Ø30
5	530	318	265	199	159	133	99	84	64	53
10	1061	637	531	399	318	265	199	168	127	106
15	1592	955	796	597	477	398	298	251	191	159
20	2122	1273	1061	796	637	530	398	335	255	212
25	2652	1591	1326	995	796	663	497	419	318	265
27	2865	1719	1432	1074	859	716	537	452	344	286
30	3183	1910	1591	1194	955	796	597	503	382	318
35	3713	2228	1857	1393	1114	928	696	586	446	371
40	4244	2546	2122	1591	1273	1061	796	670	509	424
45	4774	2865	2387	1790	1432	1194	895	754	573	477
50	5305	3183	2652	1989	1591	1326	995	838	637	531
55	5836	3501	2918	2188	1751	1459	1094	921	700	584

the type of drill bit being used, and as a broad principle, stub drills can cope with higher feed rates than long series.

There is a close relationship between the size of the drill being used and the hardness of the workpiece material. Drill manufacturers' data gives some guidance on the feed rate to be applied when drilling materials of differing surface machining rates and diameter. These feed rates are based on the use of jobber drills. Long series may require a lower feed rate, and stub drills may be able to cope with higher feed rates, than those stated.

The feed rate and the spindle speed is also effected by the condition of the drill, the stability of the workpiece as well as the material being drilled. Operators should use the figures for feed and speed as a guide, and if poor drilling conditions are observed, change the settings. This can often involve reducing the speed rather than increasing, and in difficult materials, increase the feed rate to ensure that the cutting edge of the tool bites into fresh material and prevents localised work hardening. Consideration should be given to the use of cobalt drills or reduce the speed of HSS drills by 33% when machining heat treated alloy steels, Nickel alloys and some titanium based metals to ensure effective drilling.

7.13 USE OF COOLANTS

In general it is probably more common to find more drilling machines without a facility for coolant than with. A bench drill is almost certainly not going to have a facility, and many of the simpler pillar drills will not. However, larger pillar drills and most radial arm drill have a facility for, or actively use, coolants supplied through a pipe system to the cutting tool and work piece.

The primary role of this flow of coolant is to reduce the temperature of the drill and workpiece. The intention of this process is to extend tool life, increase productivity and help prevent any heat treatment of the workpiece as a result of heat inadvertently introduced as a bye-product of the cutting process. This amount of heat can be substantial as witnessed by the 'blue-ing' a swarf chips, and in extreme cases, chips expelled from the drill being bright orange from the heat. The energy used to cut metal is efficiently converted into heat. The heat degrades the cutting edge of the tooling rapidly, therefore, its removal by using a flow of coolant rapidly reduces the temperature at the tool point.

Most drilling operations can be undertaken without using coolants, however, the use of them is more important when power feed is being used, as it significantly increases the life of the drill bit.

The cutting action of a tool involves a shearing process where the material is deformed and converted to a chip, either continuous or as a series of small swarf chips which are passed up the flute of the drill as it progresses into the workpiece. The majority of energy involved in this process is converted into heat which is focussed at the tool point and along the surface of the tool where swarf rubs over the surface of the tool. In time this process will develop thermal cracks which lead to cutting edge failure. High-speed steel (HSS) tooling degrades quickly under significant load, and it is not unusual to see HSS tool cutting edges glow bright orange due to heat if incorrect drilling technique is employed. Drills under these conditions have a short life, produces a poor hole surface finish. The addition of relatively small amounts of coolant rapidly quenches this heat build-up and substantially extends tool life.

The selection of a cutting fluid or coolant can depend on the machinability of the material being cut, the cost and the effect that it may have on the health of the operator, especially long term. However, most drilling machines found in a general machine shop use essentially one type for logistical ease and low cost, and are used in other workshop machine tools. These tend to be from a group of water-miscible fluids. These fluids are an emulsion of oil, water and an emulsifying agent. The proportion of mixture varies, but a mix of 1 part oil/emulsifying agent to 20–30 parts water is quite common for machining steels using carbide tooling. If using HSS tooling then a slightly richer oil mix of 1 part oil to 10–20 parts water is used. Machining stainless steels may again involve a richer mix of approximately 1:5, however, differing manufacturers products will have recommended mix values, and operators should consult product data sheets for the optimum mix. A broad 'rule of thumb' is that the harder or more difficult a material is to machine, the more concentrated the fluids become. This also applies to types of tooling, where HSS tooling would generally have a richer oil mix than carbide tooling machining the same material.

The advantage of this emulsion of oil and water is that water has excellent cooling properties, and the oil content provides good lubrication, with a secondary advantage of providing some corrosion protection to the machine. While there are a wide variety of oils which can be tailored to differing materials it is normal for a machine just to have one type which can be used across a wide variety of workpiece materials. The only exceptions to this would be where some materials that are vulnerable to a chemical reaction such as copper, brasses and bronze etc. may be subject to staining from chemicals within the coolant especially some sulphur or chlorine compounds. Where components that have aesthetic requirements, or high surface finish requirements, adjusting the cutting depth and feed rates to reduce cutting temperatures and not using coolant, or changing to a passive type are the options open to the machine operator.

Delivery of coolant to the workpiece and tool cutting edge is normally via a sectional/flexible pipe, however, some older designs make utilise a rigid pipe comprising an elbow joint and extending delivery section to accommodate differing orientations. Both designs have their advantages and disadvantages, in that rigid pipes are stronger, but can be more difficult to keep oriented in the right direction, while sectional pipes can be more easily oriented but are more vulnerable, when swarf wraps or knocks the pipe away. Irrespective of this the machine operator needs to ensure that coolant continues to be delivered to the drill as the operation progresses. When considering what orientation to deliver coolant to the tool and workpiece it is worth assessing how the coolant is going to get down the flutes of the drill. into the cutting edge. Drills that are choked with swarf present a problem, although long continuous chips tend not to generate this and there is room for coolant flow passing the swarf chip.

While coolant manufacturers and literature suggests that coolants should be delivered at 10–20lt/min, this represents a best case. In general it is not a requirement to deliver large quantities of coolant to the tool point which can often be a rather wet process, and depending on the type of coolant being used can be a long term hazard to the operator. However, the delivery of a stream to the tool such that there is little or no steam witnessed rising from the tool/cutting area is often a good indicator. Workpieces that are more than hand hot when checked at the end of each cut are an indicator of insufficient cooling. Poor surface finish and 'blueing' of swarf chips is also an indicator of insufficient cooling.

When machining more exotic materials such as magnesium or metals that may react with water no coolant would be used. In general this should not be a problem for the operator, as the cutting speed and feed rate should not be so high that heat does not dissipate fast enough, and the operator needs to have an awareness of these marginal conditions.

Where a pumped coolant supply is not available alternative methods can be used. Smaller machines such as bench drills and smaller pillar drills tend to cut smaller sizes and as a consequence generate less heat. Therefore, applying coolants using a spray bottle is one approach often taken, however, there are safety issues with aerosol emulsified oils. More commonly, and extremely effectively, is the regular application of cutting oils to the drill. It provides less immediate cooling than an emulsified oil/water mix, but prevents cracking of the cutting edge, and provides lubrication. To ensure that heat build-up is not too significant when using cutting oils the feed rate, and speed can be lowered.

7.14 BASIC DRILLING

The basic process for drilling a hole into a workpiece is relatively simple, however, like many processes it also has the potential to go wrong if a few basic principles are not put in place. The operator needs to identify where on the workpiece the hole is going to be drilled and this is normally indicated by marking out to identify a centre point prior to the drilling operation taking place. Marking can be a simple chalked point, through use of markers to more precise scribed line. The accuracy required is stated on the component drawing, and the method employed is normally driven by the degree of precision required.

To successfully drill any hole a centre mark is required. The included angle of the majority of twist drills is 118°, and they do not easily self-start. The provision of a punched or drilled centre provided an indent that allows the drill bit to start, and prevents it 'wandering' over the surface of the workpiece. A simple punched centre is created by placing a centre punch at the marked centre point and is struck with a hammer to make an indentation. The force required to make an effective mark is not particularly great, and a degree of circumspection when striking the punch is useful, as attempting to use a heavy blow often leads to inaccurate punching or damage to the operators knuckles. When a centre indentation has been made the drill point will commence drilling at that point. Where no centre indent has been provided and a small drill diameter is being used it is not uncommon for a drill to break as its lateral movement on the surface of the workpiece, becomes too great.

An alternative to providing a centre mark using a morse centre drill, or spotting drill. These drill bits are designed to bite into a plain workpiece surface and create a tapered hole for the twist drill to commence drilling within. Spotting drills are perhaps more commonly seen on CNC machines and historically morse centre drills have been the most common form of providing a centre. They are produced by a number of manufacturers, and commonly come in eight sizes. Figure 7.16 shows a number of sizes of morse centre drills.

Figure 7.16 Morse centre drills

The size of centre drill selected needs to be appropriate for the size of hole being drilled by the twist drill, as it is important for the mouth of the finished centre hole to be less than the diameter of the drill being used to produce the hole. Failure to achieve this will leave a chamfer at the mouth of the hole. Very small centre holes can still provide an effective start for a twist drill, however, very small morse centre drills are also relatively fragile, and it is generally good practise to use the largest centre possible, without providing an overly wide mouth. Morse centre drills should never be used to drill a hole, as once they are advanced into the workpiece beyond the angled face such that the plain diameter enters the workpiece, there is nowhere for the swarf chips to be ejected and it is common for the head of the centre drill to shear off leaving the point embedded in the workpiece.

The morse centre drill, or selected twist drill, where a centre punch has been used to create the drill location position, is inserted into the drill chuck, a firmly tightened. The correct speed for the diameter of the machine should be set either by adjusting a pulley system, setting the correct gearing or speed mechanism.

The workpiece should be placed in a work holding device such as a hand vice or machine vice. Where larger workpieces are to be drilled these need to be secured to the machine table such that some movement can be achieved to allow alignment of the workpiece under the spindle centre line. It is normal for the operator to bring the quill down to allow the workpiece to be positioned directly under the drill point, and the positioning of the workpiece by eye is normally precise enough. If excessive rotation of any quill capstan is required to position the drill point over the centre mark, then it is worthwhile raising the table. It is inevitable that the workpiece will be moved off centre when any raising or lowering of the machine table is undertaken. While tables which are wound up and down using a crank mechanism can be stable, it is especially common where the table is retained using a friction clamp and allows easy radial movement. Therefore, it is often a good idea to remove the workpiece and any work holding device off the table when adjusting height to remove the weight that has to be held during adjustment and tightening of the friction device.

For more precise alignment, mounting an optical centre finder into the chuck gives the operator direct vision onto the centre mark. These optical devices have an angled eyepiece and prism that allows the operator to view the workpiece from the spindle centre line. There is normally a pair of cross hairs that allow the operator to move the workpiece until it is perfectly aligned. By moving the eyepiece through 90° the operator can be satisfied that the required alignment has been achieved. Another method for aligning the workpiece is to use a centre finder. This is similar to a method used in milling where a tool is inserted into a chuck, which has a needle which can swing off axis. Where these devices are used in drilling any stylus that has a ball on the end is replaced with one with a needle point. An example of this can be seen in Figure 7.17.

Figure 7.17 Centre finder

This device is inserted in the chuck, the motor started, and the rotating needle carefully guided into the axis of the spindle until it is revolving without any discernible 'wobble'. The operator then moves the workpiece until the needle point of the centre finder is directly over the marked point. It is not unusual for this type of device to be used with a jeweller's loupe, however, car must be taken to ensure there is no risk of entanglement, and some types of interlocked guard may prevent this type of device being used, as it requires the machine spindle to be rotating.

Once the table height has been correctly set and the workpiece correctly position under the spindle centre line, any chuck guard should be put in place. It is normal for drilling machines to have a chuck guard fitted which can be rotated out of the way for insertion and removal of drill bits, but dropped into place during drilling to protect the operator from any swarf chip ejection, or being raked by long continuous chips.

Once the guard is in place the operator starts the machine and rotates the capstan or handwheel to advance the drill point into the workpiece. If a morse centre drill is being used the operator should check it is drilling in the marked position, and continue to press it into the workpiece until approximately half way up the conical portion of the drill. Where a drill is being used the operator should check the drill is aligned such that the drill point is engaged with the punched centre.

When drilling any hole a continuous pressure is applied. This is especially important when drilling some more difficult materials such as stainless steel, where continuous feed is an important factor in preventing localised work hardening. However, when drilling many materials, but especially a wide selection of steels it is normal for a long continuous chip to be produced. When drilling by hand and without any feed mechanism engaged, this can lead to wrapping of the swarf chip around the drill, damage to the drill guard, and where the drill protrudes below the guard can lead to a long tail of swarf that can rake over the operators hands if holding the workpiece or hand vice in place. While locating the hand vice or workpiece such that it rests against the machine pillar often removes the requirement to hold the workpiece, removing the proximity of the operators hands. However, a momentarily removing downward pressure on the workpiece breaks the swarf chip and allows short lengths to be ejected for the cutting area. If the material properties allow for this release of pressure then this is the optimum method for ensuring long lengths of swarf chip do not become a hazard to the operator.

The pressure on the capstan is retained until the depth of hole is achieved, or passes through the workpiece. The operator should ensure that the length of drill is sufficient to reach the depth required, and if not a long series drill should be used. It is entirely possible to start drilling the hole with a long series drill, however, the operator has a choice of starting the hole using a stub or jobber drill and finishing to depth. It is important that the top of the drill bit flutes never enter the hole drilled as this prevents swarf ejection and leads to breakage of the drill in the hole.

When a drill is to pass through the workpiece to generate a plain hole, there is a point where the drill point breaks through. If consistent pressure is applied to the drill bit throughout the drilling process, as the drill point breaks through there is a tendency for the drill bit to 'grab' the workpiece. This can result in the workpiece being pulled upwards out of simple frictional work holding devices. especially where the stepped edge of a hand vice, provides little frictional resistance, or for more rigidly held workpieces, can lead to the drill bit becoming stationary and the chuck rotating about the shank. To avoid this the operator can often feel the drill starting to break through the lower surface of the workpiece, and releasing the downward pressure on the drill capstan allows the drill bit to gently break through. This avoids and 'grabbing' by the drill bit and tends to provide a reduced size of burr on the workpiece lower surface.

7.15 LARGE DIAMETER HOLE DRILLING

Large diameter hole drilling can be undertaken using three approaches. Trepanning is one method and is the only realistic alternative to producing holes in excess of ∅100mm other than flame profiling and plasma cutting. The restriction to trepanning is that it is limited to comparatively thin workpiece thicknesses and cannot drill a blind hole, therefore, they are limited to producing through holes. Hole saws are another method for producing holes in a workpiece up to ∅100mm. There are two types of hole saw, one that is effectively a circular blade not unlike a wide hacksaw blade. These types of hole saw are not well suited to hard materials such as steel and have no cutting edge relief, but are effective on thin plastics, wood and thin metallic sheet materials. There are however, more complex designs, often with carbide teeth that allow holes to be cut up to ∅100mm through materials such as stainless steels and cast iron up to 25mm thick. Holes saws all have a similar design, in that they have a central pilot drill, to anchor the saw, and the thickness of the material they can drill through is limited by the internal depth of the saw cavity. As with trepanning tools, hole saws, cannot be used to open up an existing hole as the require material for the pilot drill to cut through and anchor the drill, and can only be used to drill a through hole, and cannot be used to produce a blind hole. Traditional taper shank twist drills are the remaining approach for drilling large diameter holes up to ∅100mm. The method of operation is not different from use of any other sized twist drill, but the loads are significantly greater and the approach to drilling needs to be modified to accommodate these. The drilling of blind holes is easily achieved, and the depth of hole that can be achieved at 100mm is approximately 250mm.

Large diameter twist drills are all manufactured with a taper shank This is to ensure that the torsional loads required to drill a hole are transmitted to the drill bit without slippage, and the frictional area that is provided by chucks is not sufficient. If the drill bit was designed to fit into a chuck the diameter of the shank would have to be reduced well below the drill diameter to be able to fit into a chuck that was not of excessive size, and the loads involved in drilling would probably induce a shear failure in the plain diameter. Figure 7.18 shows a typical large diameter twist drill.

Drilling large diameter holes is best achieved through drilling a centre hole and then a pilot hole. The obtuse angle and size of large drills means that the cutting point is never going to be able to grip a simple punched centre point, and while large diameter drills are extremely stiff, the reduction on load required to initially penetrate the workpiece is substantially reduced when a pilot hole is present.

The size of the pilot hole is subjective, but there are a number of factors to consider. The time being taken to drill a hole can be an issue in a production environment is often an issue, as is the power of the machine, the stability of the workpiece, and the condition of the drill point. By using a series of holes that increase in diameter is going to reduce the power required to drill each hole as the volume of material being removed for each diameter is proportionally reduced. That said only drilling on the corners of a drill is going to wear the cutting edge unevenly and takes more time due to repeated operations. Using successive drill diameters that are too close to each other introduces the problem that is experienced when opening out a hole, and the drill point can 'skate' on hole entry creating a polygonal shaped hole entrance chamfer. Having drilled a morse centre hole accurately and using a large morse centre drill, the operator should then commence drilling a pilot hole. For efficiency, the operator should drill a pilot hole using a taper shank drill bit, and the chuck can be removed from the machine spindle. If a large morse centre has been used then a similarly larger pilot hole can be drilled. The intention of the pilot hole is to guide the larger drill such that it is produced in the correct location. It does not especially matter if the pilot hole does not penetrate to the maximum depth required, and often the proportionately shorter flute length prevents this, however, operators will notice the increase in power required when the larger diameter drill reaches the bottom of the pilot

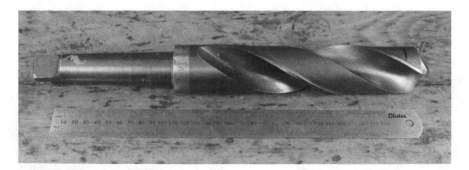

Figure 7.18 Ø32mm taper shank twist drill

hole. Where diameters in excess of ⌀25mm are going to be drilled, it is suggested that the pilot hole be in the region of ⌀12–16mm, but irrespective of this the diameter of drill bit selected for a pilot hole should have a diameter greater than the drill point web distance of the large diameter drill to allow the web of the drill to 'find' the pilot hole and engage the drill fully and follow it as a guide.

Large diameter drills producing holes in metals involve high rotational loads and high amounts of pressure to ensure effective cutting. The capacity of a machine to drill larger diameters is restricted by design, in that the size of the morse taper in the spindle will limit the diameter of drill bit that can be fitted into the machine. This generally requires the use of larger workshop machines, and the advantage of this is that the majority of machines that are capable of drilling holes larger than ⌀25mm will have a feed mechanism. Larger machines also tend to have more rigidity which is essential to ensure that the workpiece can be bolted down firmly to prevent movement when drilling.

While drilling can be achieved by hand the length of the capstan handle, whether it is geared, the amount of material being removed by the drill, and the resistance of the material can be issues when drilling, and lead to operator fatigue. Many materials will work harden, especially when coolant or cutting oils are not being used, if continuous pressure is not applied throughout the drilling process. The continuous pressure is applied to ensure that for each revolution the tool point cuts into fresh material, and for optimum results this needs to be consistently applied. The mechanical advantage of the capstan mechanism, and strength of the operator may provide a limit to the diameter that is possible to be drilled through materials, where large diameters may be easy to drill through engineering plastics or wood, but a reducing size being achievable as harder materials such as aluminium, brass, low carbon steels, through to cast irons, alloy and stainless steels. To overcome this, the use of a power feed mechanism is recommended. Feed rates can be calculated using data produced in handbooks [1], or from basic tables, however, these should be used as a guide and can be adjusted by the operator observing the cutting process. For some materials, such as those that have a tendency to work harden, there may be a minimum feed limit and if difficulty in cutting is observed, a better approach may be reducing the spindle speed rather than reducing the feed rate. In all drilling the power of the machine, drill bit material, and the surface material removal rate provide a baseline for a volume of material that can be removed per revolution. While the surface material removal rate provides an upper limit for spindle speed and feed rate, the two are proportional when considering machine power, therefore, reducing feed rate may allow a greater spindle speed and slowing spindle speed may allow a greater feed rate. This is never a particular issue when drilling smaller holes, however, large diameter drill involves high machine loads and the use of mechanical feed, and the correct proportional feed is advised in order to provide an acceptable finished hole, and a safe working condition.

7.16 DRILLING THIN WORKPIECES

Drilling thin workpieces can present a number of difficulties, especially where the workpiece material thickness is less than the distance between the drill point and the full drill diameter. Issues that can be found in drilling any workpiece tend to be amplified when drilling through thin workpieces, especially when holes of any appreciable diameter are being drilled. Difficulties are encountered on material penetration and on passing through the material. This is especially amplified when the drill point has passed through the workpiece, yet the hole has not been opened up by the drill bit lip length to the full plain diameter.

When considering what tool to use to drill a diameter, alternative techniques such as use of hole saws, or trepanning tools can often be a far better approach than attempting to use large diameter twist drills. Drilling thin sheet materials can often be problematic when using twist drills over $\varnothing$10mm as the effect of the lip length creates a surging effect before the drill bit has cut the workpiece material to its full diameter. This effect can be observed at quite small diameters. If a 1.5mm thick workpiece is being drilled the depth of the conical height of a standard 118° drill point is 1.5mm for a $\varnothing$5mm drill bit. While relatively small diameters can be controlled by the operator having a circumspect approach to drilling through thin sheets, as larger drills are used the forces involve increase exponentially, and control becomes more difficult.

Difficulties can be experienced because the point of the drill is exiting the lower side of the workpiece material before the full diameter of the drill has been cut. As the diameter of drill bit being used increases for a given material, the effect is amplified as the lip of the drill cutting point digs into the material being cut, with increasing torque as the diameter increases. In extreme cases this leads to the workpiece material being torn, rather than drilled, or the drill effectively punching through the material as it is dragged up the flutes and providing both a badly distorted workpiece and a hole that is non circular. While the example described above is for a workpiece material that is 1.5mm thick the same effect is experienced in proportion to the drill diameter and workpiece thickness.

Drilling a small pilot hole, having a well clamped down workpiece and also utilising a sacrificial backing pad, such as plywood can assist in drilling correctly formed holes in thin workpiece materials as the drill point remains embedded in a material and continues cutting until the full diameter passes through the workpiece. Drilling of pilot holes acts as a guide and removes the requirement for significant downward pressure on the workpiece. It is uncommon to drill a centre hole in sheet materials, and a punched centre is more common. Therefore, for drilling larger diameters drilling a small centre hole of a diameter equivalent to the drill point web size provides for positive engagement of the required diameter drill.

While step drilling, and producing the desired diameter hole by drilling in incremental diameters is commonly undertaken, care must be used when attempting this in thin workpiece materials. If the workpiece is not kept firmly clamped in a position such that is does not move between drill increments then it is likely that the finished hole will be off centre, and the drill will have removed material using one cutting edge only, and if a supporting sacrificial backing is not used the workpiece has the risk of an irregular or polygonal edge profile being generated. One method for drilling these types of hole is to increase the spindle speed substantially, and use minute feeds often delivered by hand to reduce the chance of workpiece snatch. Materials that work harden are less of an issue when undertaking this approach has the edge that becomes work hardened is of minimal thickness, and therefore has less effect, albeit providing proportionately more wear to the drill point cutting edge.

7.17 DRILLING DEEP HOLES

Drilling deep holes can introduce some issues that require management. These relate to swarf removal and accuracy of holes. The definition of a deep hole is often stated as any hole that is drilled that has a depth greater than four times its diameter. For practical purposes most operators would suggest that a deep hole is considerably more than this, using a more general definition that deep holes start at a depth that a jobber drill could no longer reach.

Larger diameter drills are going to be able to drill to greater depths than smaller diameters as their form is driven by proportions. However, long series drills allow much deeper holes to be drilled, but introduce constraints which relate to bending of the drill given its reduced stiffness, and accuracy of the hole. This is normally countered with reduced penetration rates, a spindle speeds, however, swarf removal a coolants become a greater issue than with stub or jobber drills.

While extra length drills can be purchased, long series drills are roughly 30–35mm longer than their jobber equivalents, giving an extended drilling depth. One problem of drilling deep holes is that the precision of a drilled hole using a standard twist drill reduces once a hole is being drilled in excess of five times its diameter. Often this is not an issue, if the hole is being drilled as a clearance hole, or a tapping hole is being drilled for a thread and the mating part has been manufactured to a standard hole clearance. In these situations the standards for clearance holes, among other criteria, accommodate errors in position from drilling. In situations where greater precision of position or form is required, alternatives need to be identified such as gun drilling, or reaming. However, gun drilling in particular, while providing a deep hole to a depth of three hundred diameters or more requires specialist drilling machines. This is due to a requirement for high pressure cutting oil, or coolant, to be fed down the drill to the tool point. This cools and lubricates the cutting edge, and removes swarf from the hole as the drill bit progresses into the workpiece.

A approach to drilling a deep hole is to undertake it in stages. While a hole can be drilled with a long series drill it is often more efficient to commence with a standard jobber drill or stub drill penetrating to the depth of twist drill flutes. This process, as with all drilling operations should be undertaken as a series of pecks to break off long continuous chips which are often experienced when drilling softer materials especially at smaller diameters. Once the initial hole has been drilled, the drill bit is changed for a long series drill, and again drilled to a depth where the top of the flutes has been reached. Once the maximum depth a long series drill is capable of drilling to has been reached, extra length drills must be used. These drills are available in quite surprising lengths, with Ø6–14mm drills all being available with flute lengths up to 300mm. Shorter lengths reducing in steps of 50mm are also available, however, care must be taken with these drills to ensure that they are proper supported and that long series drills have been used to provide the maximum length of support within the hole prior to commencing to drill with an extended length drill. A Ø6mm drill 300mm long will have a considerable amount of flex and will be easy to break if overloaded.

If deep hole drilling is commonly undertaken or the material properties preclude the use of regular pecking to remove swarf, then parabolic flute drills can be procured. These drills are available across all drill bit types and are designed to be used to a depth ten times the drill diameter, before pecking is required. They are particularly useful when drilling softer materials or those that have a tendency to produce long serrated continuous chips, presenting a hazard to the operator and potentially marking the workpiece.

7.18 DRILLING TO A DEPTH

To drill a hole to a known depth the operator needs to first establish a z-axis datum. This is normally achieved by rotating the quill capstan or handwheel without the spindle turning to bring the drill point into contact with the workpiece, and identifying the indicated distance on the vertical quill scale as can be seen in Figure 7.19.

If a relatively small centre hole has previously been drilled, it is normal for the depth of drill point intrusion into this to be ignored, as plain drilled holes to a depth are rarely tightly toleranced. It is normal for holes to be dimensioned for the length of full diameter required rather than to the point of the drilled hole, and it is usual to see a dimension for the depth of a hole

Figure 7.19 Linear quill feed scale

and there to be a separate arrow with 'permissible drill point' to recognise the further protrusion.

To set the depth of hole, the operator merely needs to identify the relative position between the start point (i.e. the drill point touching the upper surface of the workpiece and the intended depth the hole is to be drilled to), plus an allowance for the distance taken up by the drill point.

Almost all bench, and pillar drills have a linear scale and quill depth stop ring, along with a rotating nut, but rarely have a micrometer scale which is often found on turret mills with quill feeds. This normally restricts depths to be visually controlled to an accuracy no greater than 1mm. Some radial arm drills will have a linear scale but it is common for this to be a marked or calibrated ring around control wheels or a digital counter or DRO.

To drill to a defined depth, the operator adds the depth of hole required to be drilled to the reading on the scale where the drill point touches the workpiece. The machine is then started and the hole drilled as normal until the calculated depth is reached. This is perhaps the least accurate method as the operator is having to apply a downwards force to drill into the material, hold the workpiece or vice, and read the scale to determine if the depth has been reached. A more controlled method is to use a depth stop if one is fitted to the scale. This is not always present, but provides a physical stop to the quill preventing over drilling. The approach is identical to that used on a turret mill where a rotating or sliding nut is moved down the scale until the value for the depth is shown. When using this technique once the spindle has been started, the operator rotates the capstan or handwheel to press the drill into the workpiece, and merely has to continue until the mechanical stop is reached. The capstan or handwheel is then returned to the start position extracting the drill from the workpiece and the machine stopped.

The problem with either approaches described is ensuring an accurate depth of hole where the component tolerance is less than 1mm. This is especially difficult when the scale is only calibrated in mm. The use of the rotating nut, allows for much greater accuracy. By rotating the nut up the threaded shaft and locking it in place with a lock nut the quill cannot be depressed beyond the set point, as the quill depth stop ring presses up against it preventing further advancement of the drill. Setting this visually by observing the alignment between the top surface of the rotating nut and the linear scale, then locking it off is one approach but is still limited in its accuracy. A more precise method for setting a depth is to use gauge blocks (slips) trapped between the top of the depth marker and the depth stop. Again this can only be achieved where a drilling machine has a rotating or sliding stop and a fixed stop as can be found on many milling machines. If it is achievable on an operators machine then it provides an accuracy of sub millimetre, which the achievable accuracy limited only by the machine depth screw backlash.

The operator builds a pack of gauge blocks (slips) set to the depth required, including an allowance for the drill point protrusion between the quill feed stop block and rotating nut. With the gauge block pack in place the rotating nut is locked tight, and the gauge blocks then removed. This provides a depth stop for the quill feed that is accurate and provides a physical stop to a set depth.

The approach described can be employed, however, where extreme accuracy is required the operator should consider whether the operation requires a precision such that it may be better undertaken on a milling machine. While this is not always possible, it provides one approach to drilling a hole with an accuracy greater than the machine is calibrated for.

7.19 FLAT-BOTTOM DRILLING

Standard twist drills are normally manufactured with a drill point ground to a 118° included angle. This means that any hole drilled will have a bottom profile that matches this. This is normally not a problem and it is common to see component drawings dimensioned showing the depth of full diameter to be drilled and an indicated 'permissible drill point' notation indicating the hole bottom profile. However, in some cases, a flat hole bottom profile is required. Where the required feature does not allow for, or require, a through hole of smaller diameter, a counterbore cannot be used. If the hole is of any significant depth a slot drill can also not be used. When these options are excluded a flat bottom drill needs to be manufactured and used.

The normal way for producing a flat-bottomed drill is to grind the point flat of a standard twist drill. There tend to be two approaches, one being providing a completely square end profile, and a second grinding a point that has an included angle of approximately 170° to give a slightly enhanced cutting performance. Figure 7.20 shows an examples of twist drills ground to produce a flat-bottomed holes.

The process of producing a flat bottomed hole is best achieved by drilling a hole normally and allowing the depth to proceed such that the drill point reaches the maximum depth required. Using a drill of the same diameter which has a flat drill point is then moved down the hole until the corners of the drill engage with the material left behind by the 118° drill point until the maximum depth is achieved. Using a completely square end profile drill can be problematic if a residual drill point is to be avoided as the web distance and flat profile prevent cutting at the centre but will provide a truly square bottom with a small central conical drill point evident. Having a small included angle is the alternative approach and provides a compromise where the bottom of the hole is not completely square to the sides, but provides a better drilling action with less corner wear and removes, or significantly reduces, any residual centre point.

7.20 COUNTERBORES

Counterbores are holes that open out an existing hole to provide a wider diameter hole and one with a flat bottom. They are almost universally used

Figure 7.20 Drill ground square to produce flat-bottomed hole

to provide a cavity for the head of a bolt, or cap screw to be recessed and ensure there is not protrusion of a fastener above the workpiece surface. The diameter of the counterbore produced for any given size will depend on the type of fastener being inserted with counterbores to accommodate hexagonal head screws or bolts being significantly larger than those used to house the head of a cap screw, or cap screw and washer. Counterbores for hexagonal head fasteners also need to provide clearance for a socket to be inserted to allow tightening again increasing the diameter required.

The production of a counterbore can require some downforce and may be beyond the capacity of smaller belt-driven bench drills despite having a tool that fits the spindle, and operators need to assess whether the machine will have the capacity to drill the counterbore, especially if workpiece materials are of harder alloys or stainless steels.

A counterbore has a central shaft or spigot that protrudes below the cutting edges of the tool. This spigot engages into a previously drilled hole, and locates the tool to ensure the counterbore produced is concentric with the original hole. This is important as the counterbore is being produced to house the head of a bolt or fastener, and concentricity between the original hole and the counterbore must be ensured. This means that counterbores are often supplied in sets, with the spigot diameter varying depending on whether it is to be inserted in a hole drilled to a tapping size, or for a standard clearance hole for the shaft of the bolt to pass through. The cutting diameter of the tool will also have been sized for a varying head diameter to allow for socket insertion or to accommodate the head of a cap screw. This means that there are a wide selection of tools required to provide a relatively small selection of hole sizes and operators will need to ensure that they have the correct spigot/cutter diameter combination for the task. Ensuring that the spigot diameter is matched to the hole is particularly important, as if a counterbore using a spigot designed for a tapping hole is used when the drilled hole is sized for a clearance hole, the counterbore will almost certainly wander leading to a lack of concentricity, and oversized hole, or more likely one that is not round.

There are some tools available that allow for interchangeable sizes and have a standard spigot size. These can simplify tool selection and the counterbore has aa standard pilot hole drilled, the counterbore produced and the central hole opened up to the tapping or clearance size required. Most tools used are two or three flute HSS cutters, although carbide insert tools can be procured. The design of the flutes of the counterbore will vary with those used to machine brass having straight flutes compared to those used in steels. They behave in the way that milling cutters behave and good supplies of coolant or cutting oils assist with the production of a good surface finish, and extend tool life. All counterbores are drilled to a depth and td establishing the depth drilled can be from marking any depth scale with a wipeable marker to setting a fixed depth stop at the correct depth.

Speed settings should be those for a drill of the same diameter of the counterbore. To produce the counterbore the actual operation is little different than any normal drilling operation. The workpiece is retained in a vice or other work holding device, and due to the spigot being inserted into a pre drilled hole there is little wandering of the cutter. However, the larger diameters of counterbore, especially those sized to allow a socket to be inserted and engaged with a hexagonal head, will be relatively slow, but providing a high degree of torque. This means that any vice or workpiece should be positioned such that it comes up against a mechanical stop such as the machine pillar to prevent rotation. Once the workpiece has been positioned the machine is started and the operator presses the tool down into the workpiece until the desired depth is achieved. The operator should ensure that where difficult materials such a stainless steel are being counterbored, the downwards cutting pressure is maintained to ensure the cutting edge does not dwell and work hardened the face being cut into.

7.21 COUNTERSINKS

The drilling of a countersink is often surprisingly difficult. There are a number of types of countersink that can be used from ones which have just one cutting edge to those that have many. It is common for a two flute countersink to be found, however, multi-flute countersinks are better for fine chipping materials like brass. Countersinks can be tapered to fit into a spindle taper, or have a plain diameter to fit in a chuck. Speeds tend to be slower than for drilling and often the speed for reaming would be used for making a countersink.

Countersinks all come to a point and will be pressed into an existing hole in the workpiece to form an angled entry rim. However, the component drawing may dimension either the depth of the chamfer and its angle, or by the surface diameter of the chamfer and its angle. Either approach requires calculation by use of trigonometry, for precisely toleranced chamfers, but where holes have a wide tolerance and have been dimensioned by depth, it is often easy to just insert the countersink in the neck of the hole when the spindle is not rotating, identify the depth reading in the spindle quill scale with the countersink remaining engaged in the neck of the hole. The countersink is withdrawn the spindle started and the countersink pressed into the workpiece to the depth previously identified with the countersink in the mouth of the hole plus the depth of the chamfer required.

However where a more precise depth of countersink, or drilling to a known diameter is required, a calculation needs to be undertaken. This calculation is to establish the displacement that the point of the countersink needs to travel from level with the upper surface of the workpiece, to a depth that will generate the diameter (and therefore depth) of the countersink required:

$$\text{Displacement} = \frac{\text{dia of countersink req'd}}{2} \times \frac{\text{cot countersink angle}}{2}$$

This creates a right-angled triangle and gives the distance from the countersink point placed on the top surface of the workpiece to the depth it needs to be inserted to form the correct diameter of countersink mouth.

To create a countersunk mouth with a diameter of 20mm using a 90° included angle countersink:

$$\text{Displacement} = \frac{<\text{dia}> 20\text{mm}}{2} \times \tan\frac{90°}{2}$$

$$= 10 \times \tan 45°$$

$$= 10 \times 1 = 10\text{mm}$$

7.22 REAMING

Holes that have been drilled with a standard HSS twist drill tend to have a significant amount of scoring within the hole bore and often drill an oversize hole. This issues are caused by the design of a twist drill, the feed rate, speed and how well the drill has been sharpened. To provide a high quality hole of a more precise size a reamer is used. These types of tool can be plain shanked or have a morse taper allowing insertion into Jacobs type chucks, or directly into a machine spindle and retained in the morse taper.

Reamers come in two principal types: hand reamers and machine reamers. As with many cutting tools there are then sub-types with right-handed/left-handed, jobber, finishing, spiral, straight, and those fitted with a guide similar to the spigot on a counterbore all being available. An example of a spiral flute reamer and straight flute reamer is shown in Figure 7.21.

It is important when using a reamer in a drilling machine that a machine reamer is used. Hand reamers will all have a parallel shank and often have a square head to allow insertion into a tap wrench. More importantly they also have a tapered section towards the end to assist with entry into the hole. Where a reamer can pass completely through the workpiece this is not necessarily a problem as a parallel hole will be achieved, however, where a blind hole is being reamed or the reamer cannot protrude far though the workpiece die to a collision with the vice, or machine table it will leave a tapered section of hole. The tapered section will leave a portion of the hole undersize, and if the hole is being reamed to a known size to allow fitment of a component like a dowel pin, it will not be able to fully engage in the hole. Machine reamers do not have this taper as insertion is easier when firmly retained in the machine chuck, or machine spindle taper.

Where reaming is to be undertaken on machines that do not have a reversible spindle direction right handed reamers should be used. A decision needs to be made whether to use straight, or spiral fluted reamers. Spiral fluted reamers provide more of a shearing action, but as most of the cutting

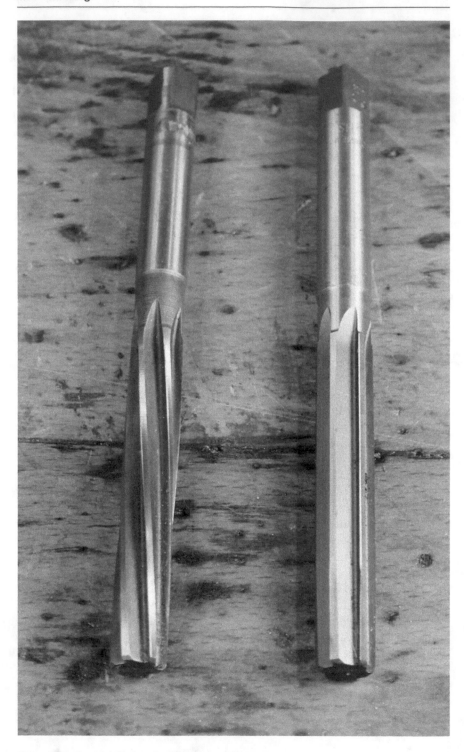

Figure 7.21 Straight flute and spiral flute reamers

is undertaken at the bottom cutting edge this is less of an issue especially for machine reamers where a tapered section is not present. Spiral reamers are extremely useful where a hole with a keyway is being reamed as the spiral profile bridges the slot and prevents chatter.

When reaming blind holes it is especially important to ensure a machine reamer is being used to ensure a parallel hole, however, all reamers have a chamfer or bevel on the bottom edge. It is the bottom edge of a reamer that does the cutting and the lead angle and lead length of the tool results in a residual bevel being left at the bottom of a blind hole being reamed.

The speed that is used when reaming is generally about half the speed that would be used for an equivalent sized HSS twist drill, but any power feed rate is normally about double that used for a drill. This is because only small amounts of material are being removed and it extends the life of the reamer, however, where a poor surface finish is observed the feed rate should be reduced.

Reamers are sized to produce an accurate diameter hole, and suffer less from the problems that twist drills do in generating oversize holes. To ensure that reamers do not suffer from producing oversize holes care needs to be taken when calculating the allowance, or how much material to leave for removal by the reamer. This allowance depends to a degree on the type of workpiece material being reamed and its size. The allowance should never be more than 5% of the hole diameter, and for softer materials is generally about 0.2mm and decreasing to 0.15mm for harder materials. By reducing the amount of workpiece material to be removed, there is less shearing, and heat production which provides a better surface finish. However, this only occurs in a band, and if too little material is left to remove the reamer can rub or push away workpiece material instead of cutting it leading to frictional heating and a poor surface finish. The operator also needs to have an understanding of the depth of grooves left by the twist drill as very soft workpiece materials may have been subject to deep scratching by the twist drills and failure to allow a big enough allowance may prevent removal of these marks.

The operator needs to decide if all holes are going to be drilled and then reamed. Best practise suggests that a hole is drilled and then reamed as the workpiece is not moved and the reamer will be on the centreline of the drilled hole. This can result in the production of a series of drilled and reamed holes taking significantly longer due to constant tool changing, but it does avoid the problem of bell mouth holes where a misalignment of reamer with a previously drilled hole results in an oversize hole at the entry point and decreasing to the correct size some way down the hole.

Having reduced the speed and adjusted any feed rate if power feeding, the operator uses a reamer in a similar manner to drilling any hole. However, good lubrication is required to prevent the workpiece material picking up or galling and cutting oils rather than emulsified oils can often be a good choice to prevent this happening. The reamer is advanced into the hole until it passes through the other side or the bottom is reached, at which point the reamer is retracted and the machine switched off. The hole should be clean parallel sided with a good surface finish. Where this has not been achieved it is normally down to a combination of factors. Poor tool sharpening, excessive feeds, poor reamer alignment with the drilled hole, incorrect speeds excessive depth of material to be removed, or too little are all contributory factors.

7.23 TAPPING AND USE OF AUTO RELEASE TAPPING AIDS

The production of threaded holes is entirely feasible using a drilling machine. There are three approaches that can be taken, however, the only consistent method is to use a reversing tapping attachment. If the workpiece is well secured, or held in a vice of reasonable mass, it is common to insert a tap into the drill chuck, engage it into a tapping hole and engage the tap into the hole by manually turning the spindle for a couple of turns, then releasing the chuck leaving the tap engaged in the mouth of the hole and in correct alignment with the axis of the hole and avoiding the production of a bell mouth when a tap is manually inserted off axis.

Operators have also often been tempted to insert a tap into the chuck, and using a slow speed allow the helix of the tap to pull itself into the workpiece. This is not especially good practice, and while can be effective for some size/material combinations often ends in failure with a stripped thread. It also requires the drilling machine to have a reverse mechanism or tap extraction will need to be manually undertaken. Straight fluted taps are designed to be used by hand and have periodic counter rotation to break of swarf. This relatively slow process allows for the swarf chips to be pushed out of the way. Power tapping does not provide for any counter rotation and chips are continually produced leading to choking, and then damage to the thread form. This is especially likely when tapping blind holes, therefore spiral taps, or for through holes, spiral point taps should be used.

As this form of tapping requires the tap to pull itself into the workpiece the resistance of the quill feed mechanism needs to be low. Small threads, and soft materials will often be immediately stripped due to the resistance of the quill feed mechanism. Similarly when tapping blind holes the drill must be stopped before the bottom of the tapping hole is reached as any over rotation will immediately strip the thread. While some operators try to overcome this by having a limited length of tap inserted in the chuck, and only tightening the chuck such that it can tap the workpiece but spin in the chuck when the bottom of a hole is reached. In harder materials this can often lead to a broken tap, and in softer materials still has the potential to strip the thread especially on smaller diameter threads. For these reasons tapping without a tapping aid is not recommended.

Reversible tapping aids are widely available, and are sized for a range of threads, and are especially effective in tapping blind holes which is what they are designed for and if intended to be used on through tapped holes, a bar of metal needs to be placed under the workpiece to allow the tap to press onto. They are designed to be used on a drilling machine, and incorporate a clutch and reversing mechanism. Machine tapping aids are normally inserted directly into the machine spindle but also require a vertical peg or part of the machine body to act as a post for a reaction arm to come up against to allow the torque mechanism to work. Most tools have an adjustable torque setting to ensure that the range thread sizes can be properly cut without damage, and the torque setting for an M3 thread would be substantially less than that required to cut an M12 thread. If the torque setting is too low the tapping attachment will stall and fail to cut the thread. If it is set too high for the thread size being cut there is a risk it will not disengage when the bottom of the hole is reached and strip the thread being cut. Manufacturers' instructions and experimentation by the operator on scrap materials will identify the combinations of torque settings, for thread size and material types. Once the workpiece has been drilled with a tapping size drill the attachment is inserted into the machine spindle and the tap inserted into the attachment chuck. Generally these are of the collet type which require a spanner to tighten, and a spiral flute tap is used. With the correct torque set the quill feed capstan or handle is rotated to press the tap into the hole. Once the tap engages, the pressure on the capstan should be reduced and the tap will drive itself down until the bottom of the hole or lower plate is reached. The tap will then stop rotating, and as the operator rotates the capstan to withdraw the tap, the chuck will automatically reverse and exit the workpiece. If the operator continues with heavy downward pressure the attachment mechanism will not correctly function and the thread may be stripped. The automatic reversing mechanism does not require the machine to reverse, but is a function of the tapping attachment internal gearing, therefore, can be successfully used on machines that only have a right-handed spindle rotation.

Spiral flute taps are normally used when power tapping as the swarf is forced up and out of the workpiece preventing chocking and over cutting which would damage the thread being cut. They are especially effective where thread are being cut on softer or more ductile materials, which tend to produce long continuous chips. Hand tapping involves occasional counter-rotation to break the chips being formed, however, this is not possible on machine tapping therefore the use of a spiral flute tap assists with ejection of continuous chips. Through holes can be cut using spiral flute taps or spiral point taps. Spiral point taps have a leading flute ground at an angle to the tap axis. The leading flutes are ground into the tap left handed and this causes the swarf chips to be pushed ahead of the tap, and as such means that the flutes up the tap body can be smaller. This has the advantage of strengthening the tap, which when tapping fine threads on harder alloys significantly reduces the potential for tap breakage in the workpiece. They work best when producing though holes and if used in a blind hole the tapping drill will need to have been drilled deep enough to accommodate all of the swarf chips pushed ahead of the tap.

Lubrication is important when power tapping, and generating excellent thread forms requires a copious supply of lubricant or cutting oils. Cutting oils are often the best solution rather than the use of emulsified oil coolant as they not only reduce heat and remove swarf, but assist in the formation of the swarf chips.

Cutting speeds when power tapping are generally considerably slower than used when drilling with HSS twist drills. This is to allow the chips to form correctly, achieve a good surface finish and to allow the swarf chips to be ejected correctly. A surface machining rate specifically used for tapping, and RPM values for a standard selection of spiral point taps are identified in manufacturers tables.

7.24 POWER FEED DRILLING

Larger drilling machines and radial arm drills commonly have a power feed facility. This is especially useful when drilling materials that are susceptible to work hardening if a drill point is left to dwell in the cavity without enough pressure being applied to cause the drill point to continually cut into the workpiece material. It is also useful when drilling multiple large diameter holes as this also reduces operator fatigue.

While a wide variety of feed rates can be calculated for drilling depending on the drill type, drill diameter, and workpiece material, it is common for a machine to have a limited selection of feed rates. The section on feeds and speeds goes into detail on the combination of feeds and speeds, however, it is rare to be able to achieve the exact required spindle revolution rate and feed rate. Therefore some compromise is required, and a broad assertion can be made that reducing the spindle RPM and using a higher feed rate can be a successful compromise especially when drilling difficult materials.

Feed rates are normally linked to the RPM rate set for the drill, and this is normally replicated on the machine, with feed set as a proportion of spindle speed. Figure 7.22 shows the feed selection values for an imperial machine with feed rate expressed in thousands of an inch/revolution.

In this example to there are three options with 0.007" (0.18mm), 0.004" (0.1mm), and 0.010" (0.25mm). To select a feed rate the control knob is rotated to one of the three options. The feed is engaged using a separate control lever, which is not always obvious, and an example is shown in Figure 7.23.

What may be a limited selection of feeds is normally not a problem even when drilling materials subject to work hardening, as suitable feed rates can

Figure 7.22 Pillar drill feed rate selection control

be achieved, and while they begin to fall behind an optimal rate at diameters over Ø20mm they are still progressing into the workpiece with each revolution and work hardening should not be an issue.

Figure 7.23 Feed engagement lever

The use of power feed, especially when drilling larger diameter holes, does require the workpiece to be firmly retained in a machine vice or bolted to the machine table. The combination of continuous feed and rotational forces combine to provide no release of load on the workpiece, and while the correct set-up of feed and spindle speed presents no risk to successful drilling, if the workpiece moves during the operation it often leads to drill breakage as quick release of the drilling pressure is not possible. Firmly retaining the workpiece in position, and correct setting of spindle speeds and feed rate is an important aspect of power feed drilling.

7.25 SWARF CLEARANCE AND CONTINUOUS CHIP ISSUES

Drilling softer and more ductile materials often produces long continuous chips, especially at smaller diameters. While the drill is cutting effectively the long chips become a safety hazard as they sweep in an arc around the drill bit, and also have a tendency to scratch the vision panels of any guarding system. The production of continuous chips is a function of the drill cutting edge, flute helix and materials type, and the simplest method to break these chips is to release the pressure when hand drilling, breaking the chip which is then discarded. When using a drill bit of larger diameters it is possible to use a thin radiused wheel to grind grooves into the cutting edge of the drill point. These act in exactly the same way that the radiused cutting edge of a milling ripper cutter works, in that the shape and effective depth of cut varies causing the chips to break in short lengths. Reducing the speed of the drill and increasing feed rates can also encourage the swarf chips to shear more often.

Drilling materials which produce copious amounts small swarf chips can be harder to deal with. The design of the drill brings swarf up to the surface of the workpiece which is then ejected from the drill bit by the rotational forces. Allowing significant piles of the swarf chips to remain in place can lead to damage or the workpiece surface finish and prevents the swarf chips from being properly ejected. It also prevents coolant or cutting oils to enter the cavity being drilled causing heat build-up. The use of significant amounts of coolant can wash away this build up, however, many small or medium sized machines do not have this facility. In this situation the operator will need to brush away the swarf chips using a small swarf brush or paint brush. This can be undertaken while the drill is being fed into the workpiece or the drill extracted, the swarf chips brushed away and drilling re-commenced.

When drilling multiple holes the operator should also take car that swarf chips do not get under and vice or work holding device that may either make it less stable and more likely to spin, or produce an off axis hole in the workpiece.

7.26 PROBLEMS WITH DRILLING

Drilling is a relatively straightforward process, however, like all material cutting processes it can still experience conditions that lead to poor results. Not only does this mean that the hole will be sub-optimal, but the drill bit can often be damaged as well. The solution to many of these issues comes back to some basic principles of utilising the correct peripheral speed for the material being drilled, and the correct feed rate where power feed is being used. Continuous feed when difficult materials are being drilled, and supporting the drill bit and workpiece correctly.

Damage to twist drill bit

Twist drills can be damaged in a number of ways, some of which are quite obvious, and some less so, although the damage can lead to incorrectly drilled holes and the evidence of the drill damage is seen in an incorrect profile of the drilled hole or in a poor surface finish.

The most visible of damage is where a parallel shank twist drill has not been sufficiently tightened in the chuck, or that the drill speed or force used to drill a hole has resulted in the drill bit stalling and the chuck continuing to rotate. In extremis a drill bit can become welded into the workpiece creating significant difficulty in extraction, however, irrespective of the cause, the rotation of a chuck around its shank causes significant scoring and pick up of material. An example of this is shown in Figure 7.24. Where the material being drilled will impart significant torsional loads that are likely to overcome the grip of the chuck on the drill shank then a solution is to utilise an taper shank drill.

In extremis this scoring can be so deep as to prevent the drill subsequently producing an accurate and parallel sided hole when reinserted in a chuck. This is due to the shank being so deeply scored it cannot be clamped on an axis aligned with the spindle. However, this is rare, and more common are a series of radial scores, along with a burr picked up where the slippage stopped. As the shank of a drill is relatively soft these burrs can be simply removed by filing off. Failure to do this will result in the drill not being properly located when subsequently mounted in a chuck.

Breakdown of the drill bit cutting edge outer corners is a common problem. This prevents the drill from properly cutting and requires the drill to be resharpened. It is caused by a number of factors such as the peripheral speed for the drill bit being set too high which burns out the corners of the cutting edge.

Corner erosion can also be caused by inadequate lubrication or coolant supply to the cutting point, and more commonly by having an interrupted feed on work hardening materials such as stainless steels that require a continuous cut to take place. If continuous feed is not maintained friction causes the workpiece to rapidly work harden to the point where the difference between the hardness of the cutting edge of the drill bit, and that of the

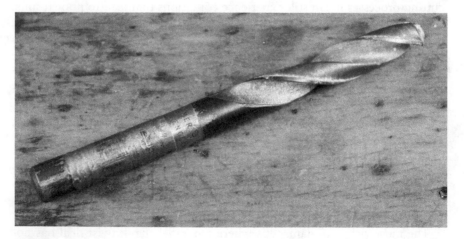

Figure 7.24 Scored drill shank

workpiece is negligible. At which point the drill fails to cut and frictional forces erode the drill corners.

A workpiece that is not properly supported is another cause of drill bit corners breaking down where irregular cutting forces impacts and wandering causing damage. This emphasises the importance of ensuring that a workpiece is well supported either in a vice or other work holding device, or clamped to the machine table, unless it is of a significant mass and prevented from rotating, by the use of table registers or the machine pillar. The final, and probably second most common cause of corner erosion, is when opening out existing holes. Opening out a hole requires a significant body of workpiece material to be removed. Where a small difference in diameter is being removed all of the cutting forces, and heat generation are focused on the corner leading to rapid erosion.

Chipping of the lips of the drill point, is a little less commonly experienced, however will result in poor drill performance. Chipping of the cutting edge occurs when the drill point lip clearance is too great, or has been quenched when re-grinding. It is common for drills to be sharpened by hand in the workshop and many workshop grinders have a small reservoir for quenching materials being ground. Where a drill has been quenched in water to cool it while being ground it can lead to cracking of the cutting edge and subsequently chipping. However, chipping can also occur where any power feed rate has been set too high and the drill surging on breakthrough, especially where thin workpiece materials are being drilled.

One type of drill bit damage that is less commonly experienced in general workshop drilling is splitting up the web. This is where a crack running up the web along the axis of the drill is observed. It is caused by a number of issues such as insufficient lip clearance. Drills that have not been ground with the correct amount of lip clearance angle increase the loads on the drill at the cutting edge and also allows for heat build-up due to heel drag. The solution is no surprise, in that regrinding of the drill will prevent this failure mode, however, if the drill is spilt a long way up the web, this may require a lot of grinding. As with chipping of the drill cutting edge lips, quenching overheated drills when sharpening is another cause of web splitting, and the operator needs to aware of heat build-up and manage progressive cooling when grinding, rather than direct quenching in water.

Having a feed rate set too high is another cause. This can be difficult to balance as optimum settings are easy to achieve in most materials, where difficult materials such as some nickel, titanium, hard alloy steels or stainless steels the peripheral speed is lowered and the feed rate increased to ensure that localised work hardening does not occur. Increasing the feed rate too far and slowing the peripheral speed can lead to excessive loads on the drill causing cracking up the drill bit web.

The final two causes web splitting relate to mechanical damage. Impact or striking of the drill point with a hard object is one cause. While this may sound strange, accidentally hitting the drill point with a hammer or spanner, when position fastening a workpiece down is not unknown. Drill bits also periodically suffer damage when left on a workbench or the drill machine table having metallic vices dropped onto the cutting point, and while in a perfect world they would have been put away, or left in storage until about to be used, this is rarely achievable in a busy workshop environment, but can lead to damage. A more common cause, especially to taper shank drill bits, is the drill point striking the machine table. It is rare for the operator to be able to grip the drill, rotate the spindle capstan, insert a drill drift and strike with a hammer all at the same time. A more common scenario is that the operator rotates the capstan, inserts a drill drift into the spindle slot, and then holds the drift while striking it with a hammer to eject the drill bit. The drill bit then freely drops onto the hard machine table causing damage. The solution is simply to put a soft wooden block underneath to allow the drill bit to drop onto.

Incorrectly drilled holes

A perfectly drilled hole will always have some scoring of the walls, hence the requirement for reaming, however, the surface finish is rarely an issue for manufacturing, and a hole that has been drilled using a well sharpened twist drill of the correct type, with the correct feed and speed rate will not be rough. Similarly a correctly drilled hole will be of a diameter close to that of the size stated on the drill.

While the diameter of the hole drilled will never be exactly the size stated on the drill as all drills will produce a slightly oversize holes. A well sharpened drill would be expected to produce a hole that is oversize by no more than 0.025–0.08mm for a drill of approximately $\varnothing$3mm and 0.08–0.13mm for a drill of approximately $\varnothing$19mm. A well sharpened drill would be unlikely to produce and over size hole of a maximum of more than 0.11–0.2mm for $\varnothing$3mm and $\varnothing$19mm drills respectively. This tendency to drill oversize is well known, however, the tolerances on drilled holes normally accommodate this, and the small errors rarely require the operator to compensate by selecting a drill bit 0.1mm smaller, as holes with that level of tolerance are normally reamed to size.

However, significantly larger oversize holes can be drilled if the drill bit is not correctly ground. Where the cutting edge lips are of unequal length resulting in one lip doing the majority of the cutting, or the chisel edge on the drill point is not central. The correct geometry of the drill point is key to preventing the drill from wandering in the hole and creating a significantly oversize hole as it progresses through the workpiece.

Another common drilling problem is the production of a hole with a bell mouth. When a twist drill is pressed on the surface of a workpiece it has a tendency to wander as it is unstable and not supported. This is especially problematic where a centre point has been punched rather than drilled and the chisel point, or web of the drill is too large to engage in the marked point. This wandering creates a significantly oversize hole that reduces as the drill progresses into the workpiece and the sides of the hole provide support. This tendency is also exacerbated where the drill point has been incorrectly ground, especially where one lip, or cutting edge, is longer than the other. A method to eliminate this is to drill a centre hole with a morse centre drill prior to drilling. This allows the point to be located and stabilised as it enters the workpiece and prevents the drill from wandering.

An alternative method is to support the drill bit with a bush. This method is very common in production drilling where the drill bit enters a bush which supports the drill bit throughout its length preventing wandering and having the secondary benefit of removing the requirement for a punched centre hole or one drilled with a morse centre. However, while an extremely effective method, in the general workshop this requires the manufacture of tooling and is generally only used in exceptional circumstances, with centre drilled holes, and a correctly ground twist drill being the 'go to' approach for the operator to prevent bell mouthing.

One other potential reason for the creation of a bell mouth, is wear in the drilling machine spindle assembly. Where there is significant wear in the spindle bearings the spindle itself the spindle bearings allow trans axial movement driven by the drill wandering. There is little that can be done about this even when using a correctly sharpened drill other than using a drill bush that is firmly located in position.

When a hole is being drilled completely through a workpiece there comes a point when the drill starts to break through the bottom surface. As the drill point breaks through, especially when drilling larger diameter holes through thin workpiece materials, the lips of the cutting edge start to have a point loading rather than a load spread across the entire length of the cutting edge. The cutting edge is then removing metal from a corner, rather than a full face. As the chisel point breaks through and the length of the cutting edge reduces the drill will pull itself through as it grabs material and the helix of the flute pulls the drill down. This effect is called surging.

The diameter of the drill and the thickness of the workpiece material can have a significant effect on this as the depth of the point increases with drill size so the ratio between drill point height and material thickness increases. Where the height of the drill point is less than the thickness of the material being drilled there is significantly less tendency for surging. The height of the drill point is a function of the point angle, and in larger twist drills this can be considerable, so drilling a 5mm thick plate with a Ø25mm drill, the height of the point will be 7.5mm if the drill has an included point angle of 118°. This means the point of the drill will be breaking through the workpiece lower surface before and rapidly opening up the hole. In extremis the drill corner grabs the workpiece pulls itself through without cutting the material correctly and draws the workpiece up the flutes of the drill while spinning it. This can lead to an injury to the operator, damage to the workpiece, or breakage of the drill. Removing the drill from the workpiece can also be difficult.

If a radial arm drill is being used with power feed and a rigidly mounted workpiece, this effect is rarely experienced. However, when drilling by hand the operator will feel the drill point breaking through the workpiece and relieving the pressure on the workpiece and progressing slowly through in series of pecks can avoid this problem. This is easily achieved when using drill bits of diameters up to 20mm, however, when larger diameter twist drills are being used and the height of the drill point exceeds the material thickness, this still needs care and a well clamped down workpiece. Where very thin sheet materials are being drilled, the operator should consider if a spot weld point drill could be used as an alternative given its small central spur and flat cutting edge. This removes the ability of the drill to surge, however, are only intended for thin sheet materials and are unlikely to be available in any significant diameter.

Drilling deeper holes creates a problem for swarf chip extraction. Drills are designed to pass chips up the helical flutes which are ejected at the exit from the hole being drilled by rotational forces, and any coolant flow. This means that holes can only be drilled to the depth of the flutes to allow swarf chip ejection and drilling beyond this point will lead to the flutes choking and if forced, drill bit breakage. The use of long series drills, or extra length drills, provides some ability to drill, as does pecking to allow swarf build up in the flutes which is release by drawing the drill out of the hole being generated. The use of parabolic flute drills which are designed to clear swarf chips from depths at over ten times the drill diameter without pecking.

Problems with swarf chip extraction and choking of the drill flutes can be experienced at much shorter depths, especially when drilling softer and more ductile materials. It is not unusual for drills to stop cutting or break due to the flutes being choking with swarf, especially when drilling some softer aluminium alloys. Extraction of a drill often shows swarf compacted, and in extreme cases, melted into a hard pack of swarf in the drill flutes which is difficult to remove. This is probably more common where drilling has been undertaken on a lathe and it is harder for the operator to assess the feed rate when winding a drill in on the machine tailstock, but does occur when using a drill press. To avoid this it is important for the operator to have a correctly sharpened drill of the correct type, use the correct RPM, for the drill and workpiece material, and ensure the depth being drilled does not exceed the flute length.

Another type of incorrectly drilled hole is one where the entrance to the hole has a lobed octagonal or pentagonal shaped mouth. This is normally cased by chatter or wandering of a drill entering the a hole that has already been drilled in order to open out the diameter. This can be experienced at any size but is more commonly seen on larger sizes of drill, and observed on sizes above $\varnothing16$mm where existing holes such as $\varnothing14$mm are being opened out. The cause of this generation of a lobed polygon is a function of the drill geometry as the material behind the cutting edge is relieved to form the flute allowing more space for wandering and the drill only cutting at corners giving a less stable drill point and a greater amount of torque loading given the distance of the point of force from the drill axis.

Where an operator is required to open out an existing hole, it is something they cannot control and if the increase in size is quite small then an alternative drill bit such as a 3 flute, of 4 flute core drill can be used to overcome the tendency to generate a non-circular mouth. Where the operation is merely increasing the diameter of a pilot hole, the operator is in full control. Drilling pilot holes provides a cavity for the full sized drill to follow, removes the volume of material to be removed in on operation, the amount of swarf generated. The actual size of a pilot hole can vary, however, too large results in corner wear of the drill bit and the generation of a lobed hole entrance. Too small a diameter leaves a significant amount of material to be removed, which may overcome the power of the drill, lead to work hardening, and fails to provide enough guidance for the drill. The broad guidance would be to use a diameter of drill that is at least a diameter equal to the chisel edge, or web thickness, and up to 50% of the full diameter required.

Appendix A

Nomogram for calculation of helical milling angles

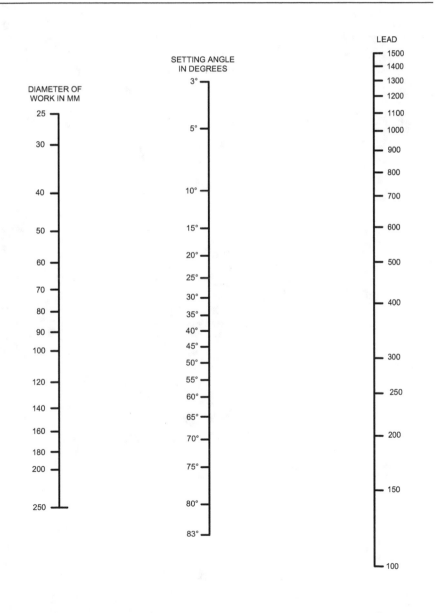

Appendix B
Simple turning exercise

Most operators began developing their skills as a lathe operator by machining a stepped bar. A stepped bar (as shown in Figure B.1) is a simple turned component that requires the operator to parallel turn, parallel turn to a length, face off, produce an undercut, generate a chamfer, centre drill, drill a hole, and part off to a known length. This exercise covers 90% of the operations commonly undertaken on a lathe and other than knurling, screw cutting, taper turning and boring is a method of getting an operator to manufacture a component to a known size and grow an operators skill and confidence when taking the first steps into effectively using a lathe.

The following list of operations is intended to be followed by someone who is taking their first steps in turning, and will identify all of the individual operations to be taken. The operator should follow these steps in order, however, it is assumed that the operator is competent to use a vernier or digital calliper, micrometer, and be conversant with the setting of speed and feed rates on their machine. A component drawing for a stepped bar is shown in Figure B.2.

Stock material to manufacture the stepped bar from is steel to BS EN 10277:2018 grade 070M20, however, any bright bar free machining mild steel is suitable. The bar size needs to be a minimum of 75mm long and give the type of exercise being undertaken Ø30mm bright bar removes the requirement to machine an outside diameter to Ø30mm and utilising the next stock bar size up. There is likely to be a degree of lack of concentricity between the plain diameter and the machined sections due to material manufacturing processes and the precision of any three jaw chuck on the machine. However, this is not significant given the nature of the exercise, and while it is normal to ensure that all errors are removed by giving the stock bar surface a light cut to 'true' it, this is not required.

The list of operations and cutting speeds assumes that the following tooling is being used:

- Carbide insert with an approach angle set to 90° of workpiece axis (or less) to be able to form or generate a 90° shoulder.

Figure B.1 Stepped bar

- Carbide insert parting tool 3mm wide.
- Medium sized HSS morse centre drill.
- Ø12mm HSS jobber drill

It is assumed that all turning tooling will be mounted on centre in an indexing toolpost.

OPERATION LIST

1 Insert the bar into the machine chuck with approximately 30mm protruding from the chuck, and tighten the chuck jaws to firmly grip workpiece.
2 Set the machine speed to approximately 1000–1300 rpm.
3 Close guarding and start machine.

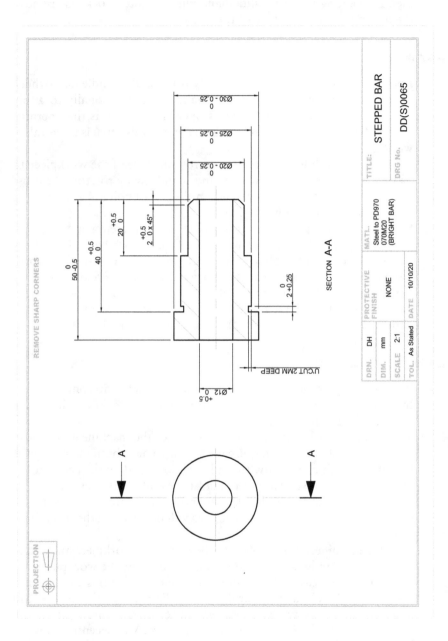

Figure B.2 Stepped bar component drawing

4 By rotating the saddle handwheel, and cross slide handwheel bring the point of the cutter into light contact with the end of the workpiece, and then by rotating the cross slide handwheel counter-clockwise remove the tool point from the face of the workpiece.

FACING OFF

5 Apply a light cut either by fractionally rotating the saddle handwheel, or by rotating the compound slide hand wheel fractionally to apply a cut. Coolant can be used for this operation or not as the operator desires. Given that there is little heat generated this often is undertaken without.

6 Using the cross slide move the tool across the face of the workpiece to its centre to form a plane surface, and withdraw by rotating the cross slide handwheel counter-clockwise until the tool point is away for the external surface of the workpiece.

7 Stop the machine and inspect the workpiece. If there remains a rough section on the end of the bar where the face has not been cleaned up, repeat the process of applying a small cut and passing it over the workpiece face. If the facing cut was too heavy a conical piece of swarf may have been generated and retained on a central pip if the tool was not exactly on centre. Where this occurs the tool again has a very light cut applied and again passed over the face which will remove it.

CENTRE DRILLING

8 Reduce the speed of the machine to approximately 750 rpm.

9 Insert a Jacobs chuck (or similar) into the tailstock, insert the morse centre drill and firmly tighten.

10 Move the tailstock forward along the bed of the machine and lock in place. This will normally require the saddle to have been moved up the bed of the machine to allow the tailstock to be slid up the bed and not require excessive extension of the tailstock quill. However it is important to check that the tool in the toolpost is not move so close to the chuck that impact with the chuck jaws will occur when the machine is started.

11 Start the machine, apply coolant to the end of the workpiece and rotate the tailstock handwheel until the centre drill enters the workpiece and forms an angled cavity. It is important to ensure that this cavity is not drilled so deeply that the parallel outer diameter of the centre drill enters the hole as this will almost certainly cause the point of the centre drill to snap of and be lodged in the workpiece. Morse centre drills are only designed to have the small point and angled face pushed into the

workpiece to form the angled face for a supporting centre to engage with. Pushing the drill into the workpiece beyond the angled section destroys the angled feature and will break the drill.

12 Unclamp the tailstock and slide along the bed to the end. Remove the morse centre drill and remove the Jacobs type chuck. Following this insert a revolving centre.

PARALLEL TURNING

13 Move the saddle along the bed of the machine towards the tailstock to allow the operator clear access to the workpiece.

14 Loosen the chuck to allow the workpiece to be slid forward in the jaws until a minimum of 60mm is protruding from the jaws, and then lightly tightened to retain in place. Rotate the saddle assembly back down the bed of the machine until the tool point is close to the end of the workpiece. Then the tailstock is also slid down the bed and the point of the centre inserted into the drilled hole. If the quill of the tailstock has not been extended enough the tailstock body will impact the saddle before the point of the centre has engaged in the workpiece. Where this happens the operator merely needs to rotate the tailstock handwheel until proper engagement in the hole is achieved. Once the centre is fully engaged, the chuck is tightened to firmly clamp the workpiece.

15 The operator should then check that the chuck can freely rotate when the right-hand side of the parting tool is moved to a length 50mm down the workpiece without impacting the tool or tool post. This is normally checked by rotating the toolpost so that the parting tool is facing the workpiece and a rule used to place the tool point 50mm down its length. The chuck is then rotated by hand. Where a collision occurs the workpiece will need to be moved further out of the chuck, necessitating loosening of the tailstock and chuck, movement of the workpiece and retightening etc. The checking process is then repeated. However the operator should ensure that the maximum of workpiece material located in the chuck as possible to provide sufficient friction to rotate the workpiece while being cut, while ensuring there is no tooling collision with the chuck. The length of stock bar cut, and the confidence of the operator are the controlling factors here, and it is not unusual for more experienced operators to be happy working close to the chuck, and understand the minimum engagement lengths required, whereas those just beginning to turn have greater clearances.

16 Once the workpiece has been set up and supported with a centre, parallel turning can commence. The first operation is to generate the Ø25mm feature which is turned to a 40mm along the length of the workpiece, forming a shoulder. There are a number of methods for

identifying the length, which include moving the tool to a point 40mm down the body and scribing a line using the tool point, placing the tool point on the end of the workpiece and then trapping a 40mm slip pack between the saddle and bed stop, or temporarily sliding the tailstock to the rear of the machine bed and using the depth function of a vernier or digital calliper to set the tool point to 40mm down the workpiece and clamping a bed stop in place. Irrespective of the method utilised a clear mark of where the material removal should stoop is essential, and the benefit of using a mechanical stop, or register means that coolant will not obscure the stop position which can occur when machining to a marked line of the workpiece.

17 The tool point is then placed in contact with the workpiece and the scale on the cross slide scale set to zero. This can be achieved with the workpiece rotating under power or when stopped. Both approaches have their merits, but if the workpiece is not rotating then care needs to be taken to damage the tool point.

18 Once the scale has been set to zero the tool is taken off the end of the workpiece adjacent to the centre, and the machine set of a rotational speed of approximately 1000–1300rpm. By rotating the cross slide handle a cut of 2.5mm is applied. Any guarding is put in place coolant applied to the tool point, the machine started and the feed handle engaged to commence the tool cutting down the workpiece. The operator should check that the tailstock centre is revolving and tighten in necessary, and then observe the movement of the carriage towards the bed stop, check for smooth cutting and ejection of swarf, or movement of the tool towards a scribed line. It's worth noting that the power of the machine feed is more than the frictional resistance of the bed stop and the feed should be disengaged as it meets the bed stop or fractionally before and the cut to the end completed by hand.

19 The machine can be stopped and once the workpiece stationary the saddle rotated back to its start position. This can sometimes result in a fine line being scored down the workpiece as the tool tip recovers from compression, or the tool can be withdrawn the saddle moved down and the tool returned to its setting. While the second approach ensures there is no marking of the work piece, it introduces the potential for error when resetting the cross slide. Normally marking is not an issue as successive cuts remove it.

20 The workpiece diameter is then measured using a micrometer or calliper. The workpiece stock size was ∅30mm, however, it is rare for a 2.5mm cut to leave exactly 2.5mm more to cut when a 5mm reduction in size is required. Therefore checking the size is important to determine the amount of material remaining. The operator then has to decide whether to remove the remaining depth of material in one cut or two. If electing to remove the remaining material in two cuts then

a cut of approximately 1.5mm followed by a finishing cut of 1mm is preferred. Depending on the nose profile of the cutting tool, very light cuts with a carbide tip with a small nose radius, while machining to the correct size, can produce a disappointing surface appearance albeit smooth.

21 Once the final cut depth is established the process of the first cut is repeated until a diameter of 25mm (or close to it) is achieved which leads to a shoulder formed 40mm along the length of the workpiece.

22 The operator then needs to form the Ø20mm feature. The process is repeated with the bed stop being located to produce 20mm length of cut or the workpiece diameter scribed. Two cuts, one of 2.5mm, followed by measurement and a finish cut to depth are applied in exactly the same manner as generating the Ø25mm feature leaving a section turned to Ø20mm by 40mm long.

23 While the operator has a degree of choice in the order that the remaining operations are undertaken, generating the chamfer is chosen next as it doesn't require changing the tool. Firstly the tailstock is unclamped and slid to the end of the bed and the centre ejected.

CHAMFER GENERATION

24 To generate the chamfer, rather than form it, the compound slide has its clamping screws loosed and the compound slide rotated counterclockwise to a position of 45° as indicated on the scale at the bottom of the compound slide assembly and aligned with a register mark scribed into the top of the cross slide assembly. The compound slide clamping screws are then re-tightened. It is common for the toolpost to require rotation as well as either the cross slide does not have enough movement to allow the tool point to get close to the central axis of the machine, or more commonly, the cross slide handle has interference with the coolant pipe. Therefore, the toolpost is loosened and rotated through 45° clockwise and re-tightened. This allows the tool to be placed close to the machine central axis and be placed adjacent to the workpiece. While the relative angle of the tool point to the workpiece feature being generated is not normal, this rarely causes a problem with chip formation or surface finish given the light cut taken.

25 The tool point is placed until close to the corner formed by the Ø20mm feature, and slightly away (toward the tailstock) from the faced off end of the workpiece using a combination of the saddle control wheel and the cross slide control wheel. The guarding is then closed and the machine started. The chamfer is generated by applying a cut using the cross slide and advancing the tool in front of the faced off end away from the operator. The compound slide control wheel is then used to pass the tool across the corner of the workpiece. Small cuts are applied

and it's not unusual for the first couple to be in fresh air, however, the operator rotates the cross slide control wheel in small increments until metal cutting commences. The application of small depth of cut applied by the cross slide control wheel, followed by the actual cut taken by moving the tool across the face of the chamfer to remove material until the length of chamfer is achieved. Measuring chamfer lengths is not easy, as it requires finding a corner and the intersection of a face. Typically this is achieved by using a ruler to determine if the chamfer is correct, and tolerances are rarely tight on this type of feature. For the stepped bar the chamfer should be 2mm long an read using a ruler. If it is noticeably short additional cuts can be applied. If it is a significant chamfer, in this exercise it will not matter unless it is a such significant size it intrudes into the entrance of the 12mm hole. Once the chamfer is of correct size, the toolpost and compound slide are rotated back to their normal position.

DRILLING

26 The next operation is to drill the Ø12mm hole. The Jacobs (or similar) chuck is inserted into the tailstock and a Ø12mm HSS jobber drill firmly tightened into it. The machine speed is set to 716 rpm, or the closest setting for the machine as this is the correct speed for a HSS drill of this size in mild steel.

27 The saddle assembly is then moved towards the headstock with any tooling having been retracted clear of the workpiece, and the operator having checked that there will be no collision with the chuck.

28 The tailstock quill is the set at either a known point against any scale on the quill (ideally zero), or a mark made with a non-permanent marker at the junction of the quill and tailstock body. the tailstock is the slid up the bed of the machine until the drill point is adjacent to the end of the workpiece and firmly clamped in place. Often operators touch the drill point into the morse centre drilled hole.

29 The machine is then started and copious amounts of coolant applied where the drill enters the workpiece allowing it to wash over the drill flutes and workpiece body. By rotating the tailstock handwheel the drill is advanced into the workpiece, and this advancement continues until a depth of 60mm is achieved. This depth allows for the fill 50mm length to be achieved and the depth taken up by the drill point allowed for, and will leave a full parallel hole in the workpiece length required. Drilling in this manner, especially where the operator smoothly advances the tool, often creates a continuous chip which moves down the bed of the lathe towards the operator. As the type of steel being drilled is not subject to rapid work hardening pausing advancement for a couple of revolutions breaks the cut and the continuous chip will

fall away with the next one being formed as the drill is advanced once again. Where the tailstock quill does not have a clear depth marked, using a rule to check the distance between the tailstock body and a position marked on the quill establishes the depth the drill has been advanced. Once the full depth of hole has been achieved the machine is stopped and the drill withdrawn by sliding the tailstock back down the bed of the machine. The drill is removed and the Jacobs chuck ejected from the quill.

GENERATION OF UNDERCUT

30 The undercut at the shoulder between the $\emptyset$30mm and $\emptyset$25mm features is the next operation. The toolpost is rotated until the parting tool is facing the workpiece, or parting tool mounted on the toolpost. The left-hand edge of the tool is then brought into contact with the workpiece shoulder and advanced until the tool point touches the surface of the $\emptyset$25mm feature. The scale on the cross slide handle is then reset to zero and then retracted off the workpiece. The dimensioning on the component drawing shows a 2mm depth of undercut rather than a diameter. This means that the tool will need to be advanced 4mm on the cross slide scale as they are normally calibrated to material removal off a diameter rather than a radius. The width of the undercut in this case will be the width of the parting tool, which is typically 3mm. The machine speed is once again set to 1000–1300 rpm, the guard set in place and the machine started. It is advisable to use plenty of coolant as the tool has no side relief and this not only cools the workpiece and tool point but washes away swarf chips preventing over-cutting. The cross slide control wheel is rotated util it comes into contact with the workpiece and then the tool is smoothly advanced into the workpiece until the scale reads 4mm. the tool is then retracted clear of the workpiece and the machine stopped.

PARTING OFF

31 The final operation is to part off the workpiece from the bar retained in the chuck. The parting tool is positioned so that the left-hand edge of the tool touches the end of the workpiece adjacent to the $\emptyset$12mm hole. The bed stop is then set using a pack of slips which is 50mm (the length the workpiece is required to be manufactured to), plus the thickness of the parting tool insert (in this case 3mm). This is because the left-hand side of the insert is being used to identify a datum position of the workpiece end, but the right-hand side of the tool needs will generate the final workpiece length. Once the bed stop has been clamped in place the tool is withdrawn from the end of the workpiece

by rotating the cross slide control wheel and the saddle rotated down the bed of the machine until it comes into contact with the bed stop. The operator then checks there will be no tool/toolpost collision with the chuck and the guarding set in place. The machine is started, and again with good coolant flow the tool is advanced into the workpiece with a slow smooth continuous action. This advancement continues until the workpiece detaches from the stub held in the chuck, and will normally drop to the base of the machine. The machine is then stopped, the saddle moved to the end of the bed and the component retrieved. When parting down to a hole it is not unusual for a thin circular shim to be found around the periphery of the hole and its height is equal to the width of the parting tool. This is removed by gripping it with pliers a pulling on it until it detaches, and the entrance of the hole cleaned up with a cranked de-burring tool to remove any residual sharp edge.

32 The component is now complete and manufactured to drawing. The remaining stud of material in the chuck is removed and disposed of, if not of a useable length, and the machine cleaned down.

Appendix C
Simple milling exercise

Operators wishing to introduce themselves to milling have a significant number of choices to make, and as a piece of machining equipment the standard workshop milling machine commonly offers up to five degrees of freedom, compared to the three of a lathe. To give operators new to milling a chance to develop their skills an exercise has been designed to introduce them to milling using simple translational movements, and a drilling operation. The component is shown in Figure C.1. Work is intended to be undertaken on a turret milling machine; however, an end milling machine with a quill feed will be able to be used.

The component drawing shown in Figure C.2 is a simple block with an open pocket, drilled hole and step that requires the operator to machine a piece of stock bar to a correct size, with all faces at 90° to each other, machine a pocket to a designated size, drill a hole that intersects the pocket wall and machine a step. Once the workpiece has been machined to a parallel sided block of the correct size, all subsequent operations can be undertaken without moving the workpiece, which ensures squareness. this requires the operations to be undertaken in a logical sequence, and with the correct setup, and this is another aspect of the exercise, as milling often requires operations to be undertaken in a sequence to avoid having to reset the workpiece, allowing for potential positioning errors to become apparent, and extend the time taken to complete the work.

OPERATION LIST

1 Take a length of aluminium or brass bar stock 50mm × 30mm × 30mm and insert it into the machine vice with the 50mm length horizontal and resting on a machine parallel. Depending on the height of the machine vice differing heights of parallel will need to be used, and it is important that the parallel be narrower than 25mm, and that a minimum of 5mm, and maximum of 15mm of workpiece sticks up above the vice jaws. The vice should be tightened until firmly gripping the

Figure C.1 Milling exercise component

workpiece. The workpiece is tapped on the top with a soft faced mallet to ensure it is seated firmly on the machine parallel and it should not slide. The vice is then firmly clamped and the parallel re-checked. If it slides easily normally a firm tap with a mallet causes it to lock up again. The workpiece is then ready for milling of the first face.

FACE MILLING

2 Using a face mill with a diameter of greater than 30mm set to the correct speed for the cutter type, close the guarding and start the machine. Use the rapid traverse feed if fitted to bring the cutter close

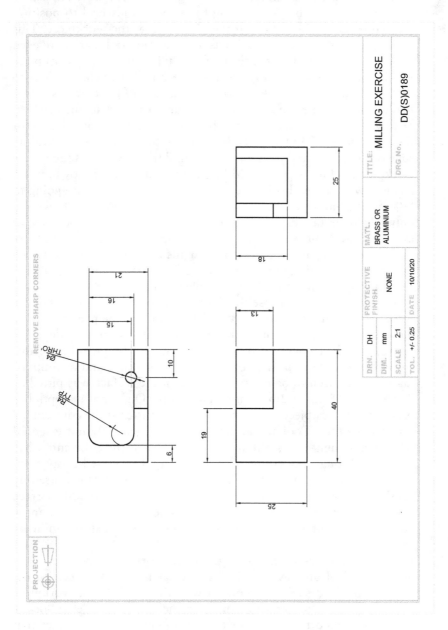

Figure C.2 Milling exercise component drawing

to the workpiece. Where no rapid feed mechanism is fitted operate the table handwheel. The intention is to bring the rotating cutter close to one edge of the workpiece, not plough it into the side. The cutter is the adjusted in the y-axis using the y-axis handwheel to position the centreline of the cutter on the centreline of the workpiece. This doesn't need to be exact, but it ensures that any tool marks are even across the surface and ensures that the full width of the workpiece will be machined, especially when the face mill diameter is close to that of the workpiece width. If the cutting edge of the cutter is below that of the workpiece upper surface the knee of the machine needs to be lowered to allow the cutter to be positioned just above the top of the workpiece. The workpiece is then moved just under the leading edge of the cutter, and the table raised until the cutter just touches the workpiece. While papers could be used to detect touch down, when facing a workpiece, this is rarely required as all surfaces are going to be machined and that degree of precision is not required. However, only a light contact between cutter and workpiece should be made to avoid plunge damage to the tooling. The table is then moved out from under the cutter on the x-axis to allow the cut depth to be set. Given the short length of component and the light cut it is not required to check for light contact across the whole face of the workpiece. Longer workpieces, especially those that have been hot rolled in manufacture can undulate, and can cause a very heavy cut in some places. Having moved the workpiece out from under the cutter a cutting depth of approximately 1mm is set by raising the knee. The intention of this cut is just to clean up the workpiece upper surface, and as 5mm of material will need to be removed in total if the surface was pitted or damaged a 1.5mm or 2mm cut would be fine. Once the cut depth has been set the workpiece is then fed under the cutter at the correct feed rate for the tool. Coolant can be used, but is generally not essential when taking light cuts on non-ferrous materials. Once the cutter has been fed through the cutter the table is rapid traversed away from the cutting head and the machine stopped. Any swarf chips carefully brushed away and the surface checked for a clean smooth surface and remove from the vice. If the surface is not fully cleaned up, apply another small cut and pass back through the cutter before removing from the vice.

3 Remove any burrs from the edge of the workpiece, clean the vice and parallel of any swarf chips and reseat the workpiece with the freshly machined surface in contact with the parallel. The workpiece is clamped as before and the operator checks the parallel is 'locked up'. This is important as the next series of cuts will form a face that needs to be parallel to the first. The height of the cutter should not need adjustment as the previous cut will have positioned the cutter

at the exact thickness of the workpiece, and if the workpiece is correctly seated on the parallel the height will not have changed. If the workpiece stock size was 30mm and the required size is 25mm, the first cut will have removed approximately 1mm leaving 4mm to be removed. The operator should then think about how many cuts they wish to remove the material at. It is entirely possible to remove 4mm in one cut however, it is possible if the feed rate is too great the workpiece will be dislodged, and there is no certainty of the component size until it is measured. Most operators will choose to set another 1mm cut, and pass the workpiece through the cutter producing a nominal workpiece thickness of 3mm above that required with to flat parallel clean surfaces. The workpiece thickness is then measured and the exact amount of material to be removed identified. Typically a roughing cut of 2.5mm will be taken followed by measurement of the exact remaining amount to be removed and the final approximately 0.5mm removed as a finishing cut. The table is traversed clear of the head swarf chips brushed away and the component thickness checked to see if is in tolerance. If so the workpiece is removed from the vice and any burrs removed using a de-burring tool and the workpiece will have two clean parallel faces 25mm in width. The vice is then cleaned ready to receive the component again.

4 Depending on the wear in the machine vice being used getting the next workpiece surface square to the two previously machined faces can be awkward. The moving face of a machine vice is always going to have a little movement, although many modern, and expensive vices are excellent, with minimal movement. The workpiece is inserted back into the machine vice with one of the remaining un-machined faces upwards and the opposite face resting on the parallel. It is important that when inserting the workpiece back in the vice, the machined face is pressed in full contact with the fixed jaw, and not rolled at a slight angle by the moving jaw. This may require packing of the moving jaw to achieve, however, the workpiece once inserted into the vice, with a machined face placed firmly up against the fixed jaw and the parallel locked up is ready to have the upper surface machined. The table will need to be dropped as the previous operations will have removed 5mm of material from the workpiece. Therefore, the table is dropped by lowering the knee by 5mm, and the cutter brought into light contact with the upper surface of the workpiece as before. The workpiece is moved out from under the cutter and once again a cut depth of approximately 1mm set and the workpiece passed under the cutter. As before the table is traversed once the cutter has passed from the workpiece, swarf removed and the workpiece if cleaned up, removed from the vice and de-burred. The operator should then check that the machined face is square to the other two faces using a tri square. If it

is not there has been a problem with setting the workpiece up in the vice and a second light cut on the face is required to correct, although this is rarely needed.

5 Assuming the face is square, the workpiece is put back in the vice with the remaining un-machined face upright, an initial 1mm cut applied and the workpiece passed under the cutter. The workpiece is then measured using a micrometer of calliper, and the remaining cuts taken as before down to size. The swarf is cleared using a small brush and the workpiece removed and de-burred. Four of the sides are cleaned up, the component is 25mm × 25mm and the next operations are to machine the ends, and reduce the stock bar to a length of 40mm. Sawn ends are rarely square to the body of a workpiece and the ends can be quite off axis. Once again though the workpiece is fitted into the machine vice, however, it needs to be placed slightly lower down so a lower parallel needs to be used, or if the base of the vice is in good condition and at least 10mm of the workpiece sticks up above the vice jaws mounting the workpiece on the base of the vice is not uncommon. While this prevents checking for parallel 'lock up' it ensures that the workpiece has maximum engagement with the frictional area of the vice jaws and is less likely to move when engaged with the cutter. The ends of the workpiece also need to be square to the sides, therefore when inserting the workpiece in the vice, it is normal to not only ensure that one machined face is in full contact with the fixed jaw, but a tri square is used to ensure the other vertical machined face is at 90° to cutting plane. This is normally achieved by lightly clamping the workpiece in the vice, placing a tri square on the parallel and moving the workpiece until the machined face is in full contact with the tri square blade edge. Following checking for full engagement of the fixed jaw, the vice is fully tightened. The table is then dropped by lowering the knee and the cutter brought into contact as before. A light cut of approximately 1mm is taken, however, if this does not clean up the surface to a smooth plane face, then an additional light cut is taken.

6 The process is then repeated for the opposite face, although the cutter height will not require re-setting. A first cut of 1mm is taken and the length of the component measured. Given that the block is sitting on a parallel or base of the vice, and the faces reduced to 25mm wide it is likely that any residual gap will be too small for a micrometer, or even a calliper, however, the depth function of a calliper can be used to measure form the top of the workpiece to the parallel it is seated on to determine the residual amount to be removed. This often requires the table to be traversed well clear of the machine head to prevent the callipers being fouled by the head and leading to an inaccurate measurement. The remaining material is then removed in a series of lights cuts down to the required size of 40mm. The swarf is brushed away

and the workpiece removed. The workpiece has now been face milled on all sides to size and all faces are square to each other.

WORKPIECE MOUNTING

7 The workpiece needs to be remounted in the vice and locked up on a machine parallel. As all operations can be undertaken from one side it makes sense to position the workpiece at a height that ensures all subsequent operations can be undertaken without having to move the workpiece. The component drawing shows that there is a step machined onto one side which has an edge 13mm down from the upper surface. As the component is only 25mm high, this means that a minimum of 13mm of the workpiece needs to protrude above the jaws of the vice. However, excessive protrusion above the vice jaws can result in low frictional forces retaining the component in place and less than 10mm engagement between the vice jaws and sides of the workpiece may result in light cuts having to be used rather than a more efficient cutting depth. Very low levels of engagement such as 5mm can lead to marking of the workpiece surface due to indentation, therefore selecting a parallel of the correct height is important. Once the workpiece has been clamped in place and the parallel underneath locked the workpiece is ready for the remaining operations to be undertaken.

DRILLING HOLE

8 One thing that separates out milling to many of its sister techniques is that often the sequence of tasks can be more crucial. The component drawing shows a Ø4mm hole passing through the component, however, approximately one-third of the diameter of the hole intersects the wall of the pocket and the step in the side. This means that any attempt to drill the hole from either size will cause the drill to deflect away from the pocket wall, and therefore needs to be drilled before the pocket, or step is machined.

9 The face mill is removed from the spindle and the operator has the choice of inserting either a Jacobs type chuck into the spindle or directly mounting a centre drill and drill into a 4mm R8 collet or similar. The approach taken will often depend on the machine and equipment available and either approach is suitable.

10 The next operation is to align the spindle of the machine over the component edges to set an x-axis datum and a y-axis datum. This can be achieved using papers of a known thickness, such as cigarette papers, where they are slid between a stationary cutter and the table position moved until there is a noticeable 'drag' on the paper. Alternatively the paper can be wetted and stuck to the workpiece with a rotating cutter,

or drill, slowly being brought in contact with the paper. When it is ejected due to rotation of the cutter the machine is stopped. An alternative is to use a centre finder with a ball stylus where the rotating ball, set on a vertical axis, is brought into contact with the edge of the workpiece, when the stylus visibly displaces the spindle axis is off the edge by half the diameter of the ball. Whichever approach taken (i.e. using a centre finder or paper method) a datum edge on the x-axis and Datum edge on the y-axis should be established and the DRO readout set to zero, and/or the table scales set to zero. Once the datum edges have been established and the operator has confidence the spindle centre is located exactly over each edge, machining the features can commence.

11 Moving the spindle away from the workpiece a centre drill is inserted into the spindle and the correct speed for the size of centre drill set. Ensuring that the drill point will not foul the workpiece the table is traversed on the x-axis until it is positioned 10mm beyond the datum edge. The drawing has been functionally dimensioned (i.e. the dimensions represent the use or operation of the finished component, rather than it being dimensioned for convenience of manufacture). By using simple arithmetic the centre position of the hole is either 19mm from one side of the workpiece or 6mm from the other, and the workpiece is positioned under the spindle on the y-axis at the correct distance from the datum edge. The machine is then started, the quill clamp untightened and the centre drill advanced into the workpiece by rotating the quill feed until a small centre drilled hole is produced. If using a morse centre drill that has the potential to produce a hole larger than Ø4mm then the operator needs to either just use the centre point of the drill or be careful not to over drill the hole and produce a tapered cavity greater than Ø4mm. Once this has been achieved the table is dropped rather than moved in an x-axis or y-axis direction to allow removal of the centre drill and insertion of a Ø4mm jobber drill. The speed is adjusted and depending on the material used ie brass or aluminium the spindle speed will be in the region of 2500rpm - 3000rpm. The table is then raised until the drill point is close to the workpiece and using the quill feed a hole is drill right through the component. The operator needs to check that the supporting parallel will not foul the drill coming through, as this will damage both the parallel and if hardened, the drill. Once the drilling operation is complete the spindle is stopped and the quill is raised back to its stop removing the drill from the cavity and the drill removed.

POCKET MACHINING

12 Following the drilling operation the pocket can now be machined. The component drawing shows that the internal corners have a radius of

4mm. This suggests that an Ø8mm slot drill is used, to generate the correct sized corners. The pocket depth is 18mm deep therefore it is important to select a slot drill that has a full flute length of a minimum of 20mm to ensure effective swarf chip ejection and prevent 'polishing of the pocket cavity side where any plain diameter comes into contact with the pocket wall. The spindle speed should then be set to the correct cutting speed for the type of cutter used.

13 As the datum edges had already been identified prior to drilling there is no need to reset the datum edges, however the z-axis datum needs to be established. This is achieved in a similar manner to finding the datum edges using papers, and a paper is either trapped between the cutter and machined face until there is a noticeable drag, or it is undertaken with the paper wet and cutter rotating when the cutter touches the paper it is ejected from under the cutter leaving the cutter at height above the face of the workpiece equal to the thickness of paper. It is common for turret milling machine DRO readouts to only have an x-axis display and y-axis display, but no z-axis. Therefore it is important to zero the scale on the z-axis handle. Once this datum has been established the cutter is moved off the surface of the workpiece ready for the first cut.

14 The pocket needs to be 18mm deep when complete which lends itself to 4 × 4mm deep cuts plus a lighter finish cut. It is generally considered that the depth of cut should not be more than half the cutter diameter, especially when milling a slot or pocket as the cutter has 180° of engagement with the workpiece. If time isn't especially important and understanding pocket generation is, then 2.5mm deep cuts can be used as this normally means the knee z-axis control handle can be given half a rotation, as one full rotation of the handle normally raises the knee by 5mm, and making keeping a mental record of the total depth of cut simpler. A cut depth of 2.5mm would mean 7 × 2.5mm cuts plus a 0.5mm finish cut. Once the peck depth for each cut and number of cuts has been decided the depth of cut is set.

15 The cutter then has to be positioned for producing the initial slot along the x-axis, and this requires the cutter to be positioned on the y-axis. The y-axis datum was set so that the centre of the spindle was directly over the datum edge. This means that the workpiece needs to be moved the thickness of the remaining wall once the pocket is machined, plus half the diameter of the cutter. A further additional distance needs to be allowed for a finish pass, as sequential pecks will leave a series of parallel lines on the pocket walls therefore it is common to leave an small amount such as 0.1mm to be removed with a full depth pass to produce a smooth unmarked pocket wall. The cutter is now to a set depth and has been set on the y-axis so that one side of the cutter will leave behind a wall of the correct thickness plus an allowance for a

finish cut, so in this case the cutter will have been moved 8.6mm of its datum (4.5mm wall thickness, 4mm for half cutter diameter, plus 0.1mm for a finish cut).

16 What now needs to be determined are the sizes for the length of the pocket and the width of the pocket. The component drawing shows using simple arithmetic that the length of the pocket is 34mm long and is dimensioned as being 16mm wide. As the cutter is Ø8mm then the length the pocket will be machined to is 34mm − 4mm (half the cutter diameter) − 0.1 (finishing cut allowance), therefore the distance the table travels to in the x-axis will be 29.9mm from the datum. Similarly as the pocket is 16mm in width and an Ø8mm cutter is being used the distance to be traversed in the y-axis will be 16mm − 8mm (diameter of the cutter) − 0.1mm (for the finishing cut allowance on one side) − 0.1mm (for the finishing cut allowance on the opposite side). Therefore, the movement on the y-axis will only be 7.8mm in total. Once the pocket size readings are determined these written down as a reminder to the operator on the limits for the table to be traversed on the x-axis and y-axis. Alternatively they can be inserted into any DRO pocket milling function.

17 Once the limits have been identified the machine is started and the workpiece fed under the cutter. It is usual for the cutter to be fed into the workpiece so that it is downcut milling and using plenty of coolant to cool the cutter and assist with swarf chip ejection. The operator feeds the work until the pocket length identified is approached. If using power feed this is disengaged just before the length required is achieved and it taken to depth by hand. The x-axis clamp can then be tightened and the workpiece fed under the cutter in a translational movement on the y-axis, until that distance is achieved. The table is then fed on the x-axis in the opposite direction to the first movement taking it out of the workpiece. The next peck depth is set and the process repeated. Where multiple pecks are to be undertaken the operator does not need to measure the depth at completion of each peck, just keep a count of the number of pecks. However, when coming close to the final peck the machine should be stopped. any burr around the top edge of the pocket removed and the pocket depth checked. As with the walls it is normal to leave 0.1mm for a finish cut and once the pocket has been roughed out to size, the machine is stopped and the pocket cleaned of any swarf chips. The x-axis and y-axis pocket limits are then re-calculated to identify the full length and width readings to be achieved (i.e. 0.1mm deeper on all sides), and the final peck of 0.1mm set on the z-axis. The pocket is then enlarged by 0.1mm using a full depth cut to leave it at the correct size.

18 The final operation is to produce the step on one side. This is comparatively simple as little material has to be removed. The workpiece

is dropped so that the cutter is at its datum position on the z-axis, and roughly centred on the section of pocket wall to be removed. In a similar manner to setting the length of the pocket simple addition identifies that the step in the wall is 21mm long. Therefore the x-axis distance to be machined to will be 21mm – 4mm (half the diameter of the cutter) – 0.1mm (finishing cut allowance) = 16.9mm. Once again the peck depth used is 2.5mm and a series of cuts undertaken. The workpiece is fed under the cutter to a distance of 16.9mm, and then the workpiece is moved backwards and forwards on the y-axis to generate a plain flat surface rather than the radiused edge formed by the cutter diameter. Once the depth has been reached and the finishing cut taken the component is complete and can be removed for the vice and the machine cleaned down.

Index

CPSIA information can be obtained
at www.ICGtesting.com
Printed in the USA
LVHW081753260522
719853LV00004B/101